Essentials of Biochemistry

Essentials of Biochemistry

Second Edition

Dorothy E. Schumm, Ph.D.
Professor of Medical Biochemistry
Ohio State University College of Medicine
Columbus, Ohio

Little, Brown and Company
Boston   New York   Toronto   London

Library of Congress Cataloging-in-Publication Data

Schumm, Dorothy E., 1943–
    Essentials of biochemistry / Dorothy E. Schumm. — 2nd ed.
        p.   cm.
    Includes index.
    ISBN 0-316-77531-2
    1. Biochemistry.   I. Title.
    [DNLM: 1. Biochemistry.   QU 4 S392e 1995]
    QP514.2.S38   1995
    574.19'2—dc20
    DNLM/DLC
    for Library of Congress   94–22956
                                                CIP

Printed in the United States of America

SEM

Editorial: Evan R. Schnittman, Rebecca Marnhout
Production Editor: Cathleen Cote
Copyeditor: Lois F. Hall
Indexer: Alexandra Nickerson
Production Supervisor/Designer: Louis C. Bruno, Jr.
Cover Designer: Hannus Design Associates
Cover Artist: Peg Gerrity

To the medical students who tolerated my experimentation with new ways of presenting material and who gave me feedback on what worked—and on what didn't. To my editors, reviewers, and professional colleagues for the invaluable advice that made this a better text than it would have been. But especially to my husband for his unfailing support.

Contents

Preface

The general order of topics has not changed much between the first and second editions of *Essentials of Biochemistry*. The second edition, however, contains a totally revised and expanded section on molecular biology and a new chapter on the tools of molecular biology added to assist the student in understanding the technologies that now impact the medical sciences. The chemistry of types of compounds (for example, carbohydrates and amino acids) has been moved from the beginning of the book to the section where the metabolism of these compounds is first discussed. All of the chapters have been updated to reflect recent advances in our understanding of biochemistry and to remove topics that time has shown to be less important to our understanding of basic biochemistry.

D.E.S.

Introduction for Students

Essentials of Biochemistry is designed as a review of the essential biochemistry needed by students in medicine, dentistry, pharmacy, and nursing. In addition, graduate students in the medically related sciences of anatomy, physiology, pharmacology, pathology, microbiology, and immunology may also find this text fulfills their need for basic biochemistry. Because the book is directed to students in the medical sciences, the concepts emphasized are those relating to eukaryotic cells and, in particular, human cells. Relevant disease states, the effects of enzyme deficiencies and excesses, and clinically important drugs are discussed. This book does not contain history, derivations of equations, or discussion of such nonmedical topics as nitrogen fixation and photosynthesis. The discussion of experimental methods is confined to the area of molecular biology in which an understanding of methods is important for medical science and practice.

As you can see from the table of contents, the order of topics is different from that in most biochemistry texts. This book begins with a chapter covering the major organelles of the eukaryotic cell. It then proceeds to molecular biology and continues with the biochemistry of proteins, carbohydrates, lipids, amino acids, and nucleotides. The final sections cover hormones and some important concepts in nutrition. These final topics should help you integrate individual metabolic pathways into the function of the whole organism.

To help you master the important concepts of biochemistry, each chapter begins with a list of objectives. These objectives are covered in this order in the chapter. There are review questions at the end of each Part with answers in the Appendix at the back of the book. The questions are in the format used by many licensing examinations and should aid you in preparing for such standardized examinations.

In looking through the text, you will note a large number of structures and metabolic pathways. At first glance, this may seem overwhelming. It is not possible to memorize all this material during the length of a typical biochemistry course or in reviewing for standardized examinations. Do not be intimidated. The purpose of the structures is to aid you in following the flow of atoms as compounds are synthesized, degraded, and interconverted. Similarly, by actually seeing the structures of various drugs and inhibitors, you will be better able to understand and remember their functions. Since structures are often featured as parts of questions in standardized examinations, you will find familiarity with them to be of value.

As you read *Essentials of Biochemistry,* keep in mind a picture of what the cell or the entire organism is trying to do—for example, synthesize carbohydrate for energy storage or degrade it for energy production, prepare for cell division, or contract muscle. Do not get bogged down in small details. If you follow this suggestion, you will find your study of biochemistry a tolerable experience. I hope many of you will find it enjoyable as well.

Part
I The Cell

1 Introduction to the Cell

The human body is composed of more than 10^{14} cells. These are organized into tissues, each with its own distinct set of functions: Muscles contract, the pancreas synthesizes digestive enzymes, the bone marrow produces erythrocytes and lymphocytes, the adrenals secrete hormones. Despite major differences in function, there are many similarities in cell composition and structure. Cells are composed of four major chemical elements: carbon, hydrogen, oxygen, and nitrogen. In addition, they contain a number of minor elements including sodium, potassium, phosphorus, sulfur, magnesium, calcium, iron, and zinc. All cells, regardless of their specialized functions, take in nutrients, oxidize fuels, and excrete waste products. These are the "housekeeping" functions. Most cells also grow and divide. Because of a common composition and many common functions, cells have a similar internal organization and contain similar chemicals. The cellular organelles partition chemical substrates into different pools, isolate enzymes, provide a solid surface on which chemical reactions occur, and protect the permanent genetic information. This chapter provides you with an introduction to cell structure and some of the chemical components of the cell.

Objectives

After completing this chapter, you should be able to

Describe the structure and major metabolic functions of the plasma membrane, mitochondria, nucleus, endoplasmic reticulum, Golgi complex, and lysosomes.

Discuss the importance of water in cellular metabolism.

Define the terms *buffer, pH, polarity,* and *hydrogen bond.*

Write the Henderson-Hasselbalch equation and be able to use it in buffer calculations.

Typical Cell

Size and Shape

Figure 1-1 depicts a typical mammalian cell and some of its subcellular organelles. The size of a typical rounded cell can vary from 20 μm in diameter for a liver cell to less than 7 μm for a mature red blood cell that lacks most of the organelles. Although the cell shown in Figure 1-1 is drawn as round and smooth, scanning electron micrographs, such as that in Figure 1-2, show that this is not the case for most cells. Fibroblasts and neurons are elongated; astrocytes are star-shaped. Peripheral neurons, the largest cells of the human body, are extremely elongated

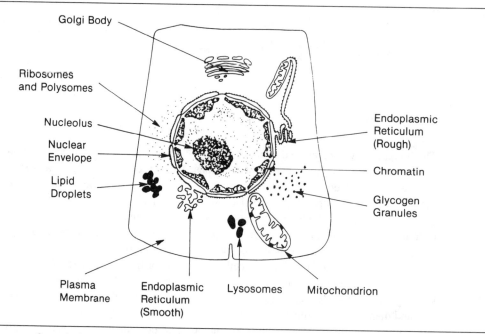

Golgi Body

Ribosomes
and Polysomes

Nucleolus

Nuclear
Envelope

Lipid
Droplets

Endoplasmic
Reticulum
(Rough)

Chromatin

Glycogen
Granules

Plasma
Membrane

Endoplasmic
Reticulum
(Smooth)

Lysosomes

Mitochondrion

Fig. 1-1. Diagram of a typical mammalian cell showing some of the subcellular structures.

and can contain extended axons many times the length of the main body of the cell.

Intracellular Attachments

Electron microscopy shows that the cell surface is covered with multiple projections, which are the remains of cell-to-cell connections. These connections permit the adhesion of cells to form tissues and facilitate the exchange of nutrients and regulatory molecules. There are three major types of cell-to-cell connections: tight junctions, gap junctions, and desmosomes. In **tight junctions,** such as in the intestinal epithelium, the outer membranes of two cells appear to be fused at several points. Nothing can pass between the two cells at the points of the tight junctions. Tight junctions also occur between epithelial cells in capillaries, where they contribute to the blood-brain barrier. In **gap junctions,** which are common in cells of the nervous system, the outer membranes are not fused but are separated by a very small space. Narrow channels connect the cells. These connecting channels allow the cells to exchange substances, such as neurotransmitters, without releasing the chemicals to the outside environ-

Fig. 1-2. Scanning electron micrograph of fibroblasts. Note the elongated shape and numerous projections from the cell surface. (From U. Bachrach and A. Shtorch. Formation of cadaverine as an effect of α-fluoromethylornithine in chick embryo fibroblasts transformed by rous sarcoma virus. *Cancer Res.* 45: 2159–2164, 1985.)

ment. **Desmosomes** are filaments that surround the outside of cells. They cement cells to one another and permit the formation of tissues.

Intracellular Components

Plasma Membrane

Surrounding the outside of all cells is the **plasma membrane,** which serves as a barrier limiting the exchange of molecules between the inside and the outside of cells. The plasma membrane is composed primarily of lipids and is selectively permeable. Whereas lipid-soluble compounds can pass through the membrane, polar and water-soluble compounds found inside and outside of cells do not pass through it easily. To exchange necessary molecules, specific proteins in the membrane provide sites for the entry and exit of polar molecules such as ions, sugars, amino acids, and nucleotides. The plasma membrane also contains receptors capable of binding hormones, neurotransmitters, and antibodies.

The plasma membrane is not symmetrical. It has a distinct outside that contains all the carbohydrate and receptor sites. The middle of the membrane is predominantly lipid. Proteins can be found on both sides of the membrane and embedded within it. However, specific proteins are found only on one side of the plasma membrane or the other but not both. The structure and function of membranes will be discussed more fully in Chapter 21.

Cytoplasm

The **cytoplasm** includes everything inside the plasma membrane except for the nucleus. It contains salts, enzymes, and a variety of substrates for cellular reactions. Some of the enzymes are entirely free in solution. These soluble enzymes include those that catalyze the initial breakdown of sugars and some of those responsible for the synthesis of proteins. Most other enzymes are attached firmly to internal membranes or are bound loosely to filaments, which form a scaffold in the cytoplasm. Binding of enzymes to a membrane or filament can orient a protein in a particular direction or facilitate an enzymatic pathway by holding together the enzymes that react sequentially with a substrate.

Filaments

Cells contain several types of filaments. These molecules maintain the shape of the cell, move organelles around the cell, and participate in the movement of chromosomes during mitosis and meiosis. The largest of the cytoplasmic filaments are the **microtubules.** They are 200 to 300 μm in diameter and are composed of a polymer of the protein **tubulin.** Microtubules are responsible for maintaining the shape of the cell and allowing material to move within the cell. They also make up the mitotic spindle, which separates chromosomes at the time of cell division. The polymerization of tubulin can be inhibited by drugs such as colchicine, which prevents the formation of the mitotic spindle and the correct distribution of chromosomes between daughter cells.

The **microfilaments,** which are found in most cells, are smaller than microtubules. The thick microfilaments are composed of the protein myosin and the thin filaments of the protein actin. Together these are the contractile proteins of muscle. In addition, microfilaments link different segments of the cell. Some are extremely stable; others are synthesized, function briefly, and then are broken down.

Nucleus

The largest of the organelles in cells is the **nucleus.** The existence of a distinct nucleus is what separates eukaryotic cells (yeast, fungi, plants, and animals) from prokaryotic cells (bacteria, mycoplasma, and blue-green algae). The nucleus is surrounded by a double membrane or envelope dotted with octagonal pores. It is through these pores that material passes between the nucleus and the cytoplasm. Ribonucleic acid (RNA) synthesized in the nucleus enters the cytoplasm by passing through the nuclear pores. Similarly, proteins destined for the nucleus are synthesized in the cytoplasm and enter the nucleus via the pores.

Deoxyribonucleic acid (DNA), RNA, and protein are found inside the nucleus, although only the RNA and DNA are synthesized there. The DNA of human cells is located in 46 chromosomes arranged in 23 pairs. Distinct chromosomes can be seen only just before the cell divides. The rest of the time, chromosomes appear as a diffuse tangle of fine threads called **chromatin.** RNA, which is synthesized in the nucleus, is found associated with protein in chromosomes and in large ribonucleoprotein complexes. In addition, protein forms a fibrous matrix inside the nucleus. Chromosomes and ribonucleoproteins are both attached to this matrix, which may be necessary for their function.

Nucleolus

The **nucleolus** is a rounded body found inside the nucleus. Unlike the nucleus and most other organelles, the nucleolus lacks a limiting membrane. The nucleolus is the site of synthesis of ribosomal RNA and of the assembly of ribosomes, the structures on which proteins are made. There may be more the one nucleolus in a nucleus, especially in cells such as those of the liver that are synthesizing large amounts of ribosomal RNA.

Mitochondria

The **mitochondrion** is the other organelle concerned with the final steps of oxidation of cellular fuels, the production of adenosine triphosphate (ATP), and the first steps in the synthesis of urea. The number of mitochondria per cell depends on the need of that cell to generate energy. A single liver cell may contain several hundred mitochondria. Mitochondria are also numerous in muscle cells but absent from mature red blood cells.

Mitochondria are usually oblong in shape. They are approximately 1 µm wide and up to 7 µm long. However, the shape can vary depending on the metabolic state of the cell. Mitochondria have a smooth outer membrane and a highly folded inner membrane with a small space between them, as shown in Figure 1-3. The folds of the inner membrane are called **cristae.** This inner membrane is highly impermeable to most ions, especially positive ions. It contains the electron transport chain and the proteins that convert oxidative energy into ATP. The enzymes responsible for oxidizing acetate and fatty acids to carbon dioxide are found in the matrix inside the mitochondrial membranes. Because of their location, electrons produced in these reactions can be easily transferred to the membrane-bound electron transport chain proteins.

Although most of the genes coding for mitochondrial proteins are found in the nucleus, mitochondria contain a small amount of DNA and the enzymes necessary to synthesize both RNA and protein. The enzymes for these reactions are distinctly different from those in the rest of the cell but resemble those

Fig. 1-3. Structure of a mitochondrion.

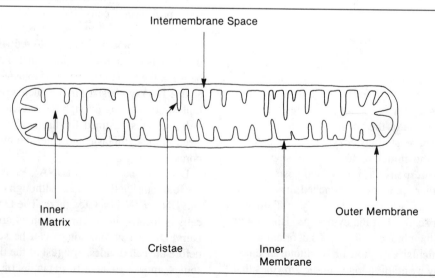

Intermembrane Space

Inner Matrix

Cristae

Inner Membrane

Outer Membrane

in prokaryotes. Because of this similarity, mitochondria are thought to have once been independent organisms. The mitochondrial proteins that are coded for by nuclear genes are synthesized in the cytoplasm and then migrate into the mitochondria. Defects in mitochondrial functions are associated with viral infections, cirrhoses of the liver, Reyes syndrome, and ammonia intoxication.

Endoplasmic Reticulum

The **endoplasmic reticulum** (ER) is a system of intracellular membranes that store, segregate, and transport substances inside the cell. The ER is continuous with the nuclear membrane and the Golgi complex. The ER can exist in a rough or a smooth form. The rough or granular ER has attached ribosomes and is one site of protein synthesis. It is plentiful in such cells as antibody-producing cells, fibroblasts, liver cells, and pancreatic cells that synthesize proteins for export.

The smooth or agranular ER lacks ribosomes and is the site of synthesis of steroids and complex carbohydrates. In liver, the smooth ER also contains the enzymes that oxidize and detoxify foreign compounds (xenobiotics). On mechanical disruption of the cell, the ER forms small membrane vesicles called **microsomes.**

Golgi Complex

Another membrane-containing cellular organelle is the **Golgi complex,** which is composed of parallel closed membrane sacs or vesicles that are continuous with the ER. Proteins destined for secretion enter one side of the Golgi and are concentrated and modified by the addition of carbohydrate or lipid. Then they exit from the other side of the Golgi bound to membranes or enclosed within membrane-bound vesicles. Because of their role in protein secretion, the Golgi complexes are numerous in cells such as liver and pancreas, which secrete large amounts of protein.

Lysosomes and Peroxisomes

Lysosomes also are membrane-bound vesicles. They contain hydrolytic enzymes that can degrade nucleic acids, proteins, and complex carbohydrates. Segregation of these digestive enzymes in the lysosomes prevents unwanted degradation of cellular macromolecules. Lysosomes can fuse with damaged cellular organelles or ingested material. Then the lysosomal enzymes can degrade the material. More than 30 inherited diseases have been associated with defects in lysosomal enzymes necessary for the degradation of carbohydrate attached to protein or proteoglycans. These diseases include Tay-Sachs disease and Hurler's and Hunter's syndromes.

Peroxisomes or **microbodies** are similar in size to lysosomes and are found predominantly in the liver and kidney. Peroxisomes derive their name from their association with hydrogen peroxide. Similar to lysosomes, peroxisomes contain a variety of oxidation enzymes. However, many of the peroxisomal enzymes produce hydrogen peroxide as one product of oxidation. Catalase, also contained in the peroxisome, then degrades the peroxide. Containment in an organelle prevents the peroxide from attacking other cellular components, especially membranes. Defects in the peroxisomes are associated with several types of adrenoleukodystrophy.

Other Cellular Structures

In addition to the structures already mentioned, cells contain ribosomes, some of which are free in the cytoplasm. These ribosomes can be either single ribosomes, called **monosomes,** or aggregates of up to 30 ribosomes, called **polysomes,** which are engaged in protein synthesis using a single messenger RNA (mRNA) molecule. Proteins from soluble polysomes are released into the cytoplasm instead of being bound to membranes.

Cells storing fats or sugars contain lipid droplets or glycogen granules. Specialized cells such as muscle, nerve, and endocrine cells contain additional internal structures. These will be discussed as we encounter them in our study of metabolism.

Chemical Interactions

Water

The structure and function of cells is produced by the chemical compounds they contain and the interactions of these chemicals with one another. The major cellular chemical is water. Water is electrically **polar**. The oxygen end is more electronegative and the hydrogen ends are more electropositive. When interacting with charged molecules, the more negative end of water is turned toward positively charged ions, and the more positive ends are turned toward negative ones. If it were not for the polarity of water, charged ions such as Cl^-, K^+, HCO_3^-, and Mg^{2+} would not exist in this form in cells.

The polarity of the water molecule also allows it to form **hydrogen bonds** with other molecules that are similarly polar. In addition to interaction with water, polar molecules can form hydrogen bonds with one another. The most important interactions for biological systems are

$$-C=O \ldots . HN-$$
$$-C=O \ldots . HO-$$
$$-C=O \ldots . HS-$$
$$-C=N \ldots . HN-$$

Such hydrogen bonds maintain the structural stability of proteins, carbohydrates, and nucleic acids.

Nonpolar hydrocarbon-like molecules that cannot form hydrogen bonds tend to cluster together away from the more polar molecules. This is referred to as **hydrophobic interaction.** Molecules of this type form droplets or layers when in an aqueous environment. Hydrophobic interaction is important for the structure of cellular membranes, the folding of proteins, and the secondary structure of nuclei acids.

pH and Buffers

Although the concentration of many cellular chemicals can vary within a wide range, the hydrogen ion concentration inside cells must remain fairly constant to be compatible with life. Most cells and body fluids maintain a **pH** (the negative log of the hydrogen ion concentration) of 7.4 despite the many reactions that utilize or produce hydrogen ions. The stability of pH is due to the function of a number of cellular **buffers** including phosphate, acetate, bicarbonate, and the amino acids that make up proteins. Water itself is a very poor buffer. In addition to

buffers, the active excretion of H^+ by the kidney helps to maintain overall pH stability in the body.

Buffers consist of mixtures of weak acids and their salts. For example, if you mixed acetic acid and sodium hydroxide, you would form a buffer containing acetic acid and sodium acetate (the conjugate base of acetic acid). Figure 1-4A shows the change of pH when sodium hydroxide is added to a solution of acetic acid (pK 4.7). Note that when the pH is at or near the pK (the negative log of the dissociation constant of the acid), the pH changes very little with the addition of more base. The solution is a good buffer at this pH. The further the pH is from the pK, the less well the solution can function as a buffer.

For some acids, such as phosphoric acid, there are multiple ionizable hydrogens. Figure 1-4B shows the titration curve for phosphoric acid. Because each hydrogen dissociates separately, there are pKs of 2.0, 6.8, and 12.2. Because of this, phosphoric acid can buffer well at several pH values. To buffer well in biological systems with a pH of 7.4, buffers should have a pK between 6 and 8. As one of the pK values of phosphoric acid is 6.8, it is a good physiological buffer.

The **Henderson-Hasselbalch equation** gives the mathematical relationship between the concentration of a weak acid and its conjugate base, the pK and the pH. It takes the following form:

$$pH = pK + \log(A/HA)$$

where A and HA are the concentration of the conjugate base and the acid, respectively. By using this equation, you can see that when the concentration of the acid and the base are equal, $\log(A/HA) = 0$ and pH = pK. The addition of a small amount of H^+ or OH^- will not change the base-to-acid ratio appreciably and therefore will not change the pH appreciably.

The concentration of a buffer also affects its ability to stabilize the hydrogen ion concentration. To see this for yourself, calculate the pH change that occurs when 10 ml of 0.01-M acid is added to 100 ml of a 0.05-M acetate buffer, pK 4.75, or to 100 ml of a 0.01-M barbiturate buffer, pK 4.0, if both buffers are originally at pH 4.0. Although the pH is closer to the pK of the barbiturate buffer, the acetate

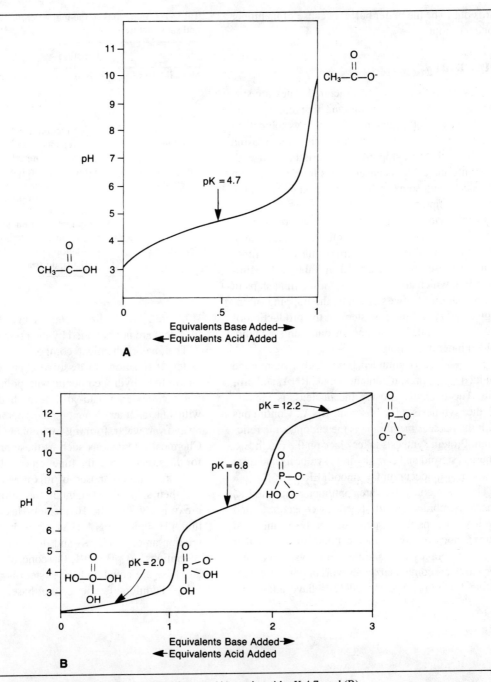

Fig. 1-4. Titration curves for the addition of a strong base to (A) acetic acid, pK 4.7, and (B) phosphoric acid, pKs 2.0, 6.8, and 12.2.

turns out to be the better buffer because of its higher concentration.

Summary

Eukaryotic cells are highly structured. They are surrounded by a plasma membrane that controls the entry and exit of compounds and allows the cell to interact with its environment. Inside the plasma membrane is the cytoplasm. It is crisscrossed with filaments and contains the enzymes for the oxidation of glucose and some of the enzymes engaged in the synthesis of proteins.

In the cytoplasm are several distinct organelles: mitochondria, nucleus, Golgi complex, endoplasmic reticulum, lysosomes and peroxisomes. The functions of these are summarized in Table 1-1. Mitochondria, which are responsible for the final steps in the oxidation of sugars and fats, the trapping of energy as ATP, and the first steps in the production of urea, are bounded by a smooth outer and a highly folded inner membrane.

The nucleus is bounded by a double membrane studded with pores. It contains DNA, RNA, and protein. The nucleolus, inside the nucleus, is the site of the synthesis of ribosomal RNA. Continuous with the nuclear membrane is the endoplasmic reticulum. Protein synthesis takes place on the rough ER, whereas steroid and carbohydrate synthesis and drug detoxification occur on the smooth ER.

The Golgi complexes are continuous with the ER. These membrane-bound structures concentrate and modify proteins to be secreted. The lysosomes and peroxisomes are two other cellular structures that have membranes surrounding them. Lysosomes and peroxisomes contain enzymes with important degradative functions. Specialized cells have additional structures, such as lipid droplets and glycogen granules, that are not bounded by membranes.

The major chemical compound of the body is water. It functions to dissolve charged compounds and to form hydrogen bonds with polar compounds. Polar compounds can also form hydrogen bonds with one another. Nonpolar compounds aggregate away from water, forming hydrophobic interactions. Chemical interactions such as these are responsible for the structure and the function of cells.

Buffers are composed of mixtures of weak acids and their salts. They maintain the cellular pH at approximately 7.4. The Henderson-Hasselbalch equation, $pH = pK + \log(A/HA)$, gives the mathematical relation among pH, pK, and the acid and base concentrations. At $pH = pK$, the concentrations of acid and base are equal and the system is a good buffer for the addition of either acid or base.

Table 1-1. Functions of the major cellular organelles and substructures

Structure	Function
Nucleus	Contains chromosomes, synthesis of RNA
Nucleolus	Site of synthesis of ribosomal RNA
Mitochondria	Final steps of substrate oxidation, production of ATP
Filaments	Maintenance of cell shape, movement of material inside cell
Endoplasmic reticulum	Synthesis of proteins, steroids, and complex carbohydrates
Golgi complex	Modification of proteins for secretion
Lysosomes	Degradation of nucleic acids, proteins, and carbohydrates
Peroxisomes	Oxidation of substrates that produce H_2O_2, destruction of H_2O_2

Part I Questions: The Cell

1. Mitochondria
   A. membranes are continuous with the endo-plasmic reticulum.
   B. are surrounded by a single membrane containing octagonal pores.
   C. are freely permeable to both positive and negative ions.
   D. are found in higher numbers in cells that require large amounts of metabolic energy.
   E. contain the enzymes responsible for the digestion of damaged cellular organelles.

2. The nucleus
   A. is surrounded by a continuous outer membrane and a highly folded inner membrane.
   B. contains degradative enzymes producing hydrogen peroxide as a reaction product.
   C. contains at least one nucleolus, the site of ribosomal RNA synthesis.
   D. is the major site of energy production in the cell.
   E. is one site of protein synthesis.

3. The plasma membrane
   A. serves as a barrier limiting the exchange of molecules.
   B. contains octagonal pores through which molecules freely pass in and out of the cell.
   C. is composed primarily of protein.
   D. is symmetrical in chemical composition.
   E. is identical in composition to the other cellular membranes.

4. Water is considered polar because it
   A. can form hydrogen bonds.
   B. is more electronegative at the oxygen and more electropositive at the hydrogen ends.
   C. forms a good buffer at physiological pH.
   D. forms hydrophobic interactions with similar molecules.
   E. is the major chemical constituent of cells.

5. Which of the following statements is true for a solution of 0.01-M weak acid, pK 6.0, and 0.02-M base?
   A. This solution will form a buffer with a pH of 6.3.
   B. This solution will be a good buffer at cellular pH.
   C. At pH 5.8, this will be a better buffer than one composed of 0.005-M acid, pK 5.8, and 0.005-M base.
   D. The addition of acid to this buffer will result in a larger pH change than the addition of an equivalent amount of base.
   E. This solution will buffer best at a pH one unit higher than the pK.

Part II Molecular Biology

2 DNA and Chromosome Structure

DNA is the permanent genetic material of human cells as well as of plants, bacteria, and some viruses. The structure of DNA contains the information for the synthesis of all the 50,000 proteins needed for general cellular metabolism and differentiated functions. It also confers our uniqueness as individuals and as human beings. Most of the DNA in cells—approximately 1×10^{-12} g/cell—is found in the nucleus. However, a small amount is also found in the mitochondria. Unlike the nuclear DNA, which is inherited from both parents, the mitochondrial DNA is believed to be inherited exclusively from the mother.

This chapter outlines the chemical structure of DNA and how information is coded in this structure. It also covers the physical arrangement of DNA in the nucleus and in the mitochondria.

Objectives

After completing this chapter, you should be able to

Describe the primary structure of DNA.

Recognize the bases found in DNA. Write the names of the corresponding nucleosides and nucleotides.

Define the terms *phosphodiester bond, complementary base, hydrogen bonding, base stacking interaction, antiparallel,* and *polarity.*

List the major characteristics of the A, B, and Z forms of DNA.

List the conditions under which DNA can be denatured and what physical properties are altered. Define the term *melting point.*

Describe the structure of a mitochondrial chromosome. Define the terms *supercoiling* and *topoisomerase.*

List the major characteristics of histones. Describe a nucleosome.

Discuss the structure of a eukaryotic chromosome. Define the terms *satellite DNA, histone, euchromatin,* and *heterochromatin.*

Differentiate among the primary, secondary, and tertiary structures of DNA.

Primary Structure of DNA

Figure 2-1 shows the structure of a small segment of DNA. It is a linear polymer of nucleotides, each consisting of a nitrogen-containing base, the sugar deoxyribose, and phosphate. Only four bases are

A G T C

found naturally in DNA: adenine, guanine, thymine, and cytosine. The first two are purines and the second two are pyrimidines. Their structures and numbering systems are shown in Figure 2-2.

The addition of a sugar to one of these bases pro-

15

duces a **nucleoside.** To differentiate between them, the carbons of the sugar are indicated by prime numbers. The sugar found in DNA is 2′-deoxyribose. The 1′ position of the sugar is attached to the nitrogen at position 9 of the purines or to nitrogen at position 1 of the pyrimidines. The names of the four nucleosides found in DNA are 2′-deoxyadenosine,

2′-deoxyguanosine, 2′-deoxycytidine, and thymidine. In contrast to the other nucleosides, which are assumed to contain the sugar ribose, thymidine is assumed to contain 2′-deoxyribose as the sugar. Therefore, thymidine does not require the 2′-deoxy notation before its name.

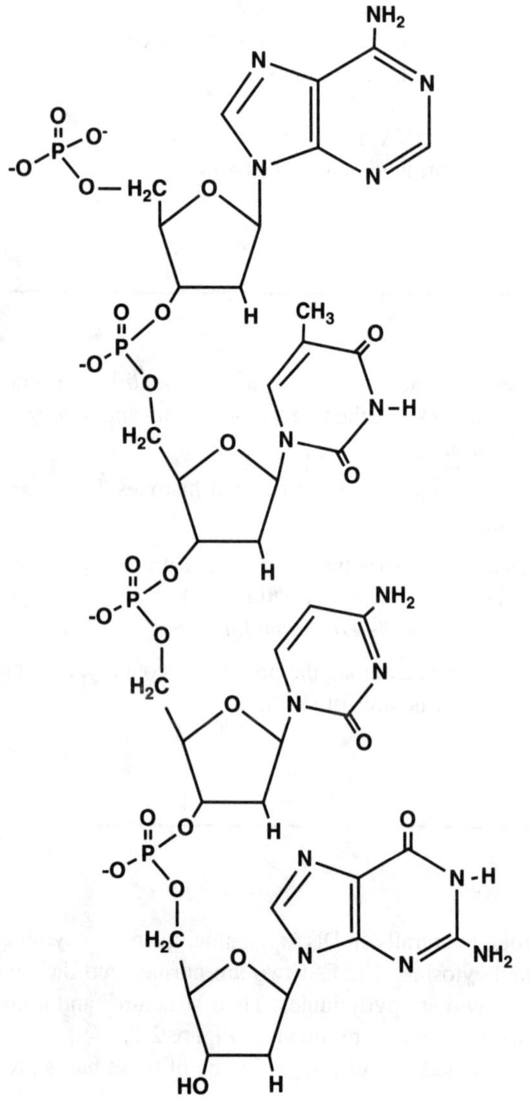

The addition of a phosphate to a nucleoside produces a **nucleotide.** Unless otherwise specified, the position of the phosphate is assumed to be at the 5′ position of the sugar. The names of the nucleotides found in DNA are 2′-deoxyadenosine monophosphate (dAMP), 2′-deoxyguanosine monophosphate (dGMP), 2′-deoxycytidine monophosphate (dCMP), and thymidine monophosphate (TMP). The structures of the nucleosides and nucleotides found in DNA are shown in Figure 2-2.

In DNA, the nucleotides are attached to one another through **phosphodiester bonds** in which a phosphate connects the 3′-OH group of one nucleotide to the 5′-OH of another nucleotide. In Figure 2-1, notice that the two ends of the DNA molecule are different. At one end (the one at the top in the figure), the 5′-OH group is not attached to any other nucleotide. However, it may be attached to a phosphate group; this is the 5′ end. At the other end, the 3′-OH group is not attached to any other nucleotide; this is the 3′ end. In moving from one nucleotide to another, you can move toward the 5′ end, in the 5′ direction, or in the opposite direction toward the 3′ end, in the 3′ direction. When a molecule has these directional properties, it is said to have **polarity.** The individual bases attached to the sugar phosphate backbone determine the information content of the DNA. It is the order of these bases or nucleotides that is the **primary structure** of DNA.

Most of the time, it is unnecessary to draw the entire structure of DNA when the only information needed is the order of the nucleotides. Several shorthand methods of drawing DNA have been developed. The same DNA that is shown in Figure 2-1 can also be drawn as follows:

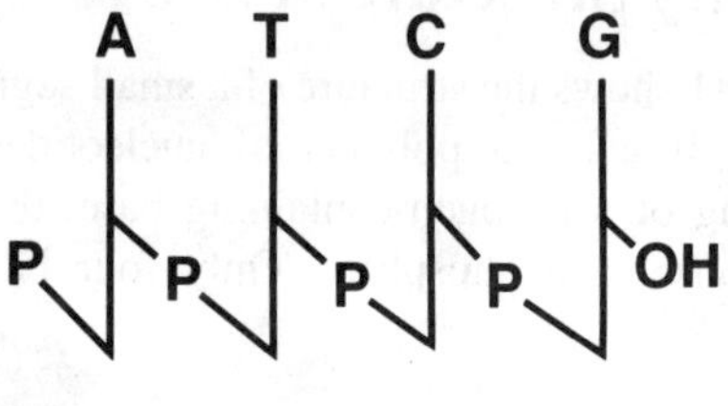

Fig. 2-1. Structure of a small segment of DNA. The 5′ end of the molecule is at the top and the 3′ end at the bottom of the structure.

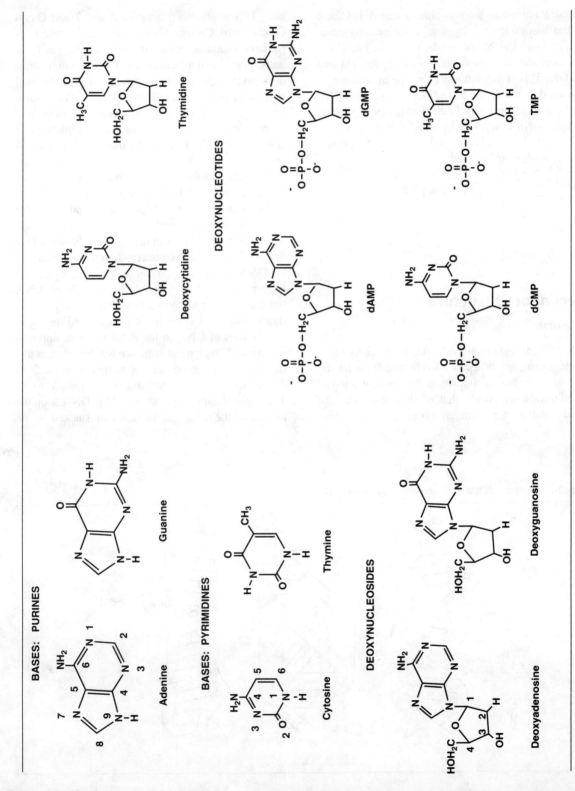

Fig. 2-2. Structures of the bases, nucleosides, and nucleotides found in DNA.

where P represents the phosphate ester; A,T, C, and G the bases adenine, thymine, cytosine, and guanine, and the straight line the deoxyribose. The 1′ position of the sugar is at the top of the straight line and the 5′ is at the bottom. This type of structure is always drawn with the 5′ end of the DNA on the left and the 3′ on the right. In all the shorthand characterizations of DNA, it is assumed that the 5′ end is on the left. Using this convention, the DNA structure can be further simplified to

pApTpCpG

or even

ATCG

Secondary Structure

B Form

The DNA in eukaryotic cells has an extensive secondary structure. When isolated from cell nuclei and analyzed for base composition, the concentration of adenine always equals that of thymidine and the concentration of guanine always equals that of cyto-

sine. This relationship between A and T and C and G, known as **Chargaff's rule,** is found in most organisms including some viruses, bacteria, yeast, insects, plants, and all mammals. There are no restrictions on the amount of any base as long as the total number of purines (A + G) is equal to the total number of pyrimidines (C + T). In neurospora, the G + C content equals 54%, whereas in tetrahymena it equals only 29%. In humans, the G + C content is approximately 41%.

Using this base content information, x-ray diffraction data, and models of hydrogen bonding between bases, Watson and Crick proposed a structure for DNA, which is designated as the *B form* of DNA. A small segment of DNA in this form is shown in Figure 2-3. This is the **secondary structure** for most of the DNA in cells. As you can see from the figure, the B form of DNA has two linear strands of DNA twisted together to form a right-handed helix . In a right-handed helix, the helix rises toward the right. The strands of DNA in the B form run in opposite directions. This type of structure is referred to as **antiparallel.** One strand runs top to bottom in the 5′ to 3′ direction, and the other strand runs runs 3′ to 5′. The sugar phosphate backbone of the DNA is on the outside of the helix and the bases are stacked on the

Fig. 2-3. Structures of the A, B, and Z helical forms of DNA.

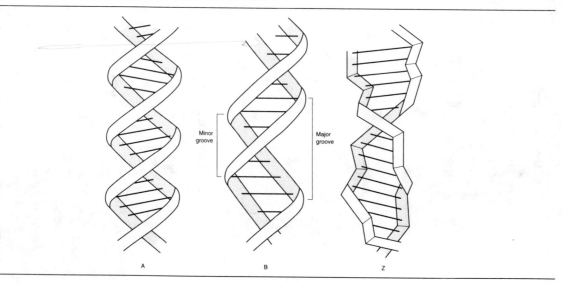

A   B   Z

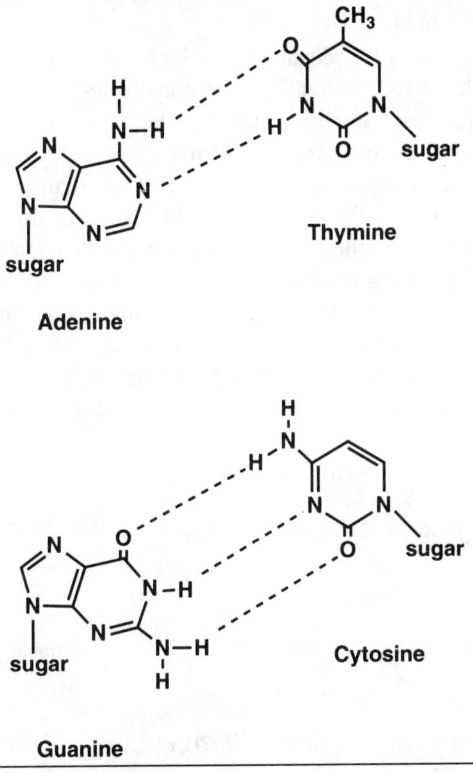

Fig. 2-4. Hydrogen bonding between the complementary bases in DNA. Adenine forms two hydrogen bonds with thymine, and guanine forms three hydrogen bonds with cytosine.

inside, away from the surrounding aqueous environment. In the B form of DNA, the bases are stacked almost perpendicular to the axis of the helix.

Hydrogen bonds between the bases and **stacking interactions** between the bases are the main forces holding the DNA helix together. Adenine can form two hydrogen bonds with thymime and guanine can form three hydrogen bonds with cytosine, as shown in Figure 2-4. This arrangement of hydrogen bonds makes the two base pairs the same size—approximately 1.08 nm across—and gives them the same angles with the sugar phosphate backbone. A and T are called **complementary bases,** as are G and C, because wherever there is an A in one strand of DNA, there is a T in the other. Likewise, a G in one strand always corresponds to a C in the other.

The base stacking interactions in the B helix consist of π bonds and hydrophobic interactions be-

tween the flat aromatic bases, which are stacked one on top of another in the helix. This interaction contributes approximately 1 kcal per base pair to the stability of the B helix, which is considerably less than the contribution of the hydrogen bonds between bases.

In the B form of DNA, there is a repeat distance of 3.4 nm, with 10 base pairs per turn of the helix. The entire structure is only 2.4 nm wide but may be several millimeters long. The B form of DNA also contains two grooves, one major and one minor. The major groove runs next to N and O atoms of the bases, and the arrangement of these atoms depends on the order of the bases. Proteins associated with specific functions of DNA, such as replication and transcription, can enter these grooves and have a direct contact with specific nucleotide sequences. In contrast, the minor groove does not show any pattern with different base sequences. It is less likely that specific base interaction occurs with proteins in the minor groove.

A Form

Although the B form of DNA predominates in cells, it is not the only form in which DNA is found. The A form, shown in Figure 2-3, can be produced by dehydration of the B form. Both the A and the B forms of DNA have the same hydrogen-bonding pattern between bases and are stabilized further by base stacking interactions. The A form is a more compact structure than the B, with 11 residues per turn. In the A form, the bases are tilted 13 to 19 degrees from the perpendicular. In addition, the grooves are altered. The minor groove is very shallow, and the major groove is extremely deep. These changes make specific protein-DNA interactions much less likely in the A form.

Z Form

The most recently discovered form of DNA is the Z (for "zigzag") form, which also is shown in Figure 2-3. Unlike the A and B forms, the Z form is a left-handed helix (helix rises to the left) with 12 nucleotides per turn. The nucleotides in the Z form rise in the helix in units of two, producing the zigzag configuration. The Z form of DNA originally was found to be the form taken by a synthetic polymer of repeating G and C nucleotides. However, cells have been shown to contain proteins capable of binding to Z DNA and other proteins that can convert the B form of DNA to the Z form. The Z form is stabilized by methylation of the bases, an event that occurs naturally in inactive genes. Table 2-1 summarizes the major properties of the three main forms of DNA.

Table 2-1. Characteristics of DNA helices

	A	B	Z
Rotation	Right-handed	Right-handed	Left-handed
Diameter (nm)	25.5	24	18.4
Nucleotides/ turn	11	10	12
Helix rise/ nucleotide (nm)	2.5	3.4	3.7
Pitch	2.5	3.3	4.5

Denaturation

Conditions for Denaturation

The secondary helical structure of DNA is stable under conditions where the hydrogen bonds are stable. If the hydrogen bonds are disrupted, the two strands of DNA can separate. A variety of conditions exist under which this can occur. These include increased temperature, extremes of pH, and the presence of compounds that disrupt hydrogen bonds or intercalate between the bases.

Each DNA has a specific temperature at which the double-stranded structure is disrupted. At this temperature, the DNA is said to melt. The specific temperature at which 50% of the helix is melted is called the **melting temperature** (Tm) for that particular DNA. The Tm is linearly related to the G + C content of the DNA. The higher the G + C content, the higher the melting temperature, because it takes more energy (a higher temperature) to break three hydrogen bonds than it does to break two.

The structure of DNA can also be disrupted by extremes of pH. DNA is stable between pH 4 and pH 10. Outside of this pH range, it is denatured due to the protonation of the ring nitrogen atoms or the attached oxygen atoms at low pH or the ionization of the OH groups at high pH. When this occurs,

the hydrogen bonds are disrupted, and the DNA is denatured.

DNA is not denatured by high concentrations of salt. In fact, strong salt solutions have the opposite effect and stabilize the structure. However, compounds such as urea and formamide that disrupt hydrogen bonds can denature DNA at temperatures well below the normal Tm. In addition, large flat compounds containing aromatic rings are capable of intercalating between the bases of DNA. This interaction can denature the DNA by stretching and untwisting the helix, which will cause the strands eventually to separate. **Ethidium bromide** and the **acridine** dyes are compounds that have this property.

Ethidium Bromide   Acridine

Physical Property Changes on Denaturation

When denaturation occurs, many physical properties of the DNA molecule are altered. Among these are the absorbance at 260 nm, viscosity, and buoyant density. Nucleotides, especially the purines, absorb light strongly at 260 nm. This absorbance is decreased by the base stacking interactions in double-stranded DNA. When DNA is denatured, the stacking is eliminated and the absorbance increases. This increase in absorbance is called **hyperchromicity.** For an average DNA, the increase in absorbance at 260 nm is approximately 40%. The absorbance increase is one of the easiest methods by which to measure the denaturation of DNA. Because adenosine monophosphate (AMP) has the highest absorbance at 260 nm, the extent of hyperchromicity is linearly related to the amount of AMP in the DNA.

Viscosity measures the length and stiffness of a molecule. When DNA is denatured, it is converted from a long stiff molecule to two random coils, and its viscosity decreases. **Buoyant density** is measured by centrifugation through a gradient of cesium chloride. Double-stranded DNA has a greater buoyant density than does single-stranded DNA. The buoyant density of DNA depends on the G + C content of the DNA sample. A DNA with a high guanine content bands at a higher density than does DNA with a lower guanine content, regardless of whether one is comparing single-stranded or double-stranded DNA samples.

Renaturation

Just as DNA can be denatured by altering temperature, pH, or urea concentrations, it can be renatured by reversing these conditions. Unlike denaturation, the process of renaturation or reannealing is a very slow one. In native helical DNA, the bases are re-aligned with A across from T and G across from C. Once this alignment is lost, it must be restored before the helix can reform. The necessary realignment occurs only by random collision of the single strands of DNA. If by chance, after a collision, the bases are again lined up properly, the structure will reform a double helix rapidly. The more concentrated a solution of DNA, the more likely it is that the complementary strands will collide and become properly aligned. The higher the **complexity** of the DNA—that is, the larger the number of different nucleotide sequences it contains—the longer it will take for any two complementary strands to collide. Thus, at the same concentration, the reaction that is slower contains the more complex DNA sample.

The rate of renaturation of DNA has been used to gain information about the number and complexity of different abundance classes in a given DNA sample. Different abundance classes may contain DNA that occurs only once per cell (unique-sequence DNA) or DNA that occurs in multiple copies (repetitive-sequence DNA). Renaturation experiments with human DNA have shown that only 50 to 60% of human DNA in each cell is unique or single-copy. This is the DNA that codes for proteins. Twenty percent of human DNA has several hundred copies per cell and codes for ribosomal RNA, transfer RNA, and the histone proteins. The remainder of the DNA has several thousand copies per cell.

Part of this highly repeated DNA includes the short sequences found at the centromeres of chromosomes. This DNA is a sequence of six nucleotides repeated in tandem hundreds of times in each chromosome. Because this DNA has a different G + C content from the bulk of the DNA in the cell, it bands on a cesium gradient at a different position from the rest of the DNA. Hence, it is called **satellite DNA.** The ends or **telomeres** of chromosomes also contain short repeated DNA sequences, which are part of the highly repeated DNA of the cell. In humans, the repeated sequence at the telomere is AGGGTT. This DNA also forms a satellite band on cesium chloride gradients. The DNA at the centromere and telomere probably has a structural function, but the function of the rest of the highly repeated DNA is unknown.

Renaturation of mixtures of DNA from different sources is one way to determine how closely the DNA samples are related in their base sequence. This method has been used in the past to construct relationships between species. Now such relationships are determined by comparing the nucleotide sequences of particular genes. The ability of DNA to form hybrid molecules with DNAs or RNAs that have the same or similar nucleotide sequences is the basis of most of the recombinant DNA techniques. These techniques, which will be discussed in Chapter 9, are important for understanding the molecular basis of disease.

Chromosome Structure

If stretched out in a single molecule in the B form, human DNA would measure approximately 1 meter long. This is obviously too large to fit into a nucleus only 10 μm in diameter. The mitochondrial DNA also is too large to fit into the mitochondria. DNA can become more compact either by twisting around itself or by winding around proteins. Mitochondrial DNA is made more compact by twisting the circular DNA around itself. Nuclear DNA is made more compact by first wrapping the DNA around protein and then twisting this structure around itself to form still more compact structures.

Mitochondrial Chromosomes

In humans, each mitochondrial DNA takes the form of a small circular chromosome of 16,569 base pairs. The DNA sequence of this chromosome has been completely determined. It contains the information for the amino acid sequence of 13 proteins, ribosomal RNA, and 22 transfer RNAs. Each mitochondrion contains 4 to 10 copies of this chromosome, and there are from 2 to more than 100 mitochondria in each cell. The circular mitochondrial DNA is found in the B form, but the DNA is structured as though it had been partially untwisted before the circle was closed. Because the B form of DNA is very stable, the DNA reforms with the expected number of helical turns (1 for each 10 base pairs). Additional twists, in the opposite direction from the helical twists, must occur in the structure as a whole. The addition or subtraction of helical turns in a circular molecule to produce one with turns in the molecule as a whole is called **supercoiling** and is shown in Figure 2-5.

For a helix of 16,560 base pairs, there are normally 10 base pairs of DNA per turn, for a total of 1656 turns. If there are more than 1656 turns, the structure is positively supercoiled and is more compact. Similarly, if there are fewer than 1656 turns, the structure is negatively supercoiled and again is more compact. Most natural DNAs that exhibit supercoiling are negatively supercoiled and contain approximately 1 superhelical turn per 20 helical turns. For mitochondrial DNA, this translates to approximately 83 superhelical turns per mitochondrial DNA molecule.

The enzymes responsible for producing and removing supercoils are **topoisomerases.** Type I topoisomerase breaks one strand of DNA, allows the DNA to untwist, and then reforms the broken phosphodiester bond. This type of topoisomerase can release stress in the DNA molecule by removing supercoils, but it cannot add superhelical turns. Type II topoisomerases, also called *DNA gyrases,* use the energy from the hydrolysis of adenosine triphosphate (ATP) to break both strands of DNA. These enzymes then add or subtract helical turns and reseal the circular molecule. We will encounter topoisomerases again in our study of DNA replication.

Eukaryotic Chromosomes

Eukaryotic nuclear DNA has a complex and very compact structure. In addition to DNA, chromosomes contain, by weight, 60% protein and 10% RNA. Human DNA is split into 46 chromosomes of varying length. Each chromosome contains a single linear piece of DNA. This DNA is complexed with positive ions, polyamines, and protein so that it is 8000 times more compact than the B form of DNA.

Under the electron microscope, salt-extracted eukaryotic nuclear chromosomes resemble beads on a string (Fig. 2-6). On analysis, the beads, called **nucleosomes,** have been shown to contain a segment of 146 base pairs of DNA and two each of four small basic proteins called **histones.** Table 2-2 gives the names, sizes, and locations of the histones. The four histones found in the nucleosome—H2A, H2B, H3, and H4—contain 100 to 135 amino acids, with a high proportion of arginine and lysine. Because these amino acids are positively charged at cellular pH, they can form salt bridges with the negative charges on the phosphate groups on DNA. They play an important structural role, and very little difference is found among the amino acid sequences of histones from different sources. For example, for histone H3 there are only two differences in amino acids be-

Fig. 2-5. Supercoiling of mitochondrial DNA. This is a negative supercoil as there are fewer total twists than would be expected for the B form of DNA with the same number of base pairs.

Table 2-2. Properties of histones

Histone	Number of amino acids	Location
H1	215	Linker
H2A	129	Nucleosome core
H2B	125	Nucleosome core
H3	135	Nucleosome core
H4	102	Nucleosome core

tween the common garden pea and the domestic cow. The substitutions that do occur are with amino acids with similar physical and chemical properties—lysine for histidine and valine for alanine.

Figure 2-7 depicts a model for a nucleosome. In this model, the eight histones create a core around which the 146 base pairs of DNA are wrapped. No specific base sequence is required for the interaction between DNA and the histone core. Even viral or bacterial DNA, which is not normally found in nucleo-

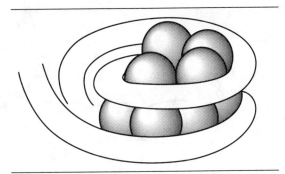

Fig. 2-7. Model of a nucleosome. The eight histone molecules form a core around which 146 base pairs of DNA are wrapped.

somes, will form this structure when incubated with histones under the correct conditions. Wrapping the DNA around the histone core requires that it be kinked at two points. Because of this, it does not maintain a B form all the way around the nucleosome.

Histone H1 is not part of the nucleosome core. It is much larger than the other four histones and shows more variation among species. H1 is located on the outside of nucleosomes and is also associated with the DNA found between the nucleosomes (linker DNA). H1 appears to stabilize the DNA strands that are entering and leaving the nucleosome. The length of DNA in the linker region between nucleosomes varies between 20 and 80 base pairs and appears to depend on the activity of the genes in that segment of DNA. Wrapping B-form DNA around a nucleosome core is the **tertiary structure** of DNA in eukaryotes and decreases the size by a factor of five to seven.

Figure 2-8 is a diagram of the higher orders of structure that make up a eukaryotic chromosome. First, the nucleosomes are packed into a 10-nm chromatin fiber, which decreases the size by a factor of five. Then this fiber is wound into a left-handed solenoid with a diameter of 30 nm. The solenoid has six nucleosomes per turn, with the nucleosomes oriented so that histone H1 is in the solenoid's center. This decreases the size by a factor of 30. The solenoid, in turn, is twisted into a supercoiled helix with a diameter of 200 nm. Finally, this helix is found in loops of 35 to 100 kilobases attached to a protein scaffold, forming a chromatid that is approximately

Fig. 2-6. Electron micrograph of salt-extracted eukaryotic nuclear chromosome showing beads-on-a-string structure. The arrow indicates a point of replication. (From D. Riley and H. Weintraub. Conservative segregation of parental histones during replication in the presence of cycloheximide. *Proc. Natl. Acad. Sci. USA.* 76: 331, 1979.)

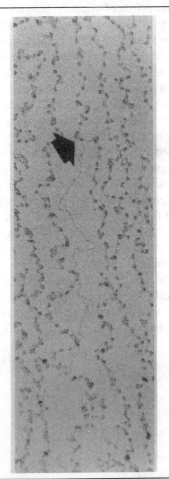

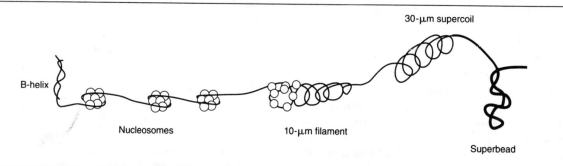

Fig. 2-8. Higher orders of structure for a eukaryotic chromosome.

700 nm in diameter. This structure, which is visible only in metaphase chromosomes, completes the final packing of nuclear DNA. One of the major proteins forming the nuclear scaffold is topoisomerase II. It is not yet clear whether the role of topoisomerase in the scaffold is enzymatic or structural. In its final form, the DNA is 8000 times more compact than the original B helical form.

Inactive DNA is dense and is not easily digested by nucleases. It is called **heterochromatin.** Genes that are active have an altered, more open structure and are easily digested by nucleases. This DNA is called **euchromatin.** Under the electron microscope, euchromatin is less dense and not easily visible. In females, the inactive X chromosome is in heterochromatin, whereas most of the active X chromosome is in euchromatin.

Summary

The primary structure of DNA is the sequence of nucleotides. The secondary structure found in cells is a right-handed helix called the *B form*, which is held together by hydrogen bonds between the bases and by hydrophobic interactions among the stacked bases. Denaturation by temperature, change in pH, or hydrogen bond–disrupting agents can change the physical properties of viscosity, buoyant density, and absorbance of ultraviolet light. The primary structure is not altered by denaturation. Short tandemly repeated DNA sequences, such as those found in centromeres and telomeres, form satellite bands on cesium chloride gradients because they have a different G + C content from the bulk of the DNA. Renaturation can occur when the denaturing conditions are removed. It is a slow process and depends on the concentration and complexity of the DNA.

Mitochondrial DNA is a circular, negatively supercoiled structure. The supercoils are formed by enzymes called *topoisomerases.* Nuclear DNA is linear and found in nucleosomes in which 146 base pairs of DNA are wrapped around a core of eight histone molecules. This, in turn, is in a 10-nm fiber that is wrapped into a 30-nm solenoid. Finally, the solenoid is in a 200-nm superhelix that is looped onto a scaffold, forming the metaphase chromosome. The final packing is 8000 times more compact than the original B helix.

3 DNA Replication

During the average human lifetime, the single original cell will divide to produce more than a billion cells. For each of these cell divisions, the DNA must be replicated with great fidelity and be equally partitioned between the daughter cells. The mitochondrial DNA must also be replicated before the mitochondria can divide. Because of their importance to the survival of the individual, these processes are tightly controlled and involve a large number of enzymatic steps.

In this chapter, we present the basic process of DNA replication and the enzymes that catalyze these reactions. We begin with mitochondrial DNA and move on to the more complicated replication of bacterial and human chromosomes. The chapter ends with a discussion of the cell cycle and its control.

Objectives

After completing this chapter, you should be able to

Describe the reactions catalyzed by DNA polymerases.

Differentiate between the DNA polymerases of prokaryotic and eukaryotic cells.

Describe the replication of mitochondrial DNA.

Define the terms *semiconservative, template, primer, helicase, single-stranded binding protein,* and *ligase.*

Describe the replication of a bacterial chromosome. Define the terms *leading strand, lagging strand,* and *Okazaki fragment.*

Describe the replication of a nuclear chromosome. Define the terms *replicon* and *telomerase.*

Outline the steps of DNA synthesis in the order in which they occur.

Match the following inhibitors of DNA replication with their site of action: cyclophosphamide, busulfan, cisplatin, etoposide, doxorubicin, vinca alkaloids.

List the major events that occur during the stages of the cell cycle and describe the rate-controlling events.

DNA Replication: Overview

The B form of DNA suggests a simple mechanism for replication. Because a base on one strand of DNA is always paired with one and only one base on

the other strand (A with T and G with C), lining up nucleotides using base pairing with the nucleotides on the parental strand should result in the formation of two double strands of DNA that are identical to the original parental strands (Fig. 3-1). This, with a few complications, is what actually happens. After replication, each of the new DNAs contains one of the original parental strands and one newly made strand. This type of replication is called **semiconservative.** If the replication has been carried out accurately, the new DNAs contain exactly the same information as the original parental DNA.

Fig. 3-1. Semiconservative replication of DNA. Each replicated DNA contains one of the original DNA strands (parental strands shown as thin black lines) and one newly made strand (daughter strand shown as thick black lines).

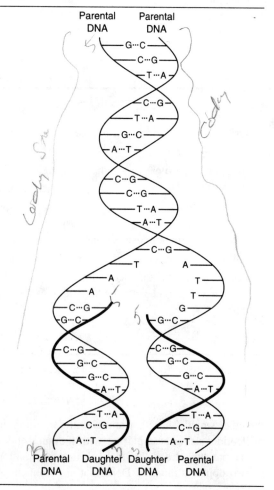

Parental DNA   Parental DNA

Parental DNA   Daughter DNA   Daughter DNA   Parental DNA

DNA Polymerases

The DNA replication is catalyzed by enzymes called **DNA polymerases.** The mechanism was first elucidated in bacteria. Fortunately, the process is similar in mitochondria and in the nuclei of cells of higher organisms. The DNA strand being synthesized, the daughter strand, is made from the 5′ end to the 3′ end using deoxynucleoside triphosphates. The strand being copied, the **template strand,** is read in the opposite direction, 3′ to 5′. The base of each new nucleotide is base-paired with its complementary base (A with T, G with C) on the template. Then, the α phosphate of the incoming nucleotide reacts with the free 3′-OH of the last nucleotide in the growing chain. The pyrophosphate lost in the reaction is immediately hydrolyzed to two molecules of inorganic phosphate by the enzyme pyrophosphatase. This reaction prevents the reversal of the synthetic reaction. The overall reaction catalyzed by DNA polymerase is illustrated in Figure 3-2.

If the selection of nucleotides were by base pairing alone, DNA would have an error rate of 1 incorrect nucleotide per 10^4 inserted. This produces far too many errors for the survival of either bacteria or humans. To reduce this error rate, many DNA polymerases possess an associated 3′ to 5′ nuclease activity that removes a nucleotide incorrectly added to

Fig. 3-2. Reaction catalyzed by DNA polymerase. The correct nucleotide is selected by base pairing with the complementary base in the template strand.

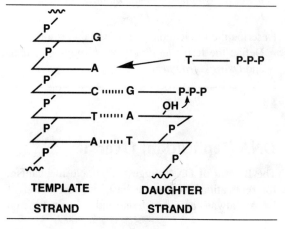

TEMPLATE STRAND   **DAUGHTER STRAND**

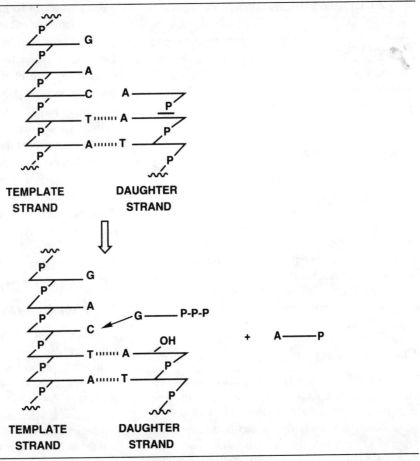

Fig. 3-3. Proofreading function of DNA polymerase. A nucleotide that has been incorrectly inserted is cleaved by a 3′ nuclease, leaving a 3′-OH and a 5′ nucleotide.

the growing DNA chain, a function known as **proofreading** (Fig. 3-3). The DNA polymerase checks each newly inserted nucleotide for fidelity of base pairing before the next nucleotide is added. If the nucleotide is incorrect, it is excised and the correct nucleotide is inserted. The addition of a proofreading mechanism produces a newly synthesized DNA molecule with only 1 incorrect base for each 10^6 inserted. This is still too high an error rate for human survival. However, there is one additional correction mechanism that functions after DNA synthesis is complete. This is part of the repair system and relies on the ability of the cell to differentiate newly made DNA from the parental DNA. We will discuss this final proofreading reaction in the next chapter.

Bacterial Enzymes

There are three well-characterized DNA polymerases found in bacterial cells. These are DNA polymerase I, DNA polymerase II, and DNA polymerase III. The properties of these enzymes are summarized in Table 3-1. DNA polymerases I and II are probably repair enzymes as cells can be deficient in these enzymes and still replicate their DNA. In each cell, there are several hundred molecules of DNA polymerase I and 50 or more molecules of DNA polymerase II. There are only 10 molecules of DNA polymerase III, which is the actual replicating enzyme in bacteria. In its most active form, DNA polymerase III contains nine individual protein subunits and can synthesize DNA

Table 3-1. Properties of bacterial DNA polymerases

	I	II	III
Molecular weight	110,000	120,000	180,000
Molecules/ccll	400	50	10
Proofreading activity (3′ nuclease)	Yes	Yes	Yes
Primer removal activity (5′ nuclease)	Yes	No	Yes
Function	Repair, primer removal	Repair	Replication

at the rate of 10^5 nucleotides per minute. All the bacterial DNA polymerases contain both a synthetic activity and a proofreading activity. In addition, DNA polymerase I has a 5′ to 3′ nuclease activity that can act on either RNA or DNA. The function of this activity will become clear as we discuss DNA synthesis in more detail.

Eukaryotic Enzymes

There are at least five DNA polymerases in eukaryotic cells: α, β, γ, δ, and ε. The properties of these enzymes are summarized in Table 3-2. DNA polymerase α and the newly discovered δ are the actual replicating enzymes for nuclear DNA. These enzymes are complex proteins with multiple subunits. DNA polymerases β and ε, like polymerases I and II of bacteria, appear to be repair enzymes. DNA polymerase γ is found in the mitochondria and is the replicating enzyme for mitochondrial DNA. In vivo, the rate of polymerization by DNA polymerase α is 500 to 5000 nucleotides per minute. This is a rate 10 to 100 times slower than for the bacterial enzyme DNA polymerase III. The slower rate may be due to the

enzyme itself or to interference by the more complex structure of eukaryotic chromosomal DNA.

Unlike the bacterial enzymes, most eukaryotic DNA polymerases have no associated nuclease activity. Only one of the polymerases isolated from mammalian cells, DNA polymerase δ, contains an associated proofreading activity. However, there is definitely a proofreading activity in eukaryotic cells; it is just not part of the polymerase molecule as it is isolated. None of the eukaryotic nuclear DNA polymerases have a 5′ to 3′ nuclease activity.

All DNA polymerases require a template, all four deoxynucleotide triphosphates, and a divalent ion (either Mg or Mn) in order to replicate a natural DNA.

Mechanism of Replication of Circular Chromosomes

Mitochondrial DNA Replication

In mitochondria, DNA replication begins at a single point on the circular chromosome. One of the strands is replicated until it is two-thirds completed; only then does the synthesis of the other strand begin (Fig. 3-4). DNA polymerase γ, like other DNA polymerases, is a copying enzyme that adds deoxynucleotides to a DNA strand which is hydrogen-bonded to the template strand. It is not capable of beginning new strands; it can only add deoxynucleotides to preexisting chains or primers that have a free 3′-OH group. How, then, does DNA synthesis get started? All new DNA chains begin with a short segment of RNA that is hydrogen-bonded to the template. RNA has a structure similar to DNA, and its bases follow similar base-pairing rules. Details of RNA structure will be discussed in Chapter 5. The priming RNA is synthesized by an enzyme called a

Table 3-2. Properties of eukaryotic DNA polymerases

	α	β	γ	δ	ε
Location	Nucleus	Nucleus	Mitochondria	Nucleus	Nucleus
Proofreading (3′ nuclease)	No	No	Yes	Yes	?
Primer removal (5′ nuclease)	No	No	Yes	No	No
Function	Replication of lagging strand	Repair	Replication of mitochondrial DNA	Replication of leading strand	Repair

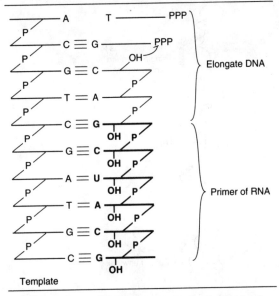

Fig. 3-5. RNA priming of DNA synthesis. Synthesis of each DNA strand begins with a 10- to 12-nucleotide segment of RNA. Then deoxynucleotides are added to this primer.

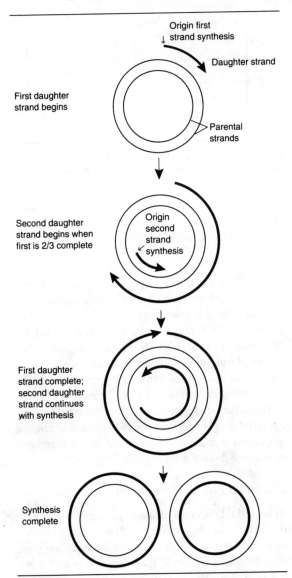

Fig. 3-4. Replication of a mitochondrial chromosome. One strand of the parental DNA is replicated from a unique start point on the circular chromosome. When this strand is two-thirds replicated, synthesis of the second daughter strand begins. The parental strands are shown as thin black lines, the daughter strands as thick black lines.

primase. After a primer segment of 10 to 12 nucleotides is synthesized, DNA polymerase can then add deoxynucleotides to the free 3'-OH of the **primer.** The priming reaction is shown in Figure 3-5.

Before the priming reaction can begin, the paren-

tal strands of DNA must be separated. This is accomplished by enzymes called **helicases.** These enzymes use the energy of ATP to denature the DNA helix, starting at a region with a high A + T content. After the DNA strands have been separated, **single-stranded DNA binding proteins** bind to the DNA to prevent it from returning to the double-helical structure (Fig. 3-6).

Simple unwinding of the double helix presents a topological problem for replication. As a segment of DNA unwinds, the helical turns are moved to other parts of the DNA, causing them to be overwound. This produces a positively supercoiled structure. Unless these supercoils are removed, further DNA unwinding and further DNA replication are not possible. **Topoisomerases,** the enzymes responsible for supercoiling of the mitochondrial chromosome, add negative supercoils to counteract the positive supercoils produced by the DNA unwinding. One negative supercoil must be added for each 10 base pairs untwisted.

Because there is no RNA in mature DNA, it is clear that the RNA must be removed at an early stage of DNA synthesis. The primer is removed by a nuclease, and the gaps left are filled in by a DNA

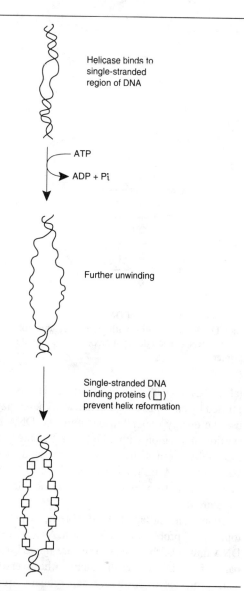

Fig. 3-6. Function of helicase and single-stranded DNA binding proteins. Helicases use the energy from the hydrolysis of ATP to untwist DNA. The single-stranded DNA binding proteins prevent the DNA helix from reforming.

Helicase binds to single-stranded region of DNA

ATP

ADP + Pi

Further unwinding

Single-stranded DNA binding proteins (□) prevent helix reformation

polymerase using the 3'-OH of the adjacent DNA as a primer (Fig. 3-7). When the gap is filled, only a single phosphodiester bond is left separating the newly synthesized segments of DNA. This final bond is formed by the action of **DNA ligase.**

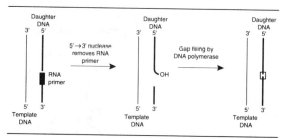

Fig. 3-7. Primer removal and gap filling in DNA synthesis. The RNA primer is removed by a 5' nuclease, one of the activities of bacterial DNA polymerase I. DNA polymerase then fills the gap using the 3' end of the next segment of DNA. The structure has a single unattached phosphodiester bond at the site where the RNA primer had been joined covalently to the DNA.

DNA ligase can join two segments of DNA only when they are separated by a single phosphodiester bond that has a 5' phosphate group and a free 3'-OH group. All DNA ligases require a source of AMP for their enzymatic activity. In prokaryotes, the AMP comes from nicotinamide adenine dinucleotide (NAD); in eukaryotes from ATP. Figure 3-8 shows the mechanism of action of DNA ligase.

When the newly replicated DNA returns to a helical form, a topoisomerase again is needed to remove the negative supertwists produced in the formation of the new double helix.

Bacterial DNA Replication

In bacteria, DNA replication begins at a single point on the circular chromosome and proceeds in both directions at the same time. Replication ends at a point directly across the circle (180 degrees) from the origin (Fig. 3-9). Bidirectional synthesis produces two regions of replication, or **replication forks,** and requires the simultaneous activity of at least four separate molecules of DNA polymerase III. Two of these enzymes synthesize in the overall direction 5' to 3', whereas the other two appear to synthesize in the opposite direction. The apparent 3' to 5' replication presents a problem as there are no DNA polymerases that are capable of synthesizing DNA in this direction. The DNA that appears to be synthesized in the overall 3' to 5' direction is actually synthesized in small segments running 5' to 3' and then combined into larger molecules, as shown in Figure

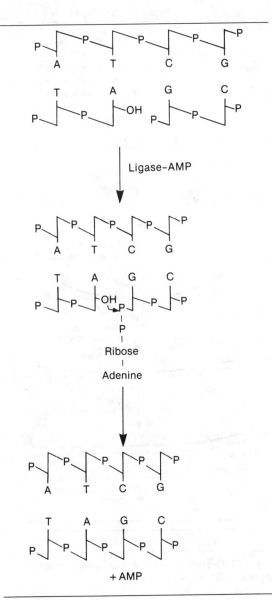

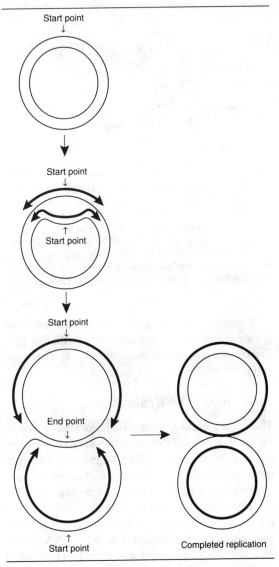

Fig. 3-8. Function of DNA ligase. The enzyme binds AMP, which then is attached to the 5′ phosphate at a single-strand break on a hydrogen-bonded DNA. The break is sealed, releasing the enzyme and AMP.

Fig. 3-9. Replication of a bacterial chromosome. Synthesis starts at a unique origin and proceeds bidirectionally around the circular chromosome. Both strands are replicated at the same time. Synthesis terminates at a point 180 degrees around the circle from the origin.

3-10. Each of these short segments is started with an RNA primer. These pieces of DNA that have a covalently linked RNA can be isolated and are called **Okazaki fragments.** Because the synthesis of these small segments must wait until a sufficiently long segment of template DNA is unwound, the DNA

strand synthesized in this way is called the **lagging strand.** The other strand is always isolated as a large piece of DNA and is called the **leading strand.** However, it too is started with a primer of RNA. In bacteria, primer removal and DNA synthesis can both be accomplished by DNA polymerase I. The 5′

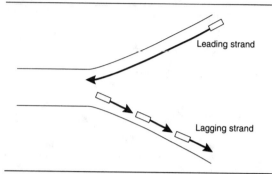

Fig. 3-10. Structure of a replication fork. In bidirectional replication, the leading strand is synthesized in the 5′ to 3′ direction as a high-molecular-weight DNA. It is primed at the 5′ end by a small segment of RNA. The lagging strand is synthesized as small segments called *Okazaki fragments,* which contain an RNA primer covalently attached to DNA. Only after removal of the primer, gap filling, and ligation is this strand made into high-molecular-weight DNA.

to 3′ nuclease activity of the enzyme removes the primer, while the polymerase activity fills the gap. The overall reaction for the replication of a bacterial chromosome is shown in Figure 3-11.

Eukaryotic DNA Replication

There are a number of additional complications in the replication of DNA in the eukaryotic cell nucleus. First, before the DNA strands can be separated by the action of helicase, the extensive tertiary and quaternary structure must be removed. In bacteria, both the leading strand and the lagging strand are synthesized by DNA polymerase III, whereas in eukaryotes, the two strands use different DNA polymerases. The leading strand appears to be synthesized by DNA polymerase δ, and the lagging strand employs DNA polymerase α. Table 3-3 compares DNA synthesis in bacteria with that in eukaryotes.

In bacteria, there is only a single replication start point on each chromosome. If there were only a single replication start point on each eukaryotic chromosome, DNA synthesis would take a week to complete. Measurements of DNA replication in a variety of eukaryotic cells show that it takes approximately 6 to 8 hours to replicate the nuclear DNA. Replication can be accomplished in this short time interval

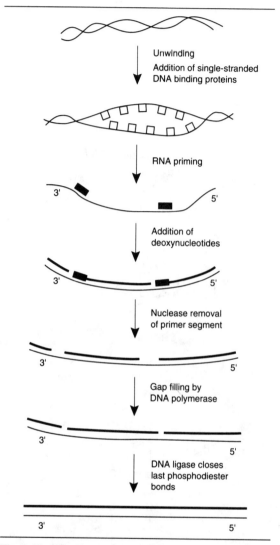

Fig. 3-11. Overall sequence of reactions necessary for replicating a bacterial chromosome.

because there are multiple replication origins on each chromosome. The origin of replication has been mapped and sequenced in mitochondria, bacteria, and many viruses. However, no specific sequence identifying the start point for DNA replication has yet been identified in the nuclear DNA of mammalian cells. Replication proceeds in both directions away from these origins and stops at a specific site some distance away (Fig. 3-12). This pro-

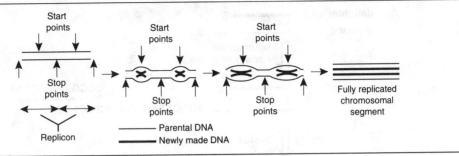

Fig. 3-12. Function of multiple replicons in eukaryotic cell chromosomes. Each replicon starts at a unique origin. Replication moves in both directions to a stop point. Adjacent replicons do not necessarily replicate at the same time

Table 3-3. Comparison of DNA replication in bacteria and the nucleus of mammalian cells

	Bacteria	Mammalian nuclei
Structure of chromosome	Circular, negative supercoil, few associated proteins	Linear, nucleosomes and higher orders of structure, many associated proteins
Number of replication origins per cell	1	10^4
Time for replication	45 min	6 to 8 hr
Replication polymerase	DNA polymerase III	DNA polymerase α for lagging strand, δ for leading strand

duces replication bubbles that eventually merge to form an entirely replicated chromosome. The segment of DNA that is replicated from a single start point is called a **replicon.**

In circular chromosomes, there is always a 3′-OH available to serve as a primer to fill in the gap left by removal of the RNA primer. When the primer is removed from the 5′ end of a newly replicated linear DNA, there is no available 3′-OH group to serve as a primer. One additional enzyme, a **telomerase,** is required to complete the synthesis of the 5′ end of linear DNA. Telomerases contain a segment of RNA as an integral part of the enzyme. This RNA can serve as a template for the synthesis of a small sequence of DNA, which is added to the parental strand of DNA. In humans, the sequence AGGGTT is added many times to the 3′ end of the parental DNA. After the addition of this sequence, the newly synthesized segment folds back on itself, forming a hairpin loop. Because it has a free 3′-OH group, it can be used as a both a primer and a template to synthesize DNA, which fills in the gap left by the removal of the RNA primer. DNA ligase then fills in the last phosphodiester bond. This reaction sequence is summarized in Figure 3-13. After both strands are fully replicated, the telomeric sequence is cleaved so that the DNA does not lengthen every time it is replicated. However, the cleavage enzyme is not entirely specific as to the amount of telomeric DNA left on the end of the DNA. The length of the telomeric sequence can vary from chromosome to chromosome and cell type to cell type. In sperm, up to 10,000 to 15,000 bases (kilobases) at the end of each chromosome are telomeric DNA.

An alternative hypothesis for the replication of the ends of linear chromosomes involves extension of the 3′ end of the parental strand by telomerase followed by normal priming of the daughter strand by RNA. DNA polymerase would fill in nucleotides and DNA ligase would close the final phosphodiester bond. Cleavage of the telomeric sequence would remove the RNA primer leaving a fully replicated DNA without any RNA.

In eukaryotes, the histones of the nucleosomes do not completely dissociate when the DNA is replicated. They tend to stay together as histone pairs and

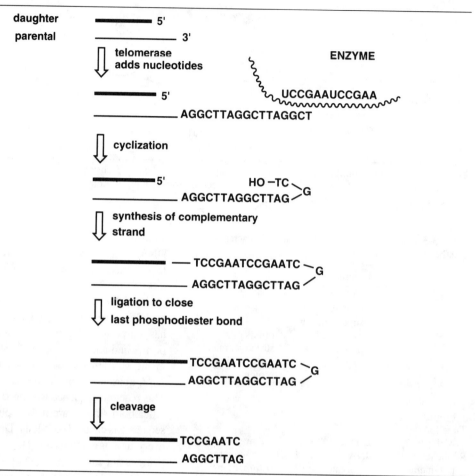

Fig. 3-13. Function of telomerases in completing the ends of linear chromosomes. The telomerase enzyme uses an RNA template, which is part of the enzyme molecule, to add a repeating sequence of nucleotides to the 3′ end of the parental strand of DNA. This forms a hairpin loop and is used as a primer and a template for synthesis of the complementary strand. After synthesis is complete, the extra sequences are cleaved.

are found associated primarily with the leading strand of DNA. Newly made histones are found on the lagging strand when it is finally ligated into a high-molecular-weight molecule.

Inhibitors

Because cancer cells replicate their DNA and divide more frequently than most normal cells, the inhibition of DNA synthesis is the target of many cancer chemotherapeutic drugs. Some such as hydroxyurea, methotrexate, and 5-fluorouracil act to decrease the concentration of deoxynucleotides. We will discuss these compounds when we cover the specific reactions they inhibit. Other drugs directly affect the replication mechanism itself. The best studied of these are the crosslinking agents. Many are related to sulfur mustard, the World War I poison gas:

$$Cl–CH_2–CH_2–S–CH_2–CH_2–Cl$$

These compounds and their nitrogen analogs readily lose Cl^-, leaving behind a positively charged carbon (carbonium ion). These can alkylate DNA primarily at the 7 position of guanine and crosslink it to other guanine residues either in the same or the opposite strand of DNA. Crosslinking also can occur between DNA and RNA or protein. The net result is DNA that cannot be properly untwisted and separated to allow for replication. Among the clinically important crosslinking agents are **nitrogen mustard, cyclophosphamide,** and **busulfan.** The structures of these compounds are shown in Figure 3-14. **Cisplatin,** an important drug for the treatment of many solid tumors, does not crosslink but does alkylate at the 7 position of guanine and disrupt hydrogen bonding in DNA.

Doxorubicin intercalates between the bases in DNA, untwisting the helix. This alters the DNA structure and its function. The drug is important for the treatment of lymphomas and cancers of the ovaries, breast, and digestive tract.

Fig. 3-14. Structure of some clinically important alkylating agents related to nitrogen mustard.

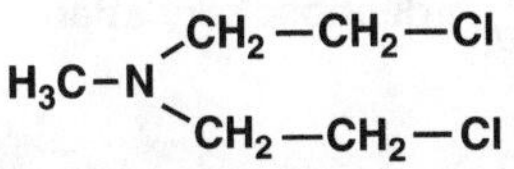

$$H_3C-N \begin{matrix} CH_2-CH_2-Cl \\ CH_2-CH_2-Cl \end{matrix}$$

Nitrogen mustard

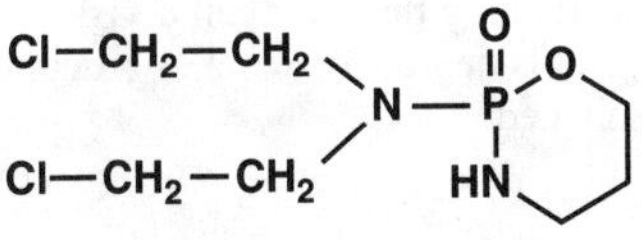

Cyclophosphamide

$$H_3C-\overset{O}{\underset{O}{\overset{\|}{\underset{\|}{S}}}}-O-CH_2-CH_2-O-\overset{O}{\underset{O}{\overset{\|}{\underset{\|}{S}}}}-CH_3$$

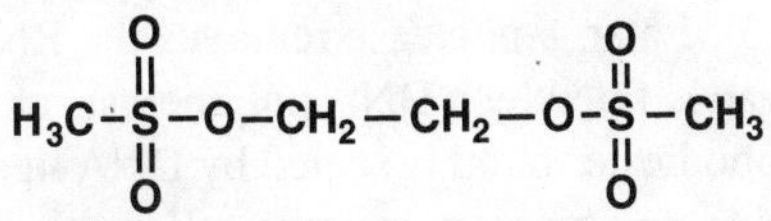

Busulfan

Inhibitors of topoisomerase II, such as **etoposide,** are being used in cancer chemotherapy as well. In the presence of these agents, the supercoils produced when DNA is being unwound for replication cannot be removed. Neither can the twists produced when the helix is reformed. This blocks DNA synthesis, and the cells die. Compounds selected to inhibit the nuclear topoisomerase have no effect on the equivalent enzyme in the mitochondria. Thus far, no clinically used drugs have been developed that target topoisomerase I, although several are in clinical trials.

Cell Cycle

In mammalian cells, DNA replication occurs only during a 6- to 8-hour period approximately 12 hours after the last cell division. This period is called **S phase** for "synthesis." The remainder of the time between cell divisions is taken up with the preparation for DNA synthesis (G_1), preparation for cell division (G_2), and the actual division itself (M). The entire period between cell divisions (G_1, G_2, S, and M) is called a **cell cycle.** For an average mammalian cell that is dividing regularly, the cell cycle lasts approximately 24 hours. Figure 3-15 shows the phases of the cell cycle and the approximate time of each phase. G_1 is the longest and most variable part of the cycle. M is the shortest. During M, the newly replicated chromosomes are attached to the mitotic spindle. The vinca alkaloids, **vincristine** and **vinblastine,** inhibit the formation of the mitotic spindle and inhibit cell division. These

Fig. 3-15. The cell cycle of eukaryotic cells. DNA is synthesized in S phase. The cell divides in M. G_1 is the gap between M and S, whereas G_2 is the gap between S and M.

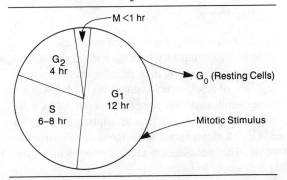

compounds are useful clinically for the treatment of leukemias and lymphomas.

Cells not undergoing regular cell division are considered to be in a phase called G_0. Under the proper stimulus, these cells can return to the cell cycle, replicate their DNA, and divide. Many fully differentiated cells, such as adult neurons, appear to be incapable of cell division. These cells remain outside the cell cycle for the remainder of their lifetime.

Progression of the cell through the cell cycle is controlled at two major points. One is prior to entry into S phase and the other is prior to M. The events at M are better understood than those at S. The major control is by phosphorylation and dephosphorylation. A protein kinase, designated p34, is dephosphorylated after binding to a second protein, cyclin B. This activates the protein kinase, which then phosphorylates a variety of substrates. These molecules, in turn, cause the condensation of the chromatin, disruption of the nuclear membrane, and alteration of the cellular architecture in preparation for cell division. After M is completed, the cyclin is degraded and the substrate molecules are dephosphorylated (Fig. 3-16). Activation of p34, necessary for the cell to proceed into mitosis, will not normally occur unless DNA synthesis has been completed.

Progression of cells into S phase requires the presence of growth factors, a low cell density, a rapid rate of protein synthesis, and the attainment of a critical cell mass. The control point is 1 to 2 hours before the beginning of S phase. The enzyme involved has not yet been identified, although there is some evidence that it may be p34 acting with different cyclins from those that function before entering M phase.

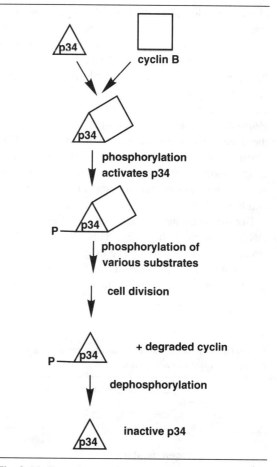

Fig. 3-16. Controls operating at the G_2 control point in the cell cycle. In late G_2, the p34 protein kinase combines with cyclin B. This allows the kinase to be phosphorylated and activated. The enzyme then phosphorylates other proteins, activating them for cell division. After the cell divides, cyclin is degraded, and p34 is dephosphorylated and inactivated.

Summary

In DNA synthesis, the separation of the parental strands is facilitated by helicase and single-stranded binding proteins. Then, a primase creates a short segment of RNA complementary to the DNA strand being replicated, the template strand. DNA polymerase adds deoxynucleoside triphosphates to the 3'-OH of this primer, with the loss of pyrophosphate. This produces an Okazaki fragment, a short segment of nucleic acid that contains covalently bonded DNA and RNA. The replicating strand of DNA is synthesized in the 5' to 3' direction. The template is read or copied in the opposite direction, 3' to 5'. Later, a nuclease removes the RNA primer. The gap is filled by a DNA polymerase, and the final phosphodiester bond is sealed by DNA ligase.

DNA polymerases require all four nucleoside triphosphates, a template of DNA, a primer of either

RNA or DNA, and a divalent cation. All DNA polymerases synthesize in the 5' to 3' direction. Prokaryotic DNA polymerases possess a nuclease activity. Most eukaryotic DNA polymerases do not. DNA polymerase α and δ are the replication enzymes for mammalian nuclear DNA, γ is the enzyme for mitochondrial DNA replication, and polymerase III for bacterial DNA. Bacterial polymerases I and II as well as eukaryotic polymerases β and ϵ are repair enzymes.

Drugs that alkylate DNA (cisplatin), intercalate into the DNA (doxorubicin), or crosslink DNA (analogs of nitrogen mustard) inhibit DNA synthesis and are used as cancer chemotherapeutic agents. Compounds, such as the vinca alkaloids, that inhibit the formation of the mitotic spindle are also useful chemotherapeutic agents.

The cell cycle is controlled at a point prior to M by the action of a protein kinase complexed with a cyclin. This allows phosphorylation of a number of proteins necessary for M phase to proceed. Control also occurs several hours before S phase begins. The enzymes functioning at this control point have not yet been identified.

4 DNA Repair and Recombination

Cellular DNA is continually being damaged by a variety of environmental and chemical agents. Before the DNA is replicated, the damage must be accurately repaired or incorrect information will be passed on to the daughter cells. To ensure the repair of the many types of damage that can occur to cellular DNA, there are a number of separate repair processes. Because of the complexity of these processes, not all the enzymes have been isolated and not all the mechanisms are clearly understood, especially in mammalian cells. However, defects in DNA repair can produce a number of medical problems, including a significantly increased risk of malignancy.

Recombination of DNA occurs by crossing over during postreplication repair and at every meiosis. Crossing over also allows the transposition of a segment of DNA from one site on a chromosome to another site, which may be on the same or a different chromosome. The production of the genes coding for specific antibodies and the integration of viruses also involves DNA recombination.

Objectives

After completing this chapter, you should be able to

List the agents that can damage DNA, the type of damage they cause, and the mechanisms that repair the damage.

Describe the events of excision repair and postreplication repair.

Describe the function of DNA glycosylase, O^6-methyl guanosine demethylase, and SOS repair.

For the inherited defects in DNA repair, match the disease with the enzyme or activity affected.

Differentiate between general recombination and site-specific recombination. Give an example of each type. Define the term *transposon*.

Types of DNA Damage

DNA is subject to damage by <u>irradiation</u> and the actions of a variety of chemicals. To keep the genetic information functioning properly, all damage must be promptly and accurately repaired. Repair must occur before the DNA is replicated, to prevent the alteration from being passed on to the daughter cells.

Table 4-1 lists a number of types of damage that can occur to DNA and the agents that cause the damage.

Living at 37°C produces a regular loss of purine bases from DNA, called **depurination.** This leaves a DNA in which the sugar backbone is intact but one base is missing. This is missing base damage. Another consequence of living at 37°C is the loss of

Table 4-1. Types of DNA damage

Type of damage	Agent
Depurination (loss of purine bases)	Heat, x-rays
Deamination (A → Hx, C → U)	Heat
Strand breaks	X-rays, chemicals
Damaged base	X-rays, alkylating agents
Dimers	Ultraviolet radiation
Incorrect base	Nucleotide incorrectly inserted during replication
Crosslinks	Chemicals

amino groups from adenine and cytosine, producing hypoxanthine (Hx) and uridine (U), respectively. This produces the incorrect base pairs, Hx:T and U:G. Because Hx base pairs with C, and U with A, if replication occurs before these bases are removed, incorrect nucleotides will be inserted into the daughter DNA across from these altered bases. Deamination will change an A:T base pair to Hx:T and then to Hx:C at the first replication. This eventually will change an A:T base pair to a G:C base pair (Fig. 4-1). Similarly, the deamination of C will change a C:G base pair to a T:A base pair. Another example of incorrectly paired bases are incorrect nucleotides added during DNA synthesis and not removed by the proofreading mechanism of DNA polymerase. As with deamination, these incorrect base pairs must be repaired before the DNA undergoes another round of replication.

Ionizing radiation in the form of x-rays or γ-rays breaks covalent bonds. This can cause several types of damage to DNA, including loss of a single base (depurination if the loss is a purine or depyrimidination if the loss is a pyrimidine), a break in the sugar phosphate backbone (single-strand break), or the more serious breakage of both strands of DNA (double-strand break). Exposure of DNA to ultraviolet light causes covalent bond formation. Specifically, it produces the formation of dimers between adjacent pyrimidine residues located on the same strand of DNA. The products can be 5,6 dimers or 6,4 photoproducts of T-T, T-C, or C-C. The structures of the thymidine photoproducts are shown in Figure 4-2.

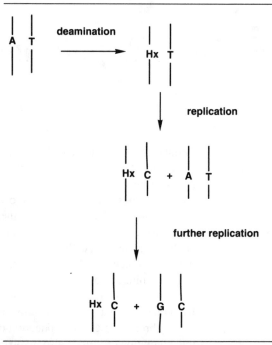

Fig. 4-1. Result of replication of DNA with unrepaired deamination of A. After the first round of replication, an Hx:T base pair is changed to Hx:C. The T base pairs normally. The next round of replication produces a G:C base pair at this site. The Hx will continue to base pair with C in further rounds of replication.

DNA also can be damaged by alkylation, with the addition of either small groups such as methyl or ethyl, or larger, more complex groups such as derivatives of benzopyrene, β-naphthylamine, or vinyl chloride. These chemicals can bind to one site on DNA, altering base pairing, or can intercalate between bases. Alternatively, chemicals can crosslink two sites on the DNA. This type of damage prevents the DNA from being replicated or transcribed into RNA. Mechanisms exist in human cells to repair all these types of damage, although the efficiency varies between active and inactive genes, between cell types, and among individuals.

Mechanisms of DNA Repair

Excision Repair

In eukaryotes, most types of DNA damage are repaired by a process called **excision repair.** A similar

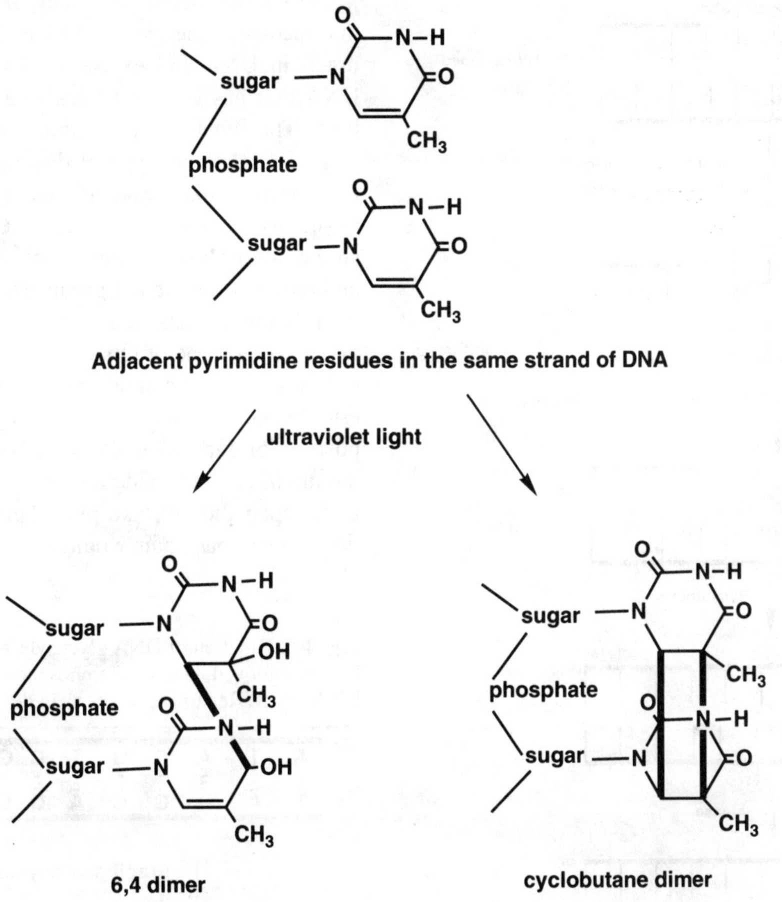

Fig. 4-2. Structure of thymidine photoproducts produced by exposure of DNA to ultraviolet light.

system operates in bacteria. In excision repair, the damage is recognized and then an endonuclease makes a nick in one strand of DNA. The nick usually is made on the 5′ side of the damage and occurs between a 3′-OH group and its adjoining phosphate. This leaves a 3′-OH group to serve as a primer during resynthesis. After the nick is made in the sugar phosphate backbone, the damaged segment of DNA is removed. This segment is only a few nucleotides in length in what is called **short-patch repair** or several hundred nucleotides long in **long-patch repair.** After removal of the damage, one of the repair DNA polymerases fills in the gap using the 3′-OH as a primer. This is a slow process because the repair polymerases dissociate from the DNA template after each nucleotide insertion. Finally, the last phospho-

diester bond is sealed by the action of DNA ligase. Damage that produces a major distortion of the DNA helix, such as a pyrimidine dimer, is repaired by long-patch repair using polymerase ε. Less distorting damage is repaired by short-patch repair using polymerase β. Figure 4-3 outlines the steps of excision repair.

DNA damage due to depurination is repaired by the short-patch excision repair system. A nick is made in the DNA on the 5′ side of the missing base. Then a few nucleotides are excised. Using the free 3′-OH, DNA polymerase β resynthesizes the DNA, and DNA ligase seals the last phosphodiester bond.

Nucleotides incorrectly added during DNA synthesis are repaired by short patch excision repair as well. The parental DNA is differentiated from the

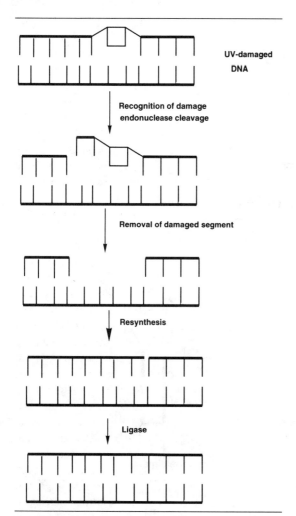

UV-damaged DNA

Recognition of damage
endonuclease cleavage

Removal of damaged segment

Resynthesis

Ligase

Fig. 4-3. Steps in excision repair. First a break is made on the 5′ side of the damage, leaving a 3′-OH. The damage is removed, the gap filled by a repair DNA polymerase, and the last bond closed by DNA ligase. (*UV* = ultraviolet light.)

newly made DNA by methylation of the 5 position of cytosine in the sequence GC in the parental strand of DNA. This modification does not change the base pairing of the cytosine. Methylation exists on the parental strand but does not occur on the daughter strand until some time after DNA replication is completed. In this way, the excision repair enzymes can recognize the correct strand of DNA to repair. This step is the final proofreading of newly replicated DNA and decreases the error rate to 1 incorrect nucleotide for each 10^8 to 10^9 nucleotides inserted.

Cells have specific enzymes, **DNA glycosylases,** that recognize the incorrect bases hypoxanthine and uracil in DNA and excise the base. This leaves a DNA that has an intact backbone and one missing base. The repair then continues as in depurination (Fig. 4-4). Another type of damage for which there is a specific repair system is methylation producing O^6-methylguanosine. A specific **O^6-methylguanosine demethylase** recognizes this altered nucleoside and removes the methyl group. This reaction returns the DNA to its unaltered form but inactivates the enzyme. In this type of repair, there are no nucleotides or bases removed and no breakage of the phosphodiester backbone. Addition of a methyl group to the O^6 position of guanosine alters the base pairing so that the modified nucleoside now pairs with T instead of C. If repair did not take place before DNA replication, a G:C base pair would be changed to an A:T

Fig. 4-4. Function of DNA glycosylases. The incorrect base is removed, leaving the phosphodiester backbone of DNA intact. Repair continues as in depurination.

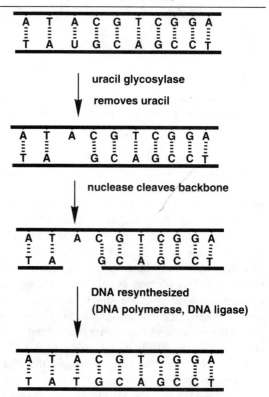

pair (Fig. 4-5). Methylation of the 6 position of G is the most frequent site of environmental methylation of DNA.

Methylation and ethylation can occur at all the N and O atoms of the bases in DNA. When such alkylations alter base pairing, they are recognized as errors and are repaired by short-patch excision repair. When small alkylations do not change base pairing, such as single methylations of the amino group at position 6 of adenosine or position 2 of guanosine, they are not recognized as errors and are not subject to repair.

A single-strand break or the loss of a single base due to x-ray damage is also repaired by short-patch excision repair. Double-strand breaks are repaired by short-patch excision repair if the breaks are not close together. Double-strand breaks directly across from one another usually are lethal. Occasionally, the break in the sugar phosphate backbone is between a 3'-OH and a 5' phosphate, then the action of a DNA ligase alone is sufficient to repair the damage.

Pyrimidine dimers are repaired in mammalian cells by long-patch repair. The damaged bases are recognized and removed from the DNA along with a large number of adjacent nucleotides. Then, the segment is resynthesized. In bacteria, a second mechanism can function for the repair of thymidine dimers. A **photoreactivating system** can split the newly made dimers and return the DNA to its original form without removing any nucleotides and without breaking the phosphodiester backbone. The enzymes of photoreactivation bind to the dimer in the dark and, when exposed to light, use that energy to split the dimer. Although many people have looked for it, there is no reproducible evidence of a photoreactivation system in human cells.

Alkylation by large groups produces a major distortion of the DNA helix and is repaired by large-patch excision repair. Many of the large alkylating agents are derived from environmental or industrial pollutants such as benzopyrene, β-naphthylamine, and vinyl chloride. These compounds must be activated by cellular metabolism, usually oxidation, before they are capable of interacting with DNA. Exposure to these chemicals is associated with an increased risk of malignancy in both experimental animals and humans. Insertion of intercalating agents, which we discussed as agents that can denature DNA, are also removed by long-patch excision repair. Although these chemicals are not covalently attached to the DNA, it requires removal of a large number of nucleotides to cause these compounds to dissociate.

Chemicals that crosslink DNA, such as nitrogen mustard, are repaired by large-patch excision repair as well. However, before repair can proceed, the first arm of the crosslink must be removed. Only then can the damaged DNA segment, which now carries the alkylating agent, be excised.

Repair of DNA is not uniform throughout the DNA helix. Euchromatin is repaired preferentially to heterochromatin. In addition, there is a preference for repairing the strand of DNA that is actually being used as a template for RNA synthesis rather than its complementary strand. This has an obvious advantage for human survival. There is also a difference in the rate of DNA repair between tissues that is not connected with the rate of gene activation. For example, cells of the liver show a much higher rate of repair than those of the brain. Because human chromosomes contain much more DNA than is ever used, unrepaired errors in the unused portions of DNA are unlikely to cause metabolic problems.

Fig. 4-5. Result of replication of DNA containing O^6-methylguanosine. Because O^6-methylguanosine base pairs with thymidine, replication changes a G : C base pair to an A : T base pair.

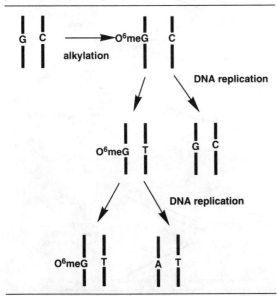

Postreplication Repair

If DNA repair is not completed by the time DNA is replicated, serious problems can develop in the transmission of genetic information to the daughter cells. In the case of the loss of a single base or an insertion, any base may be inserted across from the damage. This will produce a change in the primary structure of the DNA and a change in the genetic information. In the case of large alkylations, crosslinks, or pyrimidine dimers, synthesis may stop at the point of damage to the template strand. If synthesis stops, DNA containing gaps of nonreplicated DNA can be transmitted to the daughter cells at the time of cell division. Such gaps in the lagging strand may be fairly short, but if they are in the leading strand, the gaps can be several hundred nucleotides long. Gaps of this kind are filled by **postreplication repair.** This is a complex mechanism in which there is a crossover reaction between one parental strand and the replicating strand whose synthesis is stopped. This leaves the damage in place but allows synthesis to continue. The other newly made strand then is used as a template to fill in the gap in the parental donor strand. The mechanism is outlined in Figure 4-6. After the completion of replication, the damaged DNA is repaired by the normal excision repair process. Progress through the G_2 phase of the cell cycle is slowed under these conditions to allow the damage to be repaired.

SOS Repair

If there is extensive damage to the DNA, a different error-prone repair system is induced. This repair system, called **SOS repair,** will repair the DNA rapidly so that replication can take place. Many, but not all, the errors left by SOS repair are fixed later in G_2 by the normal excision repair process. The mechanism of SOS repair has been studied in bacteria but not extensively in humans.

Under conditions of extensive DNA damage, the cell initiates a process of programmed cell death or **apoptosis.** There is a condensation of the nucleus, shrinkage of the cell, and fragmentation of the DNA. Apoptosis occurs not only under conditions of extensive cell damage but also normally during embryonic development, hormonally induced atrophy of tissues, and lymphocyte maturation in the thymus.

Defects in Repair

There are a number of human diseases or syndromes in which affected individuals demonstrate a defect in DNA repair and have, in addition, a greatly increased risk of malignancy. All these are rare diseases that are inherited as autosomal recessive. The characteristics of the most common of these diseases are summarized in Table 4-2.

The most studied of the inherited diseases associated with defects in DNA repair is **xeroderma pigmentosum** (XP). At an early age, patients with XP show freckling of the skin in all areas exposed to the sun. Later, they develop multiple skin tumors in the exposed areas. XP patients usually die from metastasis from these skin tumors in their early twenties. Analysis of skin cells from XP patients shows that the cells are extremely sensitive to ultraviolet (UV) light. The defect in XP appears to be an inability to

Fig. 4-6. Mechanism of postreplication repair. DNA synthesis stops at the site of the damage in the template strand. The other template strand, which has a base sequence identical to the daughter strand whose synthesis has stopped, crosses over and fills in the gap. The gap in the parental strand is filled in by DNA synthesis, using the daughter strand as a template. The template strands are shown as thin black lines, the daughter strands as thick black lines.

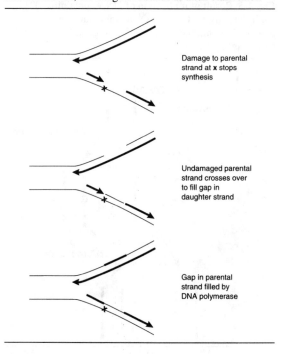

Damage to parental strand at **x** stops synthesis

Undamaged parental strand crosses over to fill gap in daughter strand

Gap in parental strand filled by DNA polymerase

Table 4-2. Some inherited human diseases that show defects in DNA repair

Disease	Increased sensitivity to	Defect	Symptoms
Xeroderma pigmentosum	Ultraviolet light	Long-patch repair of pyrimidine dimers	Photosensitivity, skin malignancies
Ataxia telangiectasia	X-rays	Short-patch repair of unrejoined strand breaks	Lymphomas, increased susceptibility to infection
Fanconi's anemia	Crosslinking agents	Removal of the first arm of the crosslink	Leukemia, anemia

recognize the UV damage and make the first endonuclease cleavage of the DNA backbone. There are at least seven different genetic defects that can produce the clinical pattern of XP. Two other syndromes, Bloom's and Cockayne's, also show an increased sensitivity to UV light and an increased cancer risk. Although showing some of the same symptoms, these diseases are genetically different from XP.

Ataxia telangiectasia (AT) is a second syndrome characterized by a defect in DNA repair and an increased risk of cancer. Affected persons are extremely sensitive to ionizing radiation such as x-rays and γ-rays. A standard chest x-ray can be fatal to someone with AT. A mammogram can greatly increase the risk of breast cancer. In addition to the sensitivity to x-rays, these individuals show a hypersensitivity to methylating agents. However, there is no increased sensitivity to UV radiation. The symptoms of AT include cerebellar ataxia, hypoplasia of the thymus, increased susceptibility to pulmonary infection, and a high incidence of lymphatic malignancy. The specific defect in AT is believed to be unrejoined single- and double-strand breaks, which normally are repaired by the short-patch repair system.

Fanconi's anemia (FA) is yet another of these defective DNA repair diseases. Patients with FA have anemia, leukopenia, hyperpigmentation of the skin, and a high incidence of leukemia. Cells isolated from FA patients respond normally to both UV and ionizing radiation. The cells appear to be hypersensitive to bifunctional alkylating agents, such as the nitrogen mustards, and are unable to remove the first arm of the crosslink.

DNA Recombination

Recombination of DNA is important for postreplication repair, meiotic crossover, and the production of the estimated 1 million possible genes coding for specific antibodies. Recombination is also the mechanism by which some viruses integrate their genes into host DNA and by which transposons, or transposable genetic elements, move from site to site.

General Recombination

In general recombination, there is a breakage and rejoining of DNA at sites that have extensive homology, such those in homologous chromosomes. Such crossing over occurs at least once for each chromosome pair at meiosis. The mechanism as presently understood is diagrammed in Figure 4-7.

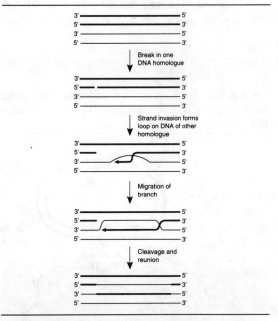

Fig. 4-7. Mechanism of general recombination. An extensive area of sequence homology between the two strands is required for general recombination to occur. One strand of DNA invades the homologous DNA and base pairs with its complementary strand. The displaced DNA loop then pairs with its complement in the other homologous DNA. The crossover migrates until breakage and reunion occur.

First, there is a nick in one strand of DNA, followed by displacement of one end of that DNA strand. This free strand then base pairs with its complementary strand in another DNA helix. The loop formed from the displacement is removed by a nuclease, and the two DNA strands are joined. Then the other free end of the nicked DNA is attached to the now free end of the second DNA strand. With an extensive area of complementary bases, the crossover can move for some distance from its point of origin. This movement requires the function of a helicase and the hydrolysis of ATP. Before the two strands of DNA can separate, the crossover must be cleaved and the resulting nicks resealed.

Site-Specific Recombination

In site-specific recombination, the complementary region is only 10 to 15 base pairs long. Here two DNAs pair at such a complementary site. Both strands are cleaved by specific enzymes that leave an overhang of 4 to 6 base pairs. The strands base pair with their complements, and the breaks are ligated. The steps in site-specific recombination are outlined in Figure 4-8. A topoisomerase is essential for the breakage and reunion of site-specific recombination.

Site-specific recombination is the mechanism used for the integration of some viruses into DNA and the movement of **transposons.** Transposons have been studied first in maize and then in bacteria, yeast, and *Drosophila.* These small segments of DNA can move from site to site, chromosome to chromosome, modifying gene expression when they integrate at new sites. This is why they are sometimes called *jumping genes.*

Human DNA contains sufficient genetic information to produce approximately 10 million different antibody molecules. These are produced by gene rearrangement in mature B cells of the immune system. To produce the light chains of immunoglobulin, 1 of approximately 500 different DNA segments coding for a variable region and 1 of 4 coding for a joining region are moved next to a segment of DNA coding for the constant region. The DNA coding for the other variable and joining regions is lost in the cell that has undergone this gene rearrangement and in all the progeny from this cell (Fig. 4-9). In other B cells, other variable and joining regions are selected.

Fig. 4-8. Steps in site-specific recombination. Only a small segment of homology is required for site-specific recombination. Both DNAs are cleaved, and recombination occurs at the homologous region.

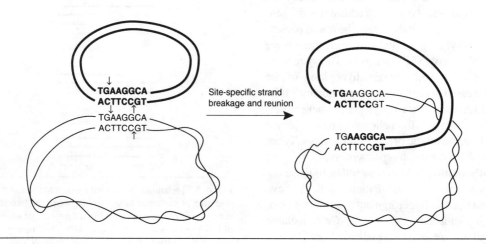

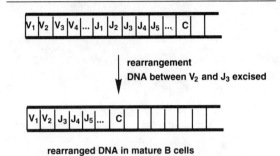

rearrangement
DNA between V_2 and J_3 excised

rearranged DNA in mature B cells

Fig. 4-9. Gene rearrangement necessary for the formation of specific antibody genes. When the rearranged gene is transcribed into RNA, one V region, all the remaining J regions, plus the C region are copied. Then all but the first of the J regions are removed in forming the mature mRNA. In this example, the mRNA will contain $V_2 J_3 C$.

Similar rearrangements occur in the production of genes coding for the heavy immunoglobulin chains, wherein one variable region is joined with one joining, one diversity, and one constant region.

Summary

Damage to DNA can be caused by heat, irradiation, or chemicals. In humans, these types of damage are repaired by excision repair in which the sugar phosphate backbone is cleaved, the damaged segment removed, and the DNA resynthesized. Large helical distortions, such as those produced by pyrimidine dimers and large alkylating groups, are repaired by the long-patch excision repair system, in which a large number of nucleotides are excised. Strand breaks and smaller distortions, such as those produced by missing bases, incorrectly paired bases, and small alkylations, are repaired by the small-patch repair system in which only a few nucleotides are removed. This is also the repair system that performs the final proofreading of newly replicated DNA.

Specific enzyme systems exist to recognize and remove base deamination products hypoxanthine and uracil from DNA and methyl groups attached to the O^6 position of guanosine.

Inability to repair DNA in patients with xeroderma pigmentosum, ataxia telangiectasia, and Fanconi's anemia produces serious clinical consequences, including an increased incidence of cancer.

General DNA recombination requires a large area of homology between two strands of DNA. This process is responsible for the crossing over that occurs at meiosis and for the strand exchange that occurs in postreplication repair. Site-specific recombination, in which there is a much smaller area of homology, is responsible for the insertion of some viruses, movement of transposons, and production of the diversity of genes coding for antibody molecules.

5 Properties of RNA

In addition to DNA, cells contain RNA, another polymer of purine and pyrimidine nucleotides. RNA transmits the genetic information stored in DNA and uses it to direct the synthesis of proteins. There are three major kinds of RNA in cells: messenger (mRNA), transfer (tRNA), and ribosomal (rRNA). mRNA carries the genetic information coded in DNA into the cytoplasm. The order of nucleotides in the mRNA determines the order of amino acids in the protein translated from it. tRNA transfers amino acids to the protein-synthesizing machinery and translates the nucleic acid "language" into amino acid "language." rRNA plus a large number of proteins make up ribosomes, the enzymatic machinery on which protein synthesis takes place. These three kinds of RNA are present in both eukaryotic and prokaryotic cells as well as in mitochondria.

There are at least two other kinds of RNA in the eukaryotic cell nucleus. Heterogeneous nuclear RNA (hnRNA) composes a large set of RNA molecules some of which serve as precursors to mRNA. Some of the small nuclear RNAs (snRNAs) function to assist in the maturation process of mRNA. The function of the other small nuclear RNAs is as yet unclear.

Objectives

After completing this chapter, you should be able to

Compare the physical and chemical properties of RNA and DNA.

List the basic properties of rRNA, mRNA, and tRNA. Define the terms *untranslated region, cap,* and *poly-A tail.*

General Properties of RNA

RNA and DNA have similar primary structures. Both consist of a linear chain of purine and pyrimidine nucleotides held together by 3′ to 5′ phosphodiester bonds. Both show polarity and have 5′ and 3′ ends. Adenine, guanine, and cytosine are common to both RNA and DNA. In RNA, uracil replaces the thymine found in DNA. In addition, the sugar in RNA is ribose instead of deoxyribose. The nucleotides found in RNA are adenosine monophosphate (AMP), guanosine monophosphate (GMP), cytidine monophosphate (CMP), and uridine monophosphate (UMP). The structures of these compounds are given in Figure 5-1.

The structure of a small segment of RNA is shown in Figure 5-2. The shorthand forms used for representing DNA can also be used for RNA. The RNA segment shown in Figure 5-2 can be drawn as

RIBONUCLEOTIDES

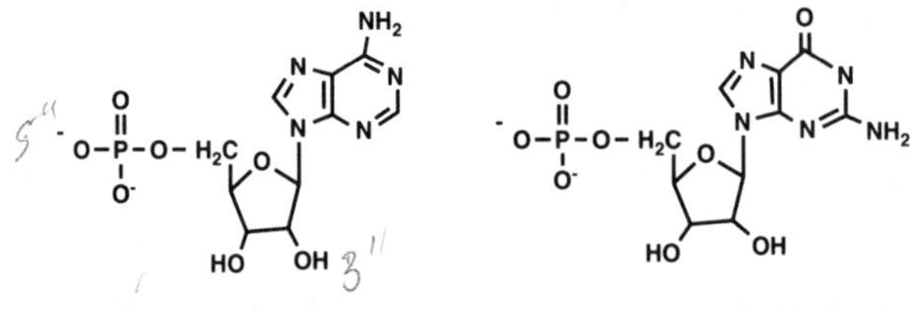

Fig. 5-1. Structure of the nucleotides found in RNA.

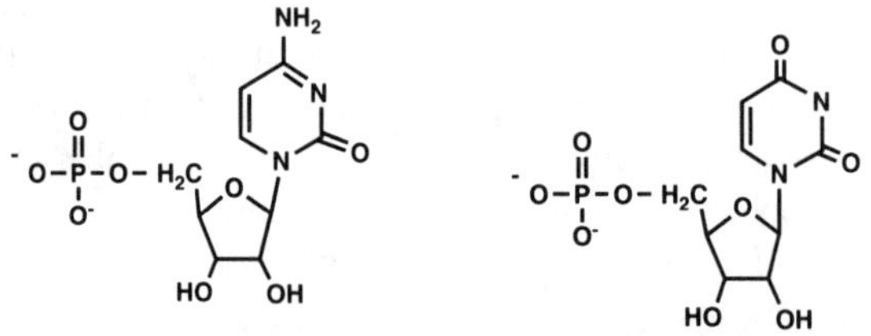

or

pApUpCpG

or even

AUCG

As we did for DNA, the 5' end is always drawn on the left. Because the hydroxyl groups of the sugar are not shown in these shorthand forms, the structure must be designated as RNA.

RNA is normally a copy of a small segment of only one strand of DNA. The strand from which RNA is copied is called the **template strand.** Its complement is called the **coding strand** (Fig. 5-3). Because RNA does not have a complementary strand in the cell, it does not form long double-stranded structures as does DNA. The secondary structure of RNA often is a random coil or is determined by its interactions with protein.

Occasionally, there are complementary regions within the same RNA molecule. When this occurs, the molecule can fold back on itself and form a base-paired stem with an unpaired loop structure, as shown in Figure 5-4. As we will see later in this chapter, this type of structure occurs frequently in tRNA. The double-stranded regions in RNA are usually short, 5 to 10 base pairs long, and are in the A helix form rather than the B helix found in DNA. This is because the B helix does not have enough room to accommodate the extra OH at position 2' on the ribose ring found in RNA. In addition to the normal base pairing, the base-paired stem in RNA can contain unusual base pair combinations such as U:G, C:C, and C:U, which do not exist in DNA.

The differences in secondary structure between RNA and DNA produce large differences in physical properties. RNA, because it does not usually exist as a large double helix, does not show much hyperchromicity under denaturing conditions. Nor does it have a well-defined melting point or show large changes in viscosity or buoyant density after

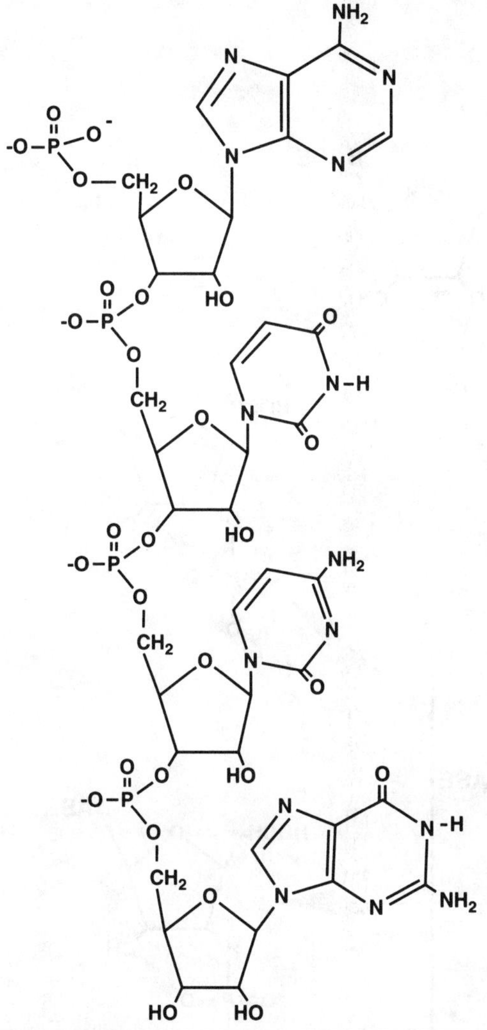

Fig. 5-2. Structure of a small segment of RNA. The 5′ end is at the top and the 3′ at the bottom.

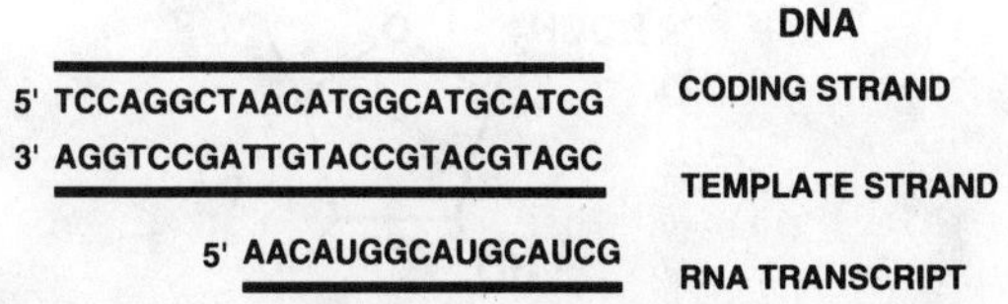

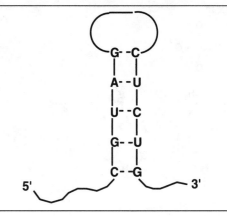

Fig. 5-4. Stem-loop structure found in RNA. A single piece of RNA folds back on itself, forming a short stem with a loop at the top. The stem may contain some unusual base-pairing combinations.

	DNA
5′ TCCAGGCTAACATGGCATGCATCG	CODING STRAND
3′ AGGTCCGATTGTACCGTACGTAGC	TEMPLATE STRAND
5′ AACAUGGCAUGCAUCG	RNA TRANSCRIPT

Fig. 5-3. Relationship of RNA to DNA. The RNA is a copy of only one strand of the DNA, the template strand. The RNA has a sequence (with U substituted for T) and orientation identical to the DNA of the coding strand.

denaturation. Because it is a copy of only one strand of DNA, the amount of adenine in RNA does not equal the amount of uracil, nor does guanine equal cytosine.

Unlike DNA, RNA can be hydrolyzed by alkali. This is possible because of the existence of the extra hydroxyl group at position 2 of the ribose. A cyclic 2′,3′-phosphate, shown in Figure 5-5, is an obligatory intermediate for base hydrolysis, and only RNA is capable of forming such a structure. The 2′,3′-phosphate can be further hydrolyzed to produce either 2′ or 3′ monophosphates.

In addition to the four common bases, RNA can contain a large number of modified bases and nucleosides. These occur at specific sites in tRNA and, to a lesser extent, in rRNA and mRNA. These modifications include methylation, ethylation, and other alkylations of the nitrogen and oxygen atoms of the purine and pyrimidine rings. The structures of a number of these modified bases and nucleotides are shown in Figure 5-6. The 2′ position of the ribose can also be methylated. Note that RNA with a methyl group at the 2′ position cannot be hydrolyzed by alkali as it cannot form the necessary 2′,3′ cyclic intermediate.

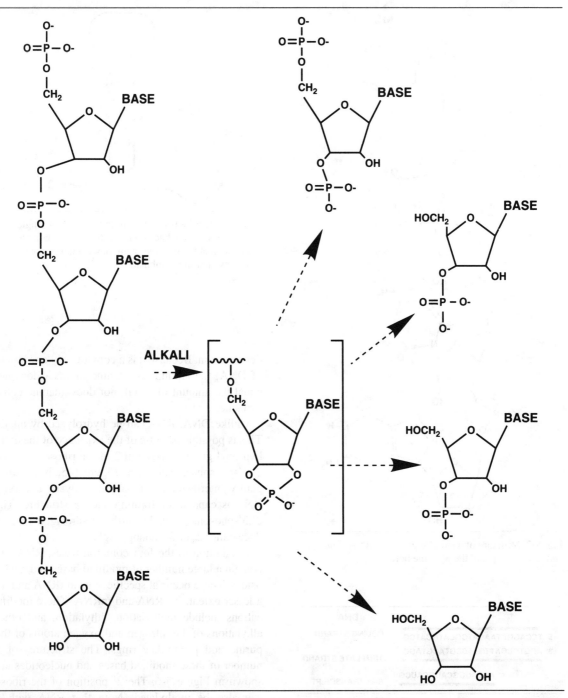

Fig. 5-5. Base hydrolysis of RNA. A 2′,3′-cyclic phosphate is a necessary intermediary for this reaction.

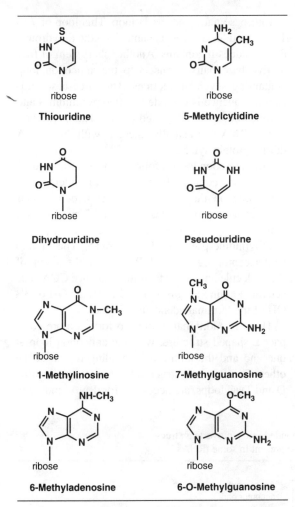

Thiouridine

5-Methylcytidine

Dihydrouridine

Pseudouridine

1-Methylinosine

7-Methylguanosine

6-Methyladenosine

6-O-Methylguanosine

Fig. 5-6. Structures of some of the unusual bases and nucleosides found in tRNA.

Major Types of RNA

Ribosomal RNA

Cells have more ribosomal RNA than any other kind. rRNA is also the most stable of the major types of RNA. It exists complexed with proteins in a large particle called a **ribosome,** which is the machinery on which protein synthesis takes place. In liver, there may be more than 10^6 ribosomes per cell, each containing approximately 50% RNA and 50% protein. The size of ribosomes, RNA, and other large molecules usually is given in **Svedberg units,** abbreviated *S.* Svedberg units are related to both the size and shape of the molecule. They are determined by measuring the rate of sedimentation in an ultracentrifuge under defined conditions. Because they are related to both size and shape, S values are not additive when nonspherical subunits are combined into larger structures.

Ribosomes in eukaryotic cells have an S value of 80. They are composed of two unequal subunits, one 60S and the other 40S. The 60S subunit contains one piece of RNA of 28S (4800 bases), one of 5.8S (160 bases), and one of 5S (120 bases), as well as 50 proteins. The smaller 40S ribosomal subunit contains a single piece of RNA of 18S (1900 bases) and 34 proteins. All human rRNAs have extensive secondary structure and contain a number of methylations on the 2′ position of the ribose. In prokaryotes, the ribosome is smaller, 70S, and the subunits are 50S and 30S. The large ribosomal subunit has a piece of RNA of 23S (2900 bases), one of 5S (120 bases), and 34 proteins, whereas the small subunit has a single RNA segment of 16S (1540 bases) and 21 proteins. The sizes and proteins of ribosomes in mitochondria are more similar to those found in prokaryotes than eukaryotes. Table 5-1 lists the composition of ribosomes in prokaryotes, mitochondria, and the cytoplasm of eukaryotes.

The small ribosomal subunit contains sites for binding the proteins involved in the initiation of protein synthesis. It also has a site for binding mRNA and decoding it by interaction with tRNA. The large

Table 5-1. Composition of ribosomes of prokaryotes, mitochondria, and eukaryotic cytoplasm

	Prokaryotes	Mitochondria	Eukaryotic cytoplasm
Large ribosomal subunit			
S value	50S	40–45S	60S
RNA	5S, 23S	16S	5S, 5.8S, 28S
Proteins	34	40–50	50
Small ribosomal subunit			
S value	30S	30–35S	40S
RNA	16S	12S	18S
Proteins	21	30–40	34

ribosomal subunit contains the enzymatic activities for actually synthesizing proteins.

Transfer RNA

Approximately 15% of the RNA in the cytoplasm of eukaryotic cells is tRNA. It has a half-life of approximately 24 hours. There are at least 20 different tRNAs found in the cytoplasm of human cells, one corresponding to each of the naturally occurring amino acids. tRNAs are small molecules containing 75 to 100 nucleotides and sedimenting between 4S and 5S. Although mitochondria have different tRNA in terms of the primary structure of nucleotides, they have structural features similar to those in the cytoplasm.

Despite their diversity of primary sequence, all tRNAs have a number of common characteristics. They all contain an amino acid acceptor region, a TψC loop, a D segment, and an anticodon loop arranged in a cloverleaf pattern, which maximizes hydrogen bonding (Fig. 5-7). The 5′ end of the molecule is part of a seven–base pair stem structure that contains a number of unusual base-pairing combinations. This stem is followed by a short stem of three

to four base pairs and the D loop. This loop of 7 to 10 nucleotides derives its name from the dihydrouridine residues it contains. Another short stem of four to five base pairs connects to the anticodon loop containing seven nucleotides. The three-base anticodon sequence is preceded by two pyrimidines and is followed by a methylated purine. This is the part of the tRNA that actually interacts with the mRNA during protein synthesis.

The anticodon loop is followed by a variable loop, which can contain from 3 to 21 nucleotides, depending on the tRNA. The final five–base pair stem connects to the TψC loop of 7 nucleotides, which always contains the nucleotide sequence ribothymidine-pseudouridine-cytidine. The amino acid acceptor region of all tRNAs is at the 3′ end of the molecule and consists of the sequence CCA. The activated amino acid is attached to the 2′-OH or 3′-OH of the terminal adenosine residue.

There is additional folding to form a more compact L-shaped structure, with the anticodon loop at one end and the amino acid–binding region at the other. The common structural elements, such as the D and TψC loops, are necessary for tRNA binding to

Fig. 5-7. General structure of tRNA. All tRNAs are hydrogen-bonded into a cloverleaf structure containing four stems and three loops. A fourth loop may be present in some tRNAs.

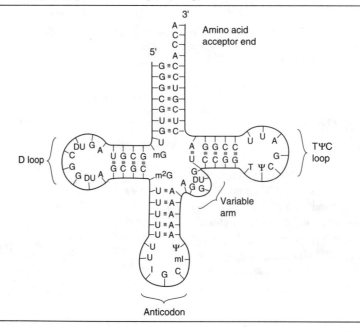

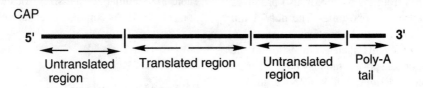

CAP

5' ────────┤─────────────┤───────────┤───── 3'

Untranslated region ⟷   Translated region ⟷   Untranslated region ⟷   Poly-A tail ⟷

Fig. 5-8. Structural features of mRNA. Most mRNAs from eukaryotic cells have a cap at the 5' end. The middle of the molecule contains the nucleotide sequence, which specifies the amino acid sequence in the protein. This is the coding or translated region. On either side of the translated region is a variable number of nucleotides that do not code for protein, the untranslated regions. The 3' end contains a 200- to 250-nucleotide sequence of adenosine residues, the poly-A tail.

ribosomes. The variable regions are necessary for binding to specific proteins and recognizing the appropriate codon in mRNA.

Messenger RNA

mRNA is the most heterogeneous in size of the major RNAs. There are thousands of different mRNAs in each mammalian cell, with sizes varying from 100 to more than 2500 bases in length. mRNA is also the lowest in concentration and the least stable. Normally, less than 5% of the RNA in the cytoplasm of a eukaryotic cell is mRNA. For some eukaryotic mRNAs, the half-life may be as short as 3 to 4 hours. However, the average half-life is closer to 24 hours. In contrast, prokaryotic mRNAs have a half-life of only a few minutes. In mammalian cells, each mRNA codes for only a single polypeptide. In prokaryotes, a single mRNA can carry the information for the synthesis of several different proteins. Such mRNAs are called **polycistronic messages.**

Despite the heterogeneity in size, half-life, and base content, all mRNAs have several structural features in common. All mRNAs in eukaryotic cells contain a **cap** at the 5' end, and most have a tail of 100 to 250 adenosine residues at the 3' end (Fig. 5-8). There is an **untranslated region** of thirty to several hundred nucleotides at both the 5' and the 3' ends. These segments may have some secondary structure, whereas the nucleotides coding for protein have little or no secondary structure. The 5' cap consists of a 7-methylguanosine attached 5' to 5' to the first nucleotide of the mRNA. The 2' position of this first nucleotide is methylated. Frequently, the 2'-OH of the next nucleotide is also methylated (Fig. 5-9). This structural modification inhibits nuclear diges-

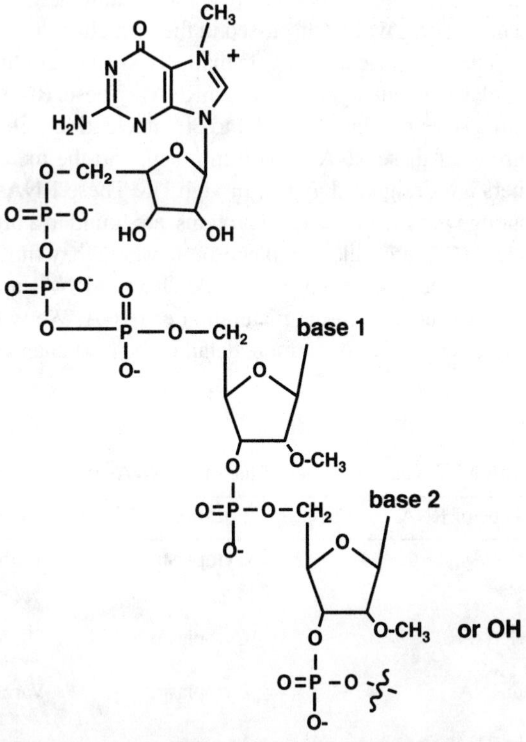

Fig. 5-9. Structure of the 5' cap found in mRNA. A 7-methylguanosine is connected 5' to 5' with the first nucleotide of the mRNA. The 2'-OH of this nucleotide is methylated. The 2'-OH of the next nucleotide of the mRNA may be methylated as well.

tion of the mRNA from the 5′ end. It also promotes binding of the mRNA to ribosomes. The **poly-A tail** at the 3′ end of most mRNAs contains up to 250 A residues and also prevents nuclease digestion of the mRNA. In addition, the poly-A tail may have other, as yet unrecognized, functions. A few mRNAs, such as those coding for the histones, do not contain poly-A tails. However, these mRNAs do have 5′ caps. Prokaryotic mRNAs contain neither caps nor tails. From the point of synthesis to degradation, eukaryotic mRNA always is associated with protein.

Other RNAs

In addition to the three major classes of RNA, there are small amounts of other types. One of these is **heterogeneous nuclear RNA** (hnRNA). These are very large RNAs that are heterogeneous in size and are found only in the nucleus. Like mRNA, hnRNAs contain 5′ caps and poly-A tails. Approximately 10% of the hnRNA eventually is converted into mature mRNA. This process will be discussed in the next chapter.

There are also a series of small RNAs found in the nucleus of eukaryotic cells (snRNA). These RNAs have sizes ranging from 100 to 300 nucleotides. One group of these RNAs is rich in uracil, and the members are designated as U1 through U6. These RNAs, along with their associated proteins, are found in a nuclear structure called a **spliceosome,** where they function in the maturation of mRNA. Other snRNAs appear to function in the maturation of rRNA. We will discuss these RNAs in more detail in the next chapter.

Table 5-2 summarizes the sizes, locations, and functions of the RNAs common to eukaryotic cells.

Summary

RNA and DNA have similar primary structures. The major differences are the sugar (ribose in RNA and deoxyribose in DNA) and one base (uracil in RNA and thymine in DNA). There are also differences in the secondary structure. Because RNA does not exist as a long double helix, it does not show hyperchromicity, decreased viscosity, or increased buoyant density on denaturation. It is hydrolyzed by base, whereas DNA is only denatured.

There are three major kinds of RNA in cells. rRNA is found as a part of ribosomes, the machinery on which protein synthesis takes place. The large ribosomal subunit is 60S and contains three different-sized rRNAs, 28s, 5.8s, and 5S, as well as 50 proteins. The small ribosomal subunit is 40S and contains the 18S rRNA and 34 proteins. tRNA is a base-paired molecule of 75 to 100 nucleotides that contains many posttranscriptionally added modified bases. Although there are at least 20 different tRNAs, they all share a common secondary structure with four stems and three loops plus the sequence CCA at the 3′ end. mRNA is a product of maturation of hnRNA and contains few modified bases. All eukaryotic mRNAs have a cap on the 5′ end, and most have a poly-A tail at the 3′ end. Some of the snRNAs function in the maturation of mRNA and rRNA.

Table 5-2. Characteristics of the major RNAs in eukaryotic cells

Type of RNA	Location	Size	Function
tRNA	Cytoplasm	4–5S	Transfer of amino acids to ribosome, interaction with mRNA to insert correct amino acid into protein
rRNA	Cytoplasm	5S, 5.8S, 18S, and 28S	With proteins, forms ribosome, machinery on which proteins are synthesized
mRNA	Cytoplasm	Variable	Processed copy of a gene, determines sequence of amino acids in protein
hnRNA	Nucleus	Variable	Some are precursors to mRNA
snRNA	Nucleus	4S	With proteins, functions in maturation of hnRNA into mRNA in spliceosomes

6 RNA Synthesis and Processing

All three major classes of RNA are synthesized in the nucleus by copying segments of DNA. Selection of the segment to be copied is the major control point in this process. As synthesis proceeds, the RNA is released and the DNA reforms a normal helical structure. All the major classes of eukaryotic RNA are modified after synthesis and before being transported into the cytoplasm, where they function in protein synthesis.

Objectives

After completing this chapter, you should be able to

Describe the reaction catalyzed by RNA polymerase.

List the different RNA polymerases found in the eukaryotic cell and identify their properties.

Discuss the events that occur during initiation, elongation, and termination of RNA synthesis. Define the terms *transcription unit, primary transcript,* and *promoter.*

Describe the events in the posttranscriptional processing of rRNA, tRNA, and mRNA. Define *intron, exon,* and *spliceosome.*

Identify the site of action of actinomycin D, cordycepin, rifamycin, α-amanitin, and proflavine.

RNA Polymerases

The synthesis of RNA using information from a DNA template is called **transcription.** The enzymes responsible for transcription are RNA polymerases. The reaction catalyzed by these enzymes is diagrammed in Figure 6-1. RNA polymerases, like DNA polymerases, add nucleotides to the 3'-OH end of a growing nucleotide chain, using base pairing with a template to select the proper nucleotide. Because RNA polymerases are capable of starting new chains, transcription does not require a primer. Another difference between RNA and DNA polymerases is that RNA polymerases contain no nuclease activity. Even though they are unable to proofread and correct mistakes in the RNA product,

Fig. 6-1. Reaction catalyzed by RNA polymerase. Ribonucleotide triphosphates are added to a growing RNA chain after selection by base pairing with the complementary base in the template.

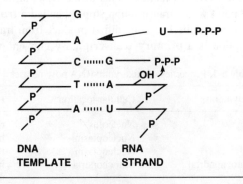

the error rate for RNA synthesis is only 1 incorrect base for each 10^4 or 10^5 bases. Because RNA is only a copy of a segment of DNA and is not part of the permanent genetic material of the cell, this higher error rate is not detrimental to cell survival.

The nuclei of eukaryotic cells contain three different RNA polymerases. **RNA polymerase I** is located in the nucleolus and is responsible for the synthesis of the precursor for rRNA. **RNA polymerase II** is found in the nucleoplasm and synthesizes hnRNA, some of which matures into mRNA. RNA polymerase II also is responsible for the synthesis of some of the small nuclear RNAs. tRNA is synthesized by **RNA polymerase III,** which also is located in the nucleoplasm. In addition, this enzyme is responsible for the transcription of the 5S rRNA, the smallest RNA found in ribosomes. The mitochondrion, like a prokaryote, contains a single type of RNA polymerase, which synthesizes all three of the major types of RNA. All these RNA polymerases require all four ribonucleoside triphosphates, either Mn or Mg as a divalent cation, and a template of DNA. The properties of the RNA polymerases are summarized in Table 6-1.

Many prokaryotic genes require a protein factor called **sigma** for proper initiation of RNA synthesis. The sigma protein binds to the RNA polymerase and remains bound to the polymerase molecule until RNA synthesis has begun. After this, the sigma factor dissociates and binds to another polymerase molecule. There is no evidence for such a cycling initiation factor in human cells, although a large number of proteins have been shown to be required for the proper initiation of synthesis by RNA polymerase II.

The section of the template DNA from the site where RNA polymerase binds to the DNA through the point where transcription stops is called a **transcription unit.** The unmodified product from transcription is a **primary transcript.** By convention,

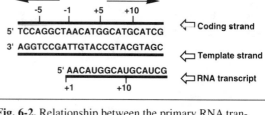

Fig. 6-2. Relationship between the primary RNA transcript and the template and coding strands of DNA. The first nucleotide in the primary transcript is number 1. The first nucleotide in the DNA on the 5′ side of the RNA synthesis start point is given the number –1.

the first nucleotide of the primary transcript is given number 1 and numbering continues with positive numbers. Moving toward more positive numbers is called *downstream.* Moving in the opposite direction from the transcription start site is called *upstream;* these nucleotides have negative numbers. Similar numbering applies to both the template and the coding strands of DNA, as shown in Figure 6-2.

Mechanism of RNA Synthesis

The general mechanism for transcription can be divided into stages: initiation, elongation, and termination. The events occurring in these stages are summarized in Figure 6-3.

Initiation

Because less than 5% of the DNA in a differentiated cell is transcribed, it is necessary for the RNA polymerase to recognize both the correct strand of DNA and the particular portion of that strand to be copied. One strand of DNA may be used as the template for the transcription of some genes, whereas the complementary strand may be used for other genes. In

Table 6-1. Properties of eukaryotic RNA polymerases

Polymerase	Cellular location	Products	Inhibitors
I	Nucleolus	rRNA precursor (28, 18, 5.8S)	Actinomycin D
II	Nucleoplasm	hnRNA (precursor to mRNA), snRNA	10^{-8}-M α-Amanitin
III	Nucleoplasm	tRNA precursor, 5S rRNA	10^{-4}-M α-Amanitin
Mitochondrial	Mitochondria	mitochondrial rRNA, tRNA, mRNA	Rifamycin

INITIATION :	RNA polymerase binds
	DNA strands separate
	First nucleotide bound
	Second nucleotide bound
	First phosphodiester bond formed
ELONGATION :	Addition of further nucleotides
	Formation of phosphodiester bonds
TERMINATION :	Synthesis stops
	RNA released
	DNA helix reforms

Fig. 6-3. Events occurring during the initiation, elongation, and termination stages of transcription.

human cells, there are no known cases in which the same segment of both strands of DNA is used in transcription.

The recognition site where the RNA polymerase binds to the template is called a **promoter.** In prokaryotes, the promoter site is located approximately 35 nucleotides upstream from the transcription start site. It contains two regions that are highly conserved. One of these consensus sequences is the nucleotide sequence TTGACC, starting at position –35. The other is the six–base pair sequence TATAAT at position –10 (Fig. 6-4A). This AT-rich region is called the *Pribnow box,* and it is melted when the RNA polymerase binds to the strand DNA that will actually be transcribed.

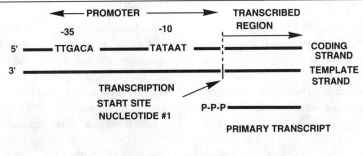

A   BACTERIAL PROMOTERS

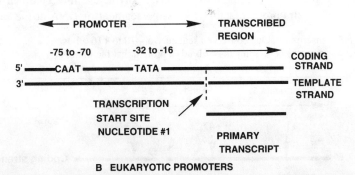

B   EUKARYOTIC PROMOTERS

Fig. 6-4. Structure of the promoter sites in prokaryotes (A) and eukaryotes (B). (A) Two conserved DNA sequences at positions –10 and –35 serve to bind and orient the RNA polymerase. (B) A conserved sequence of TATA at position –16 to –32 is found in many genes coding for proteins. A second conserved sequence CAAT at –70 to –75 is also common. Proteins binding to these sequences bind the RNA polymerase.

In eukaryotes, the promoter region for genes coding for mRNA also contains an AT-rich segment at position −16 to −32. It is called the *Goldberg-Hogness box* or *TATA box,* and it determines the actual start site of transcription. There is also a consensus sequence CAAT at position −70 to −75 in many, but not all, genes coding for protein (Fig. 6-4B). The RNA polymerase in eukaryotes does not appear to bind directly to the DNA without the assistance of a number of additional proteins. The interaction of these DNA binding proteins with DNA controls the frequency of transcription. We will discuss these controls in detail in Chapter 8.

Unlike genes coding for proteins, the promoter region for genes coding for tRNA is inside the DNA sequence that is actually transcribed. There is one conserved sequence at position −10 to +16 and a second sequence starting at +47, which serve as binding sites for RNA polymerase III and the other proteins necessary for proper initiation (Fig. 6-5). The promoter sequence for rRNA has yet to be identified clearly in human cells. However, it appears to lie in the nontranscribed spacer region between genes coding for rRNA.

Once the RNA polymerase is bound, the hydrogen bonds between the bases of the two strands of DNA must be broken and the two DNA strands separated for approximately 17 nucleotides. In this way, the ribonucleotides can be selected by correct hydrogen bonding with the bases on the template strand of DNA. Melting of the DNA and separation of the strands is accomplished by the RNA polymerase molecule, without the need for extra protein factors such as helicases and single-stranded binding proteins. However, a topoisomerase is required to remove the positive supercoils produced by unwinding the DNA helix and the negative supercoils produced when rewinding the DNA after the RNA polymerase has moved further along the DNA.

The formation of the first phosphodiester bond between two ribonucleotides can occur some distance from the site at which the RNA polymerase originally bound to the DNA. The first nucleotide of a new RNA is usually a purine, G more often than A. Like DNA replication, the DNA template is read in the 3′ to 5′ direction and RNA is synthesized in the 5′ to 3′ direction. The three phosphate residues that are attached to the 5′ end of the first nucleotide are not removed immediately from the RNA. They can serve as a marker of the true 5′ end of an RNA molecule.

Elongation

The addition of further nucleotides to the growing RNA chain is known as **elongation.** As in DNA replication, the polymerase molecule proceeds along the DNA template, joining nucleotides that are selected by hydrogen bonding to their complementary bases in the template. A DNA-RNA hybrid of 10 to 12 base pairs is an intermediate in the elongation reaction. Under cellular conditions, a DNA-DNA helix is more stable than a DNA-RNA helix. As the transcription reaction progresses, the RNA is released and the DNA helix reforms. More than one RNA polymerase molecule can be transcribing the same segment of DNA at the same time. With the

Fig. 6-5. Promoters for tRNA transcription. One conserved sequence at −10 to +16 binds RNA polymerase III. A second sequence beginning at +47 binds a protein that facilitates binding of the polymerase.

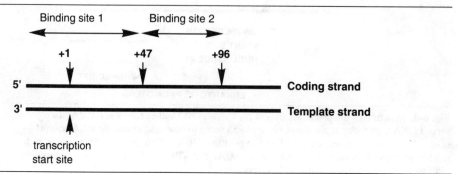

newly made pieces of RNA trailing from it, these structures resemble a tapered test tube brush when viewed under an electron microscope.

Termination

In prokaryotes, **termination** of transcription occurs when polymerization stops at a specific site on the DNA. For some genes in prokaryotes, the termination site consists of a region of RNA that contains complementary bases which can form a short-stem loop. If this is followed by a string of 4 to 8 U residues in the RNA, transcription will stop when the RNA dissociates from the DNA (Fig. 6-6). For other prokaryotic genes, termination requires the assistance of a protein factor called **rho**. This protein binds to a specific site on the newly synthesized RNA and acts as a helicase, causing a dissociation of the RNA-DNA hybrid and the termination of RNA synthesis (Fig. 6-7)

The signals for termination of protein-coding genes in eukaryotes are as yet unknown. This is because the 3′ end of the primary transcript is cleaved before the RNA can be isolated and sequenced. The DNA sequences at the 3′ ends of cloned genes do not show similar or conserved sequences.

Posttranscriptional Processing

The three major classes of RNA are all synthesized in forms larger than the final product. In addition, all contain modifications of bases. For the most part, these changes occur after synthesis has been completed. The steps of posttranscriptional processing include removal of nucleotides (from ends and from internal sites), addition of nucleotides, and modification of nucleotides.

Ribosomal RNA

The 5.8S, 18S, and 28S rRNA of eukaryotes are transcribed together in a single piece of RNA. This RNA is 13,700 nucleotides long and has an S value of 45. The genes coding for rRNA are present in multiple copies arranged in tandem in the nucleolus. In humans, there are 250 to 300 copies of the genes for rRNA in each cell. These are located on several different chromosomes.

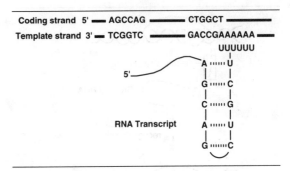

Fig. 6-6. Termination of RNA synthesis by the formation of a stem loop. Formation of a stem of complementary bases in the RNA will cause termination of transcription if it is followed by a series of U residues in the RNA (A residues in the template).

Synthesis of the 45S rRNA precursor occurs only when cells need more ribosomes for protein synthesis. After synthesis, a small number of adenine and guanine residues in the primary transcript are methylated. Many of the ribose residues are methylated as well. All the methyl groups added to the 45S precursor molecule are preserved and found in the mature rRNA. If this methylation is prevented, the 45S RNA cannot be cleaved properly to form the final rRNA products.

Figure 6-8 gives the maturation scheme for rRNA. In this process, 50% of the 45S molecule is removed and degraded. The posttranscriptional processing of rRNA is one area of metabolism in which we know more about humans than about bacteria. In eukaryotes, but not in bacteria, it is possible to isolate and examine discrete intermediates. First, the primary transcript is cleaved at the 5′ end, producing a molecule of 41S. Then there is a split between the 18S and the 5.8S sequences to produce one molecule of 20S and another of 32S. The 20S is trimmed to produce mature 18S RNA. The 32S is trimmed at the 5′ end and split between the 5.8S and the 28S sequences. Finally, the ends of the 5.8S and 28S are shortened to their final size.

The 5S RNA found in ribosomes is not transcribed as part of the 45S ribosomal precursor molecule. The genes coding for the 5S RNA are found as multiple copies elsewhere in the genome. In humans, there are approximately 1500 copies of the 5S ribosomal gene located in tandem groups on several

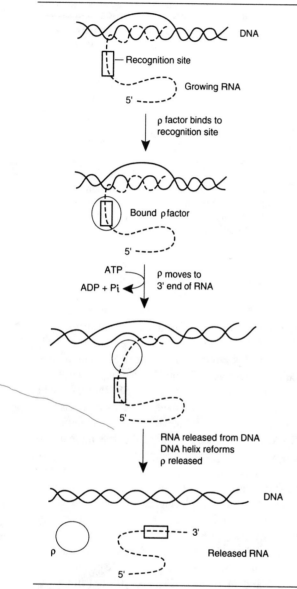

Fig. 6-7. Termination of RNA synthesis by the protein factor rho (ρ). The rho protein binds to a specific sequence in the primary transcript. It acts as a helicase to release the RNA from the DNA.

chromosomes. The 5S RNA is transcribed as a molecule longer than the 121-nucleotide final product. Unlike the posttranscriptional processing of rRNA, all the cleavages of the 5S RNA take place at the 3′

end. The 5′ end retains the three phosphate residues found on the initiating ribonucleotide.

During the maturation process of rRNA, proteins are gradually added to the RNA. When maturation is complete and the 5S has been added to the large ribosomal subunit, the ribosomes are ready for export to the cytoplasm through the nuclear pores. In eukaryotes, transport from the nucleus occurs as 40S and 60S subunits, not as 80S ribosomes.

Transfer RNA

Like rRNA, tRNA is transcribed from multiple copies in a size larger than the mature product. In humans, there are 1300 copies of the genes coding for tRNA, and these are located in tandem groups of up to 7. Cleavage of the primary transcript for tRNA occurs from both ends of the molecule as well as from the middle. In the posttranscriptional processing, the ends are cleaved first and the sequence CCA added to the 3′ end.

The next step is the removal of the intervening sequence, called an **intron.** There is only one intron in the primary transcript for tRNA, and it is always located on the 3′ side of the anticodon sequence. The mechanism for the removal of the intron is shown in Figure 6-9. First, the intron is excised, leaving a 5′-OH and a 2′,3′-cyclic phosphate in the tRNA. An RNA ligase adds a phosphate to the 5′ position and then adenylates it. The cyclic phosphate is broken, leaving the phosphate on the 2′ position. Then the ligation is completed and the 2′-phosphate removed.

The last step in maturation of tRNA is the modification of bases and nucleotides common to tRNA. The methyl donor for methylations is S-adenosylmethionine (SAM).

$$\overset{+}{H_2N}-\underset{|}{\overset{\overset{\displaystyle COO^-}{|}}{CH}}-CH_2-CH_2$$
$$\underset{H_3C}{\overset{|}{S+}}\diagdown adenosine$$

S-Adenosylmethionine

Methylations common to tRNA include the N6 of adenine, N7 of guanine, C5 of uracil, and N3 and N5 of cytosine. The structures of these modified bases

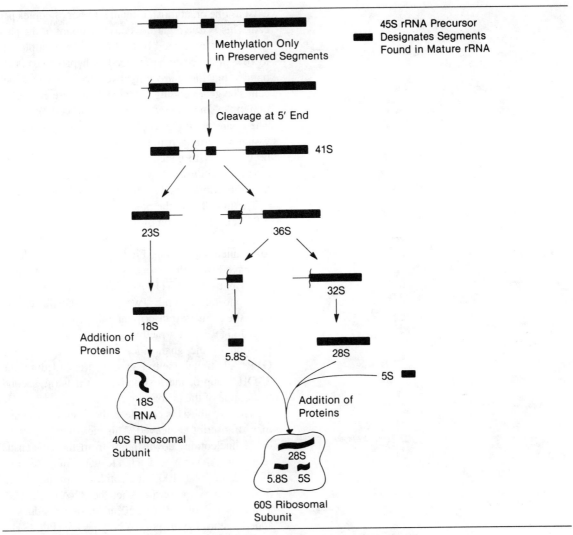

Fig. 6-8. Maturation scheme for rRNA. The 45S primary transcript is first methylated and then cleaved at both ends and in the middle to form the 18S, 28S, and 5.8S rRNA segments.

and nucleosides were given in Chapter 5, Figure 5-6. The 2′-OH group of the ribose may be methylated as well. In addition to methylation, uridine may be modified by reducing the double bond, producing **dihydrouridine,** or by rotating so that the sugar is attached to carbon 5 instead of nitrogen 1, producing **pseudouridine.** Other modifications include the addition of acetyl, isopentyl, and oxyacetyl groups. In other cases, one base may be exchanged for another.

An example is the exchange of the modified base, quenine, for a guanine in the tRNA.

Messenger RNA

Like the other two major types of RNA, mRNA is synthesized as a precursor that is considerably larger than the final product. Unlike the other two major RNAs, mRNA is not transcribed from genes present in multiple copies. The genes for individual mRNAs

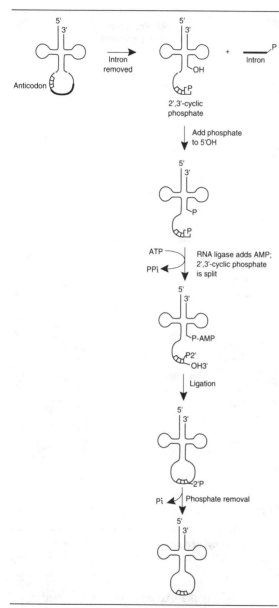

Fig. 6-9. Mechanism for removal of the intron from tRNA. The intron, which is always located on the 3′ side of the anticodon sequence, is cleaved out, leaving a 5′-OH and a 2′,3′-cyclic phosphate in the tRNA. RNA ligase first adds a phosphate to the 5′-OH and then AMP. The 2′,3′-cyclic phosphate is opened so that the phosphate is on the 2′ position. A phosphodiester bond is formed closing the gap left by removal of the intron. This releases the RNA ligase and the AMP. Finally, the phosphate is cleaved from the 2′ position. (PP$_i$ = pyrophosphate; P$_i$ = inorganic phosphate.)

exist as one or, at most, a small number of copies per cell. This is true regardless of the amount of the particular protein needed by that cell. For example, up to 10% of the protein produced by hepatocytes is albumin, but there are only two albumin genes per cell. The genes coding for the histones are an exception to the single-copy rule for protein-coding genes. There are multiple copies of the histone genes, which are arranged in tandem. This repetition may be necessary to generate sufficient amounts of histone protein during S phase, which is the only time that transcription and synthesis of histones occur.

The first step in the posttranscriptional process of mRNA is the addition of the 5′ cap (Fig. 6-10). This occurs after the growing RNA chain is only about 20 to 30 nucleotides long. First, one of the phosphates is cleaved from the 5′ end of the transcript. Guanosine triphosphate (GTP) then is added to form a 5′ to 5′ triphosphate bond. Two of the phosphates are from the primary transcript and the third is from the added GTP. Then a methyl group is added to the 7 position of the guanosine, which came from the GTP. Additional methyl groups are added to the 2′-OH group of the first and sometimes the second nucleotide of the primary transcript.

When synthesis is complete, the 3′ end of the primary transcript is cleaved. This cleavage occurs 15 to 30 nucleotides downstream from the consensus sequence AAUAAA. Such cleavage may remove 500 to more than 1000 nucleotides from the 3′ end of the primary transcript. After the RNA is cleaved, a poly-A tail of 200 to 250 adenosine residues is added by the enzyme poly-A polymerase (Fig. 6-11).

The regions of the mRNA molecule destined to be part of the mature mRNA are not continuous. They are divided by intervening sequences or **introns,** which must be excised in order to make mature mRNA. The part of the molecule that remains, the **exons,** then are spliced together to form the mature mRNA. The exons often correspond to functional domains of the protein. Introns can be very large and numerous. For example, the primary transcript for tropomyosin, an important protein for muscle contraction, contains 11 introns. Although introns are common in eukaryotic mRNA precursors, they are absent from prokaryotic and mitochondrial protein-coding genes.

By analyzing the primary transcripts of many genes, it has been possible to determine which nu-

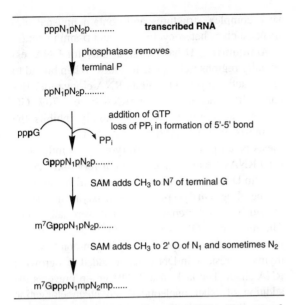

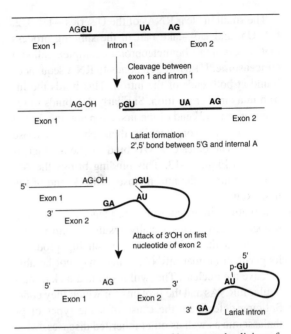

Fig. 6-10. Formation of the 5′ cap on hnRNA. First, one of the phosphates on the 5′ nucleotide of the growing primary transcript is cleaved. Then GTP is added 5′ to 5′, forming the basic cap structure. Finally, methyl groups are transferred from S-adenosylmethionine (SAM) to the 7 position of the terminal guanosine and the 2′ position of the first, and often the second, nucleotide of the primary transcript.

Fig. 6-12. Reactions in removal of introns and splicing of mRNA. Cleavage between the first exon and the first intron occurs, leaving a 3′-OH on the exon and pGU at the 5′ end of the intron. An adenosine in a conserved sequence internal to the intron attacks to the 5′ position on the intron, forming a 2′,5′ bond. This produces a lariat structure in the intron. Finally, the 3′-OH of exon 1 attacks the first nucleotide of exon 2, ligating the two exons and releasing the intron as a lariat.

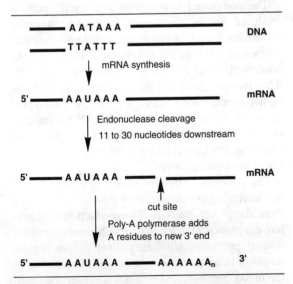

Fig. 6-11. Cleavage of the 3′ end of hnRNA and formation of the poly-A tail. The hnRNA is cleaved 11 to 30 nucleotides downstream from the consensus sequence AAUAA. Then, poly-A polymerase adds up to 250 adenosine residues to the 3′ end of the RNA.

cleotides are needed in the splicing process. An AU sequence usually is found at both the 3′ end of the exon and the intron. The sequence GU-purine-AG-pyrimidine is found at the 5′ end of the intron. In addition, there is a required AG sequence inside the intron. The reactions involved in the process of splicing mRNA are outlined in Figure 6-12. First, the 3′ end of the first exon is cleaved from the intron, either by direct cleavage or by attack by the adenosine of the conserved AG sequence in the intron. This leaves a 3′-OH on the end of the exon. The GU at the 5′ end of the intron is in a 2′,5′-phosphate bond with the attacking A of the intron. This forms the lariat structure characteristic of introns. Finally, the 3′-OH of the first exon attacks the 5′ nucleotide of the second exon, resulting in ligation of the two exons and release of the lariat intron.

The small nuclear RNAs of the U series—U1, U2, U4, U5, and U6—take part in the splicing process that occurs in a ribonucleoprotein complex called a **spliceosome.** U1 hydrogen bonds to RNA sequences found at both ends of the intron. This holds the intron in a circular position. Similarly, U5 binds to sequences at the 3′ end of the first exon and the 5′ end of the second exon, keeping them physically close together. The other U RNAs bind to one another as shown in Figure 6-13. This binding brings the AG sequence inside the intron close to the 3′ end of the first exon.

Mutations in the consensus sequences can prevent proper splicing. This usually results in introns remaining in the RNA. Defective splicing produces longer-than-normal mRNA, which may not be able to leave the nucleus. This will result in a lack of particular mRNAs and the proteins for which they code. Splicing defects are the cause of some types of β-thalassemia as well as other inherited diseases.

Inhibitors

There are several important compounds that exert their biochemical effects at the level of RNA synthesis or processing. You should be familiar with actinomycin D, cordycepin, rifamycin, α-amanitin, and proflavine. The structures of these compounds are given in Figure 6-14. Most of these compounds have very complex structures and contain unusual elements such as heterocyclic rings and D-amino acids.

Actinomycin D has a high affinity for DNA, especially regions rich in GC residues. When bound to DNA, actinomycin D inhibits RNA elongation. Because the genes coding for rRNA have a 70% GC content, actinomycin D preferentially inhibits the synthesis of rRNA. At high concentrations, the synthesis of all prokaryotic, eukaryotic, and mitochondrial RNAs are inhibited. Although very toxic, actinomycin D has found some use as an anticancer drug.

The drug **cordycepin,** or 3′-deoxyadenosine, is treated by cellular enzymes as if it were adenosine. The missing 3′-OH is ignored. In its active triphosphate form, cordycepin can hydrogen-bond with a thymidine residue in DNA and be added to a growing RNA chain. The lack of a 3′-OH group prevents the addition of other nucleotides and chain elongation ceases. Because it is an analog of adenosine, the greatest effect of cordycepin is on the growth of the poly-A tail in hnRNA. In the absence of the poly-A tail, hnRNA does not undergo proper posttranscriptional processing and no mature mRNA is produced. Ultimately, protein synthesis slows because of the lack of mRNA in the cytoplasm. At high concentrations, the synthesis of all RNA is inhibited by cordycepin.

The compound **α-amanitin** is a cyclic peptide produced by the death cup mushroom, *Amanita phalloides.* Its effect is directly on the enzymes that synthesize RNA in the eukaryotic cell nucleus. At concentrations of 10^{-8} to 10^{-9} M, α-amanitin inhibits RNA polymerase II and the synthesis of mRNA. At higher concentrations (e.g., 10^{-4} M), RNA polymerase III is also inhibited, and tRNA synthesis ceases. Neither RNA polymerase I nor the mitochondrial RNA polymerase is inhibited by α-amanitin at any concentration.

The poisonous effect of α-amanitin comes from the decreasing concentration of proteins synthesized from short-lived mRNAs. The production of digestive enzymes is affected first, which causes gastrointestinal problems. Ultimately, liver failure occurs, resulting in an increase in the level of ammonia in the blood. Neurological symptoms ensue, and death occurs after several days of acute distress.

Rifamycin affects the RNA polymerase of prokaryotes and mitochondria. It does not inhibit RNA synthesis in the eukaryotic cell nucleus. Rifamycin

Fig. 6-13. Hydrogen bonding between the U series of snRNAs and the mRNA during splicing. U1 hydrogen bonds to the ends of the intron. U5 binds to the 3′ end of exon 1 and the 5′ end of exon 2. U2 binds to the branch point consensus sequence, and U6 binds to U2. This brings the branch point near the 5′ end of the intron. The exact location of U4 during splicing is unclear.

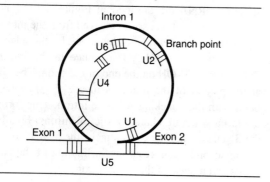

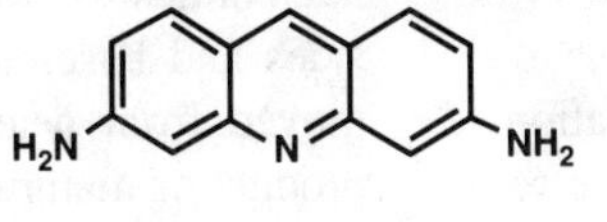

Proflavine

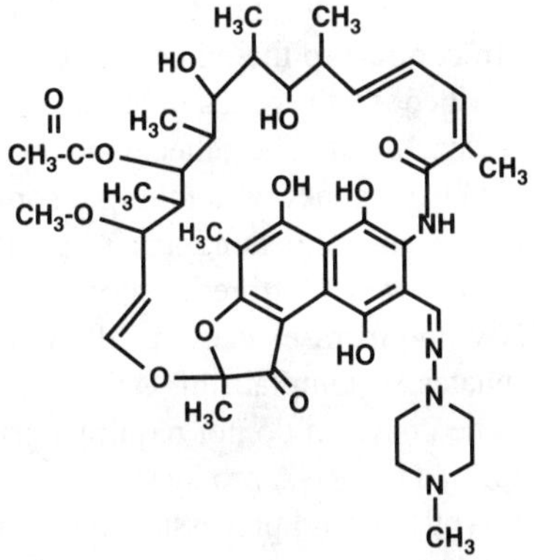

Rifampicin

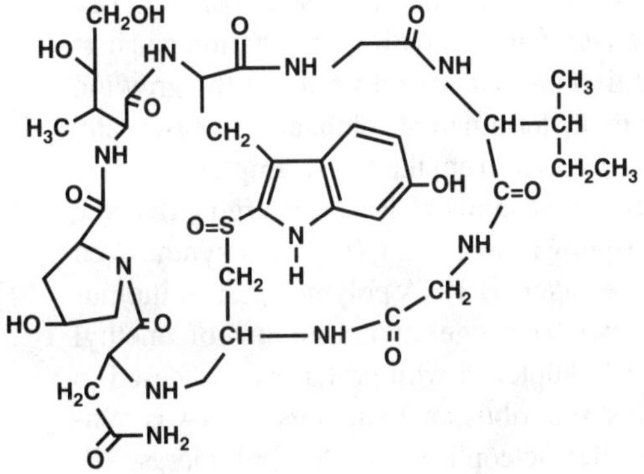

α-Amanitin

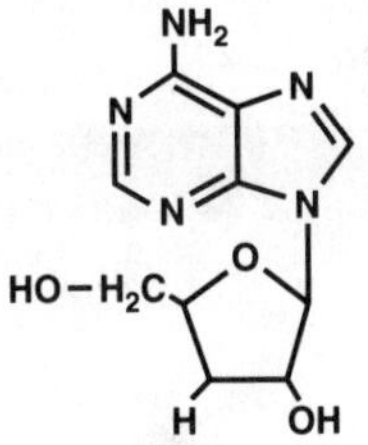

Cordycepin
(3'-deoxyadenosine)

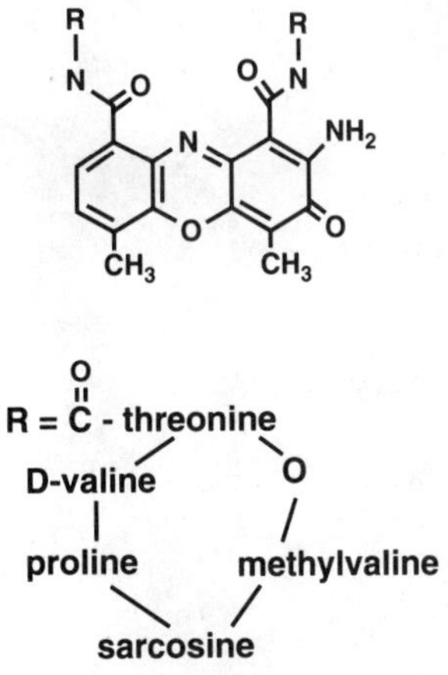

R = C - threonine
D-valine
proline   methylvaline
sarcosine

Actinomycin D

Fig. 6-14. Structure of several inhibitors of RNA transcription and processing. Rifampicin is a synthetic derivative of rifamycin.

and its synthetic analog, rifampicin, inhibit the initiation step of RNA synthesis. Because the drug shows specificity for prokaryotic metabolism, rifamycin and its synthetic derivative rifampicin are used clinically for treating bacterial infections, especially tuberculosis.

The intercalating agents such as **proflavine** have RNA processing as their site of action. These compounds disrupt double-stranded regions of RNA and the splicing complex and prevent proper processing of hnRNA into mRNA. They also affect the maturation of tRNA.

Summary

RNA synthesis occurs in three stages: initiation, elongation, and termination. The first step is the selection of the correct segment of DNA to be copied. After the polymerase molecule has bound and melted the DNA, nucleotides can be aligned by base pairing with complementary nucleotides in the template DNA. Formation of the first phosphodiester bond is the last step in initiation. Elongation consists of the addition of further nucleotides to the growing RNA chain. In termination, elongation ceases and the RNA is released from the DNA template.

All three major kinds of RNA undergo extensive posttranscriptional processing. rRNA is synthesized as a 45S precursor by RNA polymerase I in the nucleolus. After cleavage and addition of methyl groups, it is complexed with protein and released to the cytoplasm as ribosomal subunits. tRNA is synthesized in the nucleoplasm by RNA polymerase III, which also synthesizes the 5S rRNA. After synthesis nucleotides are removed from both ends, nucleosides and bases are modified, and an intron is removed from near the anticodon site. mRNA is a product of maturation of hnRNA and is synthesized by RNA polymerase II. In the maturation process, a cap is added to the 5′ end. After cleavage of the 3′ end, a tail of adenosine residues is added. Later, introns are spliced out and the exons are ligated together. In contrast to the other major RNAs, mRNA is transcribed from genes that have only a few copies in each cell. The other major types of RNA are copied from genes with multiple copies.

In bacteria and mitochondria, a single RNA polymerase synthesizes all three major kinds of RNA. All RNA polymerases require all four nucleoside triphosphates, a template of DNA, and a divalent cation. The enzymes do not require a primer and do not proofread the RNA product.

RNA synthesis and processing can be inhibited by actinomycin D, rifamycin, α-amantin, and proflavine. Only the first two of these have any clinical use.

7 Genetic Code and Protein Synthesis

Protein synthesis requires the functioning of all three major classes of RNA. The directions are given by mRNA with each three-base sequence serving as the codon or code word for a single amino acid. tRNA and the amino acyl tRNA synthetases serve as the translators of the language of amino acids and that of nucleotides. Ribosomes provide the enzymes and the structure on which the entire process takes place. Many of the commonly used antibiotics exploit the differences between the protein-synthesizing systems of bacteria and the human cell cytoplasm.

Objectives

After completing this chapter, you should be able to

Discuss the characteristics of the genetic code. Define the terms *wobble, codon,* and *degeneracy.*

Define the following types of mutations: *silent, transitional, transversional, missense, nonsense,* and *frameshift.*

Outline the events that occur during the activation, initiation, elongation, and termination phases of protein synthesis. Identify the points of energy usage. Define *polysome.*

Identify the site of action of chloramphenicol, tetracycline, streptomycin, erythromycin, cycloheximide, puromycin, and diphtheria toxin.

Genetic Code

The genetic code is given in Table 7-1. It is a triplet code found in the mRNA and composed of 64 words read in the 5′ to 3′ direction. This is the same direction in which mRNA is synthesized. For most amino acids, there is more than one **codon** or triplet. Because of this, the genetic code is said to be **degenerate.** However, it is not ambiguous. Each code word has a specific meaning. Sixty-one codons code for amino acids. Three codons—**UAA, UAG,** and **UGA**—call for stop. There are two initiation or start-here codons, GUG and **AUG.** However, only AUG appears to function as a start codon in mammals. At the beginning of a message, both GUG and AUG code for an initiating methionine. When these

codons occur in the middle of a message, GUG codes for valine and AUG for methionine.

The genetic code has always been believed to be universal. The same codons in mRNA signal for the same amino acids in plants, bacteria, fish, frogs, monkeys, and humans. However, mitochondria have a different interpretation of several codons, as shown in Table 7-2. In mitochondria, AUA calls for methionine, UGA for tryptophan, and AGA and AGG for stop. This gives mitochondria two codons for methionine (AUA and AUG), two codons for tryptophan (UGA and UGG), and four stop codons (UAA, UAG, AGA, and AGG). These four mitochondrial coding changes are the only exceptions to the universality of the genetic code that have been found.

Table 7-1. Genetic code

First base	Second base				Third base
	A	G	C	U	
A	lys	arg	thr	ile	A
	lys	arg	thr	met	G
	asn	ser	thr	ile	C
	asn	ser	thr	ile	U
G	glu	gly	ala	val	A
	glu	gly	ala	val	G
	asp	gly	ala	val	C
	asp	gly	ala	val	U
C	gln	arg	pro	leu	A
	gln	arg	pro	leu	G
	his	arg	pro	leu	C
	his	arg	pro	leu	U
U	stop	stop	ser	leu	A
	stop	trp	ser	leu	G
	tyr	cys	ser	phe	C
	tyr	cys	ser	phe	U

Table 7-2. Differences between codons in mitochondria and eukaryotic cell cytoplasm

	Interpretation	
Codon	In cytoplasm	In mitochondria
AUA	ile	met
AGA	arg	stop
AGG	arg	stop
UGA	stop	trp

GGU, GGC, GGA, and GGG all code for the amino acid glycine. The first two bases alone specify the amino acid. In contrast, the third position of the codon can contain any of the four normal bases, A, U, G, or C. This relative nonspecificity of the third base of the genetic code is referred to as **wobble.** Wobble occurs in the codons for many of the amino acids.

Wobble occurs because of some unusual base pairing between bases in the anticodon in tRNA and the codon in mRNA. These unusual base pairs, which are shown in Table 7-3, can occur because they are always between the third base of the codon in mRNA and the first base of the anticodon in tRNA. Because the mRNA and tRNA base pairing is antiparallel, as is DNA, these odd combinations of

Table 7-3. Wobble base pairs

Base at position 1 of anticodon	Acceptable complementary base in mRNA
A	U
G	C or U
I	A, G, or U
C	G
U	A or G

bases are always at the end of the three–base pair interaction between tRNA and mRNA. If the first base of the anticodon is G, it can pair with either its normal partner C or with U. We have seen G:U base pairs before in our discussion of tRNA. If U is the first base of the anticodon, it can pair with A or G. If I occurs in the anticodon in the first position, it can pair with A, G, or U. Because only three base pairs are involved, purine-purine base pairs can occur at the ends of this short double-stranded structure. C and A in the anticodon only base pair with their normal counterparts G and U, respectively. Remember that the codon and the anticodon run antiparallel. Wobble always involves the third base of the codon and the first base of the anticodon.

mRNA is read one codon at a time in the 5′ to 3′ direction. There are no breaks between code words and no forms of punctuation except for start and stop. The start codon sets the three-base reading frame. Because of this, it is possible to read an mRNA in different reading frames by starting at a different base. The nucleotide sequence given here can be read to produce three different amino acid sequences depending on the starting base.

AGG/GCU/UAC/UC/AGA
arg   ala   tyr   phe   arg

A/GGG/CUU/ACU/UCA/GA
gly   leu   thr   ser

AG/GGC/UUA/CUU/CAG/A
gly   leu   leu   gln

In higher organisms, mRNAs are read in only one of the three possible reading frames. However in viruses, such as bacteriophage φX174, all three possible reading frames are sometimes used.

Mutations

In a gene or an mRNA, the change of one pyrimidine for another (for example C for T) or one purine for another (G for A) is a **transitional mutation.** This is what would be expected from a deamination of A or C that was not corrected before DNA replication occurred. The exchange of a pyrimidine for a purine, or the reverse, is a **transversional mutation.** We do not understand the mechanism producing transversional changes. Because of the degeneracy of the genetic code, these types of mutations may produce no change in the resulting protein. This is called a **silent mutation.** For example, a change from CUU to CUC (transition) or CUG (transversion) would produce no change in the protein as all these codons call for leucine. This type of mutation can be detected only by sequencing the DNA or the mRNA.

A **missense mutation** substitutes one amino acid for another. For example, a change from CUU, a codon for leucine, to AUU, a codon for isoleucine, would produce a missense mutation. If the amino acids are similar, as they are in the example, this may produce a functional protein. If the amino acids are dissimilar, as they would be if glutamine (CAA) were substituted for glutamic acid (GAA), the protein probably would be inactive.

A third possibility is a mutation that produces no protein at all. This can occur if the mutation converts a codon for an amino acid into a codon for termination, which is called a **nonsense mutation.** Changing the codon UCA (serine) into UAA (stop) would be a nonsense mutation.

Insertion or deletion of a single nucleotide after the reading frame has been set scrambles the rest of the message. For example, consider the message

AUG/GCU/ACU/UCA/GAG/UGA
met   ala   thr   ser   glu   stop

If there is an insertion of a U after six bases, the message will read

AUG/GCU/UAC/UUC/AGA/GUG/A
met   ala   tyr   phe   arg   val

This is called a **frameshift mutation.** The addition of two bases or the deletion of one or two has a similar effect. Only the addition or deletion of multiples of three bases will restore the reading frame and leave the protein relatively intact.

Mechanism of Protein Synthesis

Like RNA synthesis, protein synthesis or translation can be divided into stages: activation, initiation, elongation, and termination. We will discuss the mechanism in terms of the eukaryotic system. The differences between prokaryotic and eukaryotic systems will be pointed out as we come to them.

Activation

The activation phase of protein synthesis involves the binding of an amino acid to a specific tRNA. The reaction is catalyzed by enzymes called **amino acyl tRNA synthetases.** These enzymes must recognize both a specific amino acid and its corresponding tRNA. Considering the structural similarities of amino acids and of tRNAs, these enzymes must be very specific in their interactions. Because there are 20 amino acids that occur naturally in protein, there must be at least 20 different amino acyl tRNA synthetases.

In the first step of the activation reaction, the synthetase enzyme attaches the amino acid to the AMP portion of ATP with the hydrolysis of pyrophosphate. In the second reaction, the amino acid is transferred to either the 2′ or the 3′-OH of the adenosine on the 3′ end of the appropriate tRNA (Fig. 7-1). This process is referred to as *charging of the tRNA.* Once

Fig. 7-1. Steps in the activation of amino acids for protein synthesis. The amino acid first reacts with ATP before being attached to the amino acyl tRNA synthetase. The amino acid then is transferred to the 2′ or 3′ position of the terminal adenosine of the appropriate tRNA.

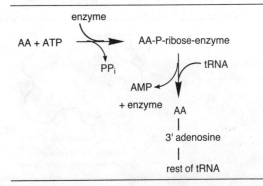

charged, only the anticodon on the tRNA recognizes the codon on the mRNA. The amino acid plays no further role in recognition.

To ensure that only the correct amino acid is added to the tRNA, many synthetase enzymes contain a proofreading function that can remove an amino acid incorrectly added to the tRNA. As seen in Figure 7-2, there are two active sites in the synthetase molecule. In the first site, the amino acid is attached to the tRNA. If an amino acid has the wrong structure or is too large, it will not fit into the first site and will not be added to the tRNA. However, an amino acid that is smaller than the correct amino acid can be attached to the tRNA. This type of error can be corrected by action at the second active site on the enzyme. This site is too small to accommodate the correct amino acid, but a smaller amino acid can enter this site and be cleaved from the tRNA.

Initiation

In initiation, all the components of the protein-synthesizing system are assembled and properly aligned to begin the process of translation. Proteins are made from the N-terminal end to the C-terminal end, with methionine as the first amino acid. To begin the process, methionine is attached to a specific initiator tRNA by the activation process previously described. In prokaryotes and mitochondria, a formyl group then is added to the α-amino group of the methionine, producing N-formyl methionyl tRNA (fMet-tRNA). Formylation occurs only after the me-

thionine is attached to the initiator tRNA, and only methionine attached to this specific tRNA can undergo the formylation reaction.

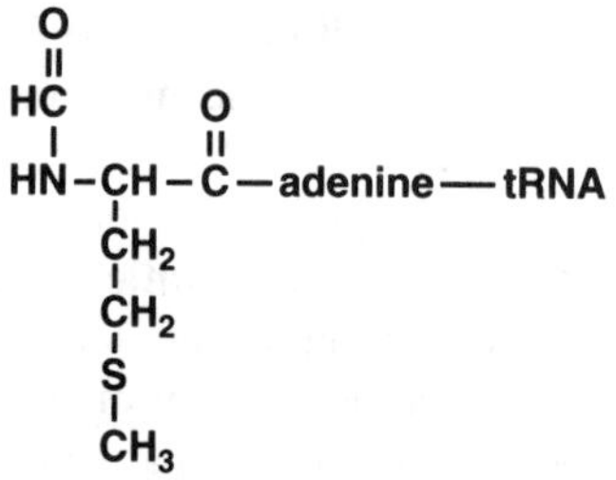

$$
\begin{array}{c}
O \\
\parallel \\
HC \qquad\qquad O \\
\mid \qquad\qquad \parallel \\
HN-CH-C-\text{adenine}-\text{tRNA} \\
\mid \\
CH_2 \\
\mid \\
CH_2 \\
\mid \\
S \\
\mid \\
CH_3
\end{array}
$$

N-Formyl methionyl tRNA

It is the fMet-tRNA that recognizes the AUG or GUG start codon and sets the reading frame that starts the process of protein synthesis. In eukaryotes, there is also an initiating tRNA that is specific for methionine. However, the methionine bound to this initiator tRNA is not formylated.

The first stage of initiation (Fig. 7-3) involves the formation of a complex among the 40S ribosomal subunit, Met-tRNA$_i$, GTP, and initiation factors. The initiation factors for eukaryotes are indicated by *e* for "eukaryote," *I* for "initiation," and then *F* for "factor," followed by a number to indicate the specific factor. Factor eIF6 is required to separate the ribosomal into 40S and 60S subunits. Then a complex is formed combining factors eIF2, eIF3, eIF4c, the initiating tRNA, and the 40S ribosomal subunit. eIF2 can function only in this complex when it has a bound GTP. Next the mRNA is added. This requires

Fig. 7-2. Proofreading of amino acyl tRNA synthetase. Amino acids that are larger than the correct amino acid do not fit into the catalytic site and are not attached to the tRNA. Amino acids that are smaller than the correct amino acid fit into the proofreading site and are cleaved from the tRNA.

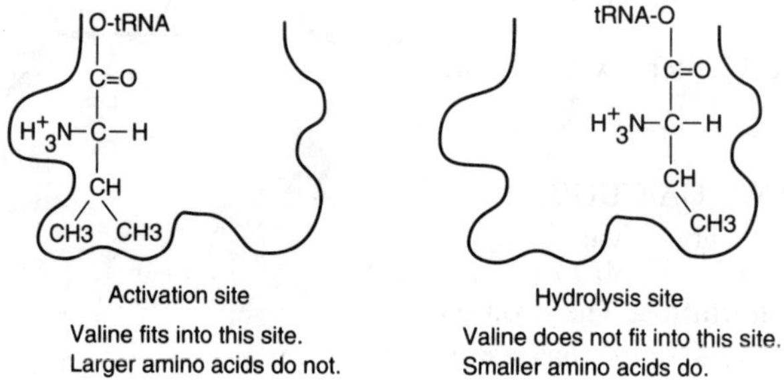

Activation site

Valine fits into this site.
Larger amino acids do not.

Hydrolysis site

Valine does not fit into this site.
Smaller amino acids do.

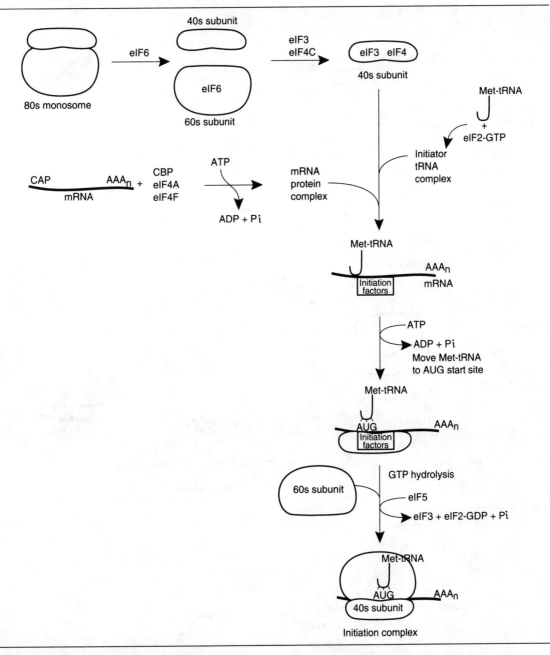

Fig. 7-3. Steps in the initiation stage of protein synthesis.

a cap binding protein (CBP) and two more initiation factors, eIF4a and eIF4. ATP is hydrolyzed as the mRNA is positioned, so that the initiating tRNA is at the AUG start codon. Finally, the large ribosomal subunit is added. This requires eIF5 and the hydrolysis of the GTP bound to eIF2. Then, eIF3 and eIF2, which is carrying guanosine diphosphate (GDP) produced by the hydrolysis, are released.

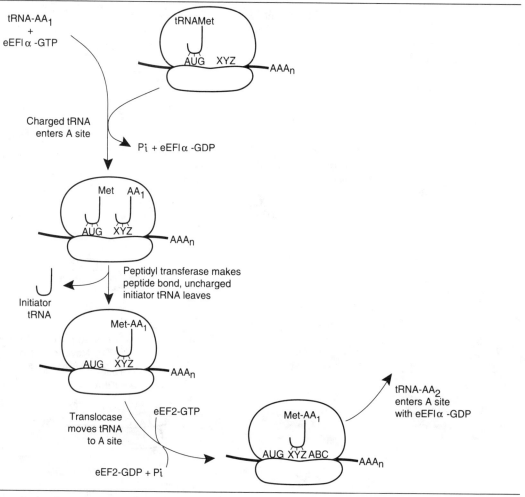

Fig. 7-4. Steps in the elongation stage of protein synthesis.

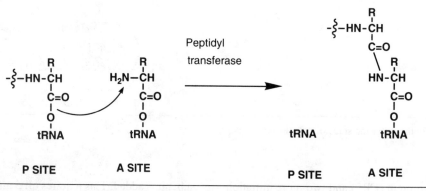

Fig. 7-5. Reaction catalyzed by peptidyl transferase. Amino acid 1, which is attached to the tRNA at the P site, is transferred from that tRNA to amino acid 2, which is attached to the tRNA at the A site. This forms a peptide bond between the carboxyl end of amino acid 1 and the amino terminal end of amino acid 2.

Ribosomes contain two binding sites for tRNA. The peptidyl, or P, site contains the growing peptide chain, and the amino acyl, or A, site contains a charged tRNA. The initiator Met-tRNA is the only charged tRNA that binds to the P site. All others bind to the A site.

Elongation

As illustrated in Figure 7-4, charged tRNAs form a complex with elongation factor eEF1α, which contains a bound GTP. This complex then enters the A site of the ribosome. If the hydrogen bonding between the codon in the A site and anticodon on the charged tRNA is correct, the charged tRNA is bound to the A site, GTP is hydrolyzed, and the elongation factor is released along with the GDP. This process changes the conformation of the ribosome so that the charged tRNA now is tightly bound to the A site.

Next, **peptidyl transferase,** a protein on the large ribosomal subunit, transfers the methionine from the initiator tRNA to the amino group of the amino acid attached to the tRNA at the A site (Fig. 7-5). This leaves the initiator tRNA uncharged. Uncharged tRNAs have a low affinity for the ribosome and dis-sociate. A translocase, using elongation factor eEF2 and the hydrolysis of GTP for energy, move the tRNA with its two attached amino acids to the P site. This opens up a new codon in the mRNA at the A site. eEF2 then dissociates with the bound GDP.

Elongation proceeds as a repetition of these reactions. A new, charged tRNA is bound to the A site by hydrogen bonding between the codon in the mRNA and the anticodon in the tRNA. The growing peptide is transferred to it. Then the entire system moves to the P site, and a new, charged tRNA enters the A site.

The initiation and elongation factors that contain bound GDP must release the GDP and bind GTP before they can function again in protein synthesis. This includes eIF2, eEF1α, and eEF2. For eEF1α, this process of exchanging GDP for GTP requires the assistance of another elongation factor, eEF1β, as shown in Figure 7-6.

Termination

When the codon in the A site is one of the three stop codons—UAA, UAG, or UGA—release factor (eRF) binds to the ribosome. Using peptidyl transferase and the hydrolysis of GTP, the protein is re-

Fig. 7-6. Exchange of GDP for GTP in cycling of initiation and elongation factors.

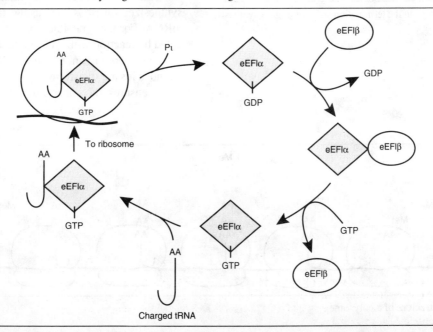

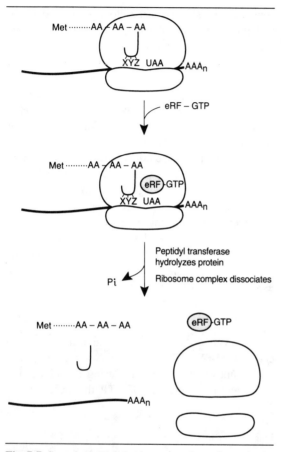

Fig. 7-7. Steps in the termination stage of protein synthesis.

leased from the tRNA. The uncharged tRNA dissociates, and the entire process can begin again (Fig. 7-7). Like other factors that enter the protein-synthesizing system with bound GTP and dissociate with bound GDP, eRF must exchange the GDP for GTP before it can function again in protein synthesis.

Polyribosomes

More than one ribosome can translate the same mRNA at the same time. In both eukaryotes and prokaryotes, this forms a large structure called a **polyribosome** or **polysome.** This structure is composed of a single mRNA and multiple ribosomes with the attached growing protein (Fig. 7-8).

Energy Requirement

Two ATPs are needed to activate each amino acid, and two GTPs are required for each elongation reaction. This means that, on the average, four high-energy phosphate bonds are hydrolyzed for each amino acid added to a growing protein chain.

Nonribosomal Synthesis of Glutathione

The tripeptide glutathione (γ-glutamyl-cysteinyl-glycine) is a major reducing agent in human cells. Glutathione and several other small peptides are not synthesized on ribosomes using directions from an mRNA. For these peptides, each amino acid is activated by reaction with ATP and then is transferred to a thio group on an enzyme. All these activated amino acids are transferred again to another enzyme, which ensures the proper order of amino acids in the

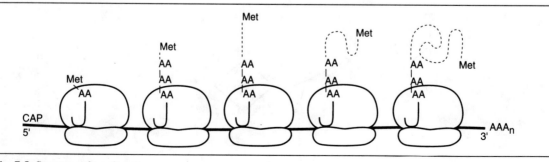

Fig. 7-8. Structure of a polysome.

final peptide. Finally, the amino acids are attached to one another to form glutathione (Fig. 7-9).

Inhibitors

Many of the familiar clinical antibiotics act at the level of protein synthesis. Among those you should recognize are chloramphenicol, tetracycline, streptomycin, and erythromycin. All these compounds inhibit protein synthesis in bacteria but not in the eukaryotic cytoplasm. At sufficiently high concentrations, all can affect protein synthesis in mitochondria.

Chloramphenicol is a broad-spectrum bacteriostatic agent. It inhibits elongation by preventing the function of peptidyl transferase. Because peptide bond formation is prevented, both the A and the P sites on the ribosome are occupied but no protein synthesis takes place. **Cycloheximide** is a compound that has a similar effect on eukaryotic ribosomes. Because chloramphenicol is associated, in rare cases, with aplastic anemia, it is used infrequently now in clinical practice.

Tetracycline is another wide-spectrum antibiotic that also affects elongation. It functions by preventing the attachment of the charged tRNA to the A site on the ribosome. Because tetracycline causes a per-

manent discoloration of developing teeth, it is not prescribed for pregnant women, infants, or young children.

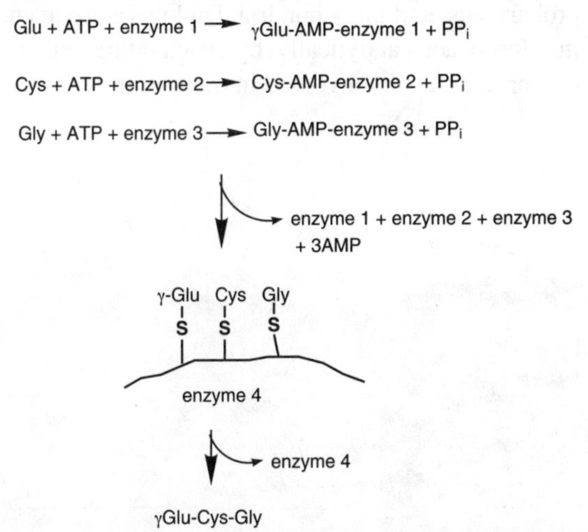

Streptomycin binds to one of the proteins of the small ribosome subunit. This distorts the subunit, which inhibits initiation and causes faulty base pairing between the codon in mRNA and the anticodon in tRNA. Incorrect amino acids are added to the growing peptide chain because of misreading of the codons. Neomycin, another antibiotic, also causes mistranslation of the genetic code due to binding to the small ribosomal subunit.

Erythromycin inhibits the ribosomal translocase on the large ribosomal subunit of a wide variety of bacteria. It is the antibiotic of choice for treating legionnaires' disease, a fast-progressing pneumonia. It also is used to treat mycoplasma infections.

There are other drugs that are used experimentally to inhibit protein synthesis in the cytoplasm of eukaryotic cells. Use of these compounds has produced much valuable information about the mechanism of protein synthesis.

To a cell, **puromycin** looks like the 3′ end of a charged tyrosyl tRNA (Fig. 7-10). Puromycin binds to the A site of the ribosome, and the growing peptide chain is transferred to it. Because puromycin is

Fig. 7-9. Synthesis of glutathione. Only small peptides such as glutathione are not synthesized on ribosomes, using the information in mRNA to determine the order of amino acids.

Glu + ATP + enzyme 1 ⟶ γGlu-AMP-enzyme 1 + PP$_i$

Cys + ATP + enzyme 2 ⟶ Cys-AMP-enzyme 2 + PP$_i$

Gly + ATP + enzyme 3 ⟶ Gly-AMP-enzyme 3 + PP$_i$

⟶ enzyme 1 + enzyme 2 + enzyme 3 + 3AMP

γ-Glu Cys Gly
  S   S   S

enzyme 4

⟶ enzyme 4

γGlu-Cys-Gly

Fig. 7-10. Structural similarity between puromycin and tyrosyl tRNA.

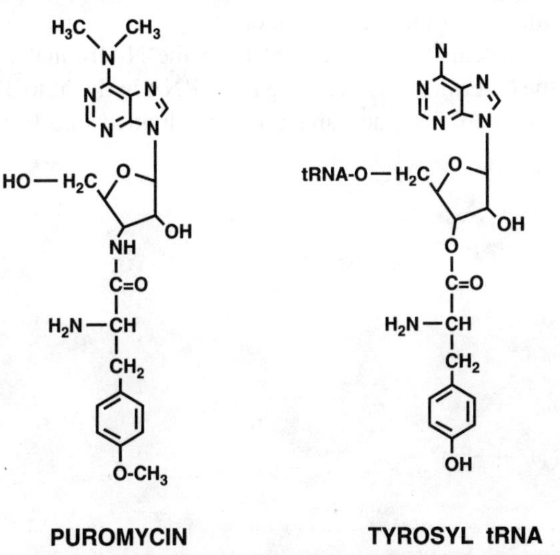

PUROMYCIN   **TYROSYL tRNA**

much smaller than a charged tRNA, it dissociates from the ribosome and carries the newly synthesized peptide with it. This produces premature chain termination. Puromycin affects protein synthesis in prokaryotes, eukaryotes, and mitochondria.

Diphtheria toxin modifies and inactivates one of the elongation factors in eukaryotes, preventing translocation. Because the toxin acts catalytically on an enzyme, only a few molecules are sufficient to halt protein synthesis and eventually kill the cell. **Interferon** also acts catalytically by inhibiting the initiation step of protein synthesis in eukaryotes. Interferon activates a protein kinase, which then phosphorylates one of the initiation factors. This inactivates the factor and stops protein synthesis.

Summary

The genetic code is triplet, degenerate, and commaless but not quite universal. There are 61 codons for amino acids, two of which also serve as start signals. Three codon words call for stop.

Changes in the base sequence of an mRNA generally change the resulting protein. A missense mutation changes the codon for one amino acid to that for another. A nonsense mutation changes the codon for an amino acid to one for stop. These changes can be due to transitions or transversions. A frameshift mutation, due to the insertion or deletion of bases, changes the reading frame and all the amino acids after the point of the mutation.

Proteins are synthesized from the N-terminal to the C-terminal end, reading the mRNA in the 5′ to 3′ direction. The mechanism can be divided into four stages. Activation involves attaching the amino acid to AMP and then transferring it to the 3′ end of the appropriate tRNA. In initiation, the small ribosomal subunit, mRNA, the Met-tRNA$_i$ and initiation factors bind together with the hydrolysis of GTP. This positions the initiating tRNA at the AUG start site on the mRNA. Then the large ribosomal subunit is added so that the initiating tRNA is in the P site.

In elongation, a charged tRNA enters the A site along with elongation factors, and the first peptide bond is formed by peptidyl transferase. Then the entire complex is translocated one codon so that the A site is free and aligned with the next codon. This reaction requires the hydrolysis of GTP. The process of tRNA binding, peptide bond formation, and translocation is repeated until a termination codon is encountered in the A site. This causes release factors to bind and hydrolyze the protein from the tRNA. Then both the protein and the uncharged tRNA leave the ribosome, which dissociates into subunits to begin the process again. Protein factors that carry GTP to the ribosome and dissociate with GDP must exchange the GDP for GTP before functioning again in protein synthesis.

Chloramphenicol, tetracycline, streptomycin, and erythromycin are all clinically important antibiotics that act to inhibit protein synthesis in bacteria. Cycloheximide is an experimental agent that inhibits protein synthesis in the eukaryotic cell cytoplasm. Puromycin is structurally similar to tyrosyl tRNA and inhibits protein synthesis in eukaryotes, prokaryotes, and mitochondria. Diphtheria toxin and interferon act catalytically by inactivating eukaryotic protein factors involved in translation.

8 Control of Gene Expression

With the exception of mature cells of the immune system, all cells of the human body contain the same genetic information in their DNA. However, they do not make the same proteins. In adults, only red blood cells and their precursors make hemoglobin. Only hepatocytes make albumin and then only in the adult. The fetus makes α-fetoprotein. Much is known about the mechanism bacterial cells use for selection of the portion of DNA to be transcribed. However, much less is known about eukaryotes. This is an area of intense research and the coming years will likely reveal much of the missing information.

Objectives

After completing this chapter, you should be able to

Define and give an example of a *constitutive enzyme,* an *inducible enzyme,* an *operon, catabolite repression, enzyme repression,* and *attenuation.*

Describe the structural motifs common in proteins that bind to DNA. Define the terms *response element, enhancer, silencer, oncogene,* and *tumor suppressor gene.*

Discuss the changes that occur to RNA in posttranscriptional processing and that are subject to control.

Discuss the events that can occur in posttranslational processing of proteins.

Describe the events preceding protein phosphorylation when initiation is by cyclic AMP, calcium, and protein kinase C. Define *G-protein.*

Control in Prokaryotes

Enzyme Induction

Some proteins are made in the same amount by nearly all cells at all times. These are called **constitutive proteins** or **constitutive enzymes.** For example, the protein products of the genes coding for the enzymes of glycolysis and other housekeeping functions are required by all cells and are constitutively expressed. Most other genes are controlled so that the protein product is synthesized only when needed.

In prokaryotes, control is exerted predominantly at the level of transcription. If a protein is not needed, the gene coding for it will not be transcribed. Because the half-life of mRNA in bacteria is on the order of minutes, control at the transcriptional level allows bacteria to adapt quickly to changing environmental conditions.

Inducible enzymes are those that are present only when the substrate for them is present. In 1961, Jacob and Monod proposed the operon model to explain the control of the enzymes that metabolize lactose. These enzymes are responsible for the transport of lactose into the cell and hydrolysis of lactose into galactose and glucose. The structural genes coding for β-galactosidase (z), galactoside permease (y), and a transacetylase (a) are linked to one another and to two regulatory sites. One of these is the promoter (p) where the RNA polymerase binds. The other is an operator (o) site, which lies between the promoter and the start of the structural genes. The operator controls the ability of the RNA polymerase to transcribe the structural genes. In addition, there is a regulator gene (i) which may be located anywhere in the genome. A group of structural genes and the genes that control them is called an **operon.** The genes for lactose metabolism are called the **lac operon.** Because the three structural genes are controlled by the same regulatory signals, the enzymes are coordinately expressed.

<u>i</u>   <u>p o z y a</u>

The i gene is constitutively expressed. The mRNA transcribed from this gene is translated into a protein product called a **repressor.** The repressor protein binds tightly to the operator site with a dissociation constant of approximately 10^{-12} M. The tight binding of the repressor to the operator site inhibits the RNA polymerase from transcribing the three structural genes. However, a small amount of transcription of these genes does take place even in the presence of the repressor.

When lactose is present, it can be converted by β-galactosidase, which is always present in small amounts, to allolactose. This metabolite of lactose can bind to the repressor. The binding of allolactose alters the repressor so that it can no longer bind to the operator. Under these conditions, the operator site is open and the RNA polymerase can transcribe the structural genes. This results in a 10- to 100-fold increase in the levels of the lactose-metabolizing enzymes. These events are outlined in Figure 8-1.

When lactose has been completely metabolized, there will no longer be allolactose in the cell to bind to the repressor. The repressor can then bind to the operator again and turn off transcription.

Two distinct kinds of constitutive mutants have been observed in operons regulated by enzyme induction mechanisms. Operator mutants exist in which the operator has been altered so that it cannot bind the repressor. Similarly, there exist i-gene mutations that produce a repressor incapable of binding to the operator. Both of these mutations lead to the production of the enzymes of the lac operon, regardless of the amount of lactose in the cell. However, if a

mutation occurs in the i gene such that the repressor cannot bind to allolactose, the lactose-metabolizing genes are permanently turned off. Mutations of these kinds have been found in the analysis of altered genes responsible for inherited human diseases.

Catabolite Repression

Bacteria prefer to metabolize glucose rather than lactose if both are present. Hence, there must be a mechanism for repressing the lac operon when both lactose and glucose are present. This happens by **catabolite repression** using cyclic AMP (cAMP) as a positive regulator. In the absence of glucose, the concentration of cAMP is high. The cAMP binds to a cAMP acceptor protein (CAP), which in turn binds near the promotor. Binding of CAP causes RNA polymerase to bind more tightly to the promotor site and increases the level of transcription of the structural genes. When glucose is present, the cAMP level is low. Then the promotor is not activated, and the RNA polymerase binds loosely and transcribes the structural genes infrequently. Little enzyme is produced, even though lactose is present and the repressor is not bound to the operator site.

Enzyme Repression

Operons for biosynthetic genes are controlled in a manner opposite that of catabolic genes. The enzymes for biosynthesis are required except when the final products of the enzymatic pathway are present. Then the genes are turned off. This type of control is called **enzyme repression,** and, in it, the regulator gene makes a protein product (apo-repressor) that cannot bind to the operator. The operator is open, and RNA polymerase can bind to the promoter site and transcribe the structural genes.

When the final product is present, such as the amino acid tryptophan for the trp operon, it binds to the repressor as a co-repressor. This combination of apo-repressor and co-repressor is able to bind to the operator, which inhibits RNA polymerase binding and the transcription of the structural genes, as shown in Figure 8-2.

Just as in enzyme induction, constant production of the enzymes occurs when a mutation in the operator prevents repressor binding. Constitutive expression also occurs when a mutation in the repressor

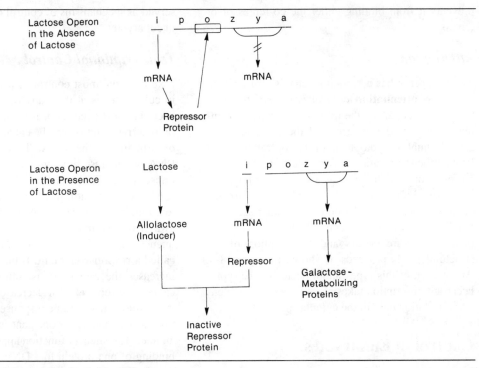

Fig. 8-1. Mechanism of enzyme induction in the lactose operon.

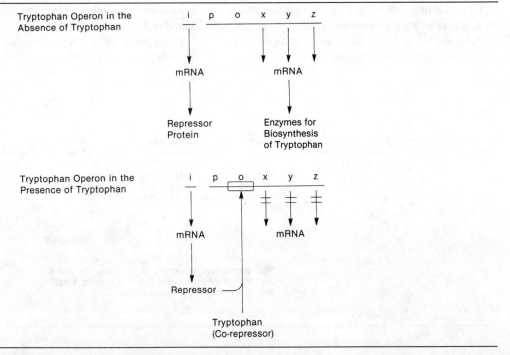

Fig. 8-2. Mechanism of enzyme repression in the tryptophan operon.

prevents it from binding either the co-repressor or the operator.

Attenuation

The trp operon has a second mechanism of control, known as **attenuation** or *transcriptional termination*. The first part of the trp operon contains an attenuation or leader sequence. If the cell contains a large amount of tryptophan, this region can be translated. When translation occurs, the mRNA forms a hairpin loop, which serves as a stop signal for RNA synthesis. Further transcription of the tryptophan operon ceases. If translation of the leader sequence is inhibited by a lack of tryptophan, a different secondary structure occurs and transcription of the structural genes proceeds as shown in Figure 8-3. Attenuation of this type cannot occur in eukaryotes because transcription takes place only in the nucleus and translation only in the cytoplasm.

Control in Eukaryotes

Most gene expression in bacteria is regulated at the transcriptional level by an on-off switch. In eukaryotes, control can be exerted at multiple levels that can increase or decrease the level of gene expression or function as on-off switches. Transcription, processing, mRNA stability, translation, gene amplification, and gene rearrangement have all been observed as mechanisms of control of gene expression in eukaryotes.

Transcriptional Control

Like bacteria, most control of protein-coding genes in eukaryotes is at the transcriptional level. Unlike bacteria, most genes appear to be in the off configuration. Transcription can be enhanced by the binding of proteins to the DNA. This binding facilitates RNA polymerase binding and increases the rate of transcription. The mechanism has some similarity to catabolite repression in bacteria. However, the protein-binding sites can be thousands of nucleotides away from the transcription start site. The DNA sequence to which the transcription factors bind is called a **response element.** If binding of the protein increases the rate of transcription, the element is called an **enhancer.** If it decreases gene expression, it is a **silencer.** The same response element can function as an enhancer for one gene and as a silencer for another. The silencer function appears to occur when binding of one protein to a DNA sequence prevents the binding of a stimulatory protein to an overlapping DNA sequence. Binding of transcription factors to the DNA bends the DNA, so that proteins bound at a distance can interact with the RNA polymerase bound at the promoter site.

Three structural motifs are common in DNA binding proteins that function as transcription factors.

Fig. 8-3. Regulation of transcription of the tryptophan operon by attenuation.

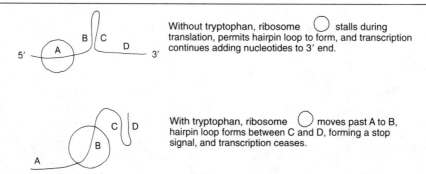

Without tryptophan, ribosome ◯ stalls during translation, permits hairpin loop to form, and transcription continues adding nucleotides to 3′ end.

With tryptophan, ribosome ◯ moves past A to B, hairpin loop forms between C and D, forming a stop signal, and transcription ceases.

These are *helix-turn-helix, leucine zipper,* and *zinc finger.* The first two structural motifs are related. The helix-turn-helix proteins have a positively charged N-terminal end that is in a protein α-helix structure. This is the portion of the molecule that interacts with the DNA by binding in the major groove. The helical part of the molecule is followed by a nonhelical segment. Then, there is a second helical segment that contains a high concentration of hydrophobic amino acids. This segment of the molecule can interact with other similar molecules, forming dimers that have a Y shape, as shown in Figure 8-4. Both homodimers and heterodimers are possible. Depending on the combination of molecules, different response elements can be activated. In the leucine zipper, the second helical segment contains leucine every seventh amino acid residue. Molecules with this structural motif interact with the leucine residues on the two arms of the dimer, fitting together like the teeth on a zipper. In the zinc finger, zinc atoms are bound by four cysteine or two cysteine and two histidine residues. This interaction holds the molecule in a specific configuration and allows it to interact with DNA. The proteins that interact with the steroid hormones are zinc finger proteins.

A number of transcription factors function as **oncogenes** if overexpressed or modified so that they are always in the active form. In the leucine zipper family of factors, fos and jun are known to stimulate the transcription of genes controlling the cell cycle. If there is an excess of these transcription factors, the cell cycle genes will be overexpressed and cells will continue to cycle even in the presence of stop signals. Other transcription factors are known to regulate genes that inhibit the cell cycle. A lack of these transcription factors prevents these genes from being transcribed and also allows uncontrolled cell cycling. Genes coding for these transcription factors can act as **tumor suppressor genes.** The retinoblastoma gene is an example of a tumor suppressor gene that is a transcription factor.

Posttranscriptional Control

The rate at which heterogeneous nuclear RNA is converted into mature mRNA is another level at which genes are regulated in eukaryotes. In addition,

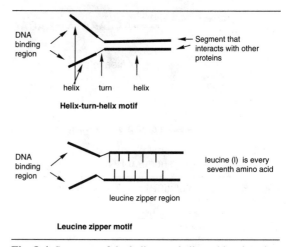

Fig. 8-4. Structure of the helix-turn-helix and leucine zipper, two types of eukaryotic transcription factors.

the choice of the poly-A addition site and splicing sites can produce different mRNAs from a single primary transcript. Figure 8-5 shows the result of choosing one of two poly-A addition sites on the mRNA produced. Because the 5' end of both mRNAs will be the same, the N-terminal portion of resulting protein will be the same. However, the C-terminal end of the protein may be different if sequences in the longer mRNA are translated.

Alternative splicing can produce a large change in the resulting protein. By retaining or splicing out segments of the RNA, different mRNAs can be produced from the same primary transcript. The result of such alternative splicing is shown in Figure 8-6. If the first exon is not translated, there may be no relationship between the proteins translated from mRNAs that have been differentially spliced. Alternative splicing is a common mechanism of control of gene expression in cells producing hormones. However, a given cell will process a primary transcript in only one way, whereas a different cell type may process the same primary transcript in another way, producing a different mRNA and a different protein.

mRNA stability is another area in which gene expression can be controlled. The average eukaryotic

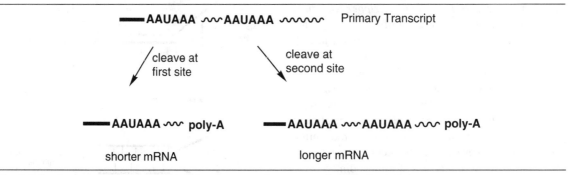

Fig. 8-5. Production of different mRNAs from the same primary transcript by poly-A choice.

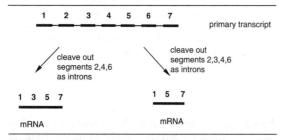

Fig. 8-6. Production of different mRNAs from the same primary transcript by differential splicing.

mRNA has a half-life of approximately 24 hours. However, this can vary from several minutes to several days. How long mRNA survives in cytoplasm is one factor that determines the number of times it will be translated into protein. Hormonal stimulation has been shown to increase the half-life of specific mRNAs by severalfold, though the mechanism for this increase in half-life is unclear.

Gene amplification is not a common mechanism for the control of gene expression in normal mammalian cells. However, it occurs in tumor cells and has been studied extensively in cells resistant to the drug methotrexate. In these cells, the gene coding for the enzyme dihydrofolate reductase, the target enzyme of the methotrexate, is amplified several hundred times. With this amplification, a large amount of the enzyme can be produced. Under these conditions, methotrexate is not effective in inhibiting the growth of these cells.

As seen earlier in Chapter 4, **gene rearrangement** is a mechanism for the control of gene expression in cells of the immune system. By rearranging the DNA, only one type of antibody can be produced by a given cell.

Posttranslational Control

Most proteins do not function in the same form in which they are first synthesized. The most obvious change is the removal of the initiating methionine from the N-terminal end of the protein. Other modifications include the addition of methyl, sulfate, phosphate, hydroxyl, and acetyl groups to specific amino acids. Prosthetic groups, lipids, and carbohydrate can be added, disulfide bonds formed, large segments of the protein cleaved, or the protein combined with other proteins to form large multisubunit structures.

Folding of the protein occurs during synthesis, often with the assistance of other proteins called *chaperones*. Disulfide bonds are formed at this time as well. However, most modifications of amino acids and the addition of prosthetic groups or coenzymes usually does not occur until after synthesis is completed. Prosthetic groups such as flavin adenine dinucleotide (FAD), NAD, or heme are required for the activity of proteins that function in oxidation-reduction reactions or in the transferring of chemical functional groups. Heme is added to four individual protein chains to produce hemoglobin. Pyridoxine, a derivative of vitamin B_6, is essential for the activity of several enzymes of amino acid metabolism. NAD from nicotinamide, another B vitamin, is necessary for many electron transfer reactions. As you can see, coenzymes and prosthetic groups are frequently derived from vitamins. These will be discussed in detail when we cover the specific reactions involved.

Insulin undergoes two different modifications after it is synthesized. The first protein formed is called *preproinsulin,* and it is folded so that only particular disulfide bonds can be formed. Formation of the disulfide bonds produces proinsulin. An internal segment of the protein is cleaved, producing insulin which contains two separate chains held together by disulfide bonds. It is only in this conformation that insulin is biologically active and can act as a hormone.

Cleavage of a portion of the peptide chain is a common activation step for other proteins as well. The digestive and blood-clotting enzymes are produced as inactive forms called **zymogens.** For example, the enzyme pepsin is formed from the zymogen pepsinogen, trypsin from trypsinogen, and thrombin from prothrombin. Activation by removal of a protein segment and refolding of the molecule, revealing the active site, protects the cell synthesizing the enzyme from its biological activity. In addition, a large amount of active protein can be created quickly without the delay required for new protein to be synthesized.

Proteins destined for particular cellular compartments contain specific targeting or localization sequences. Proteins destined for the mitochondria contain their targeting sequences in the N-terminal end of the protein. These sequences frequently are cleaved after the protein arrives at its destination. Specific localization sequences have been identified that direct proteins to the endoplasmic reticulum, peroxisomes, and the nucleus. These signals involve specific amino acid sequences which, in contrast to those for mitochondria, are not at the N-terminal end and usually are not cleaved. Modification of carbohydrate residues serves as the localization signal for proteins destined for lysosomes. Proteins destined for membranes frequently have long segments containing hydrophobic amino acids. These segments span the membrane and anchor the protein firmly to it. Soluble proteins and those to be secreted from the cell lack any localization signal.

Proteins destined for export are transported to the Golgi complex, where carbohydrate residues are added and modified. Each portion of the Golgi complex is responsible for the addition or modification of particular types of carbohydrate residues. Reactions occur sequentially as the protein moves through the Golgi. Then the protein is packaged in a vesicle and either released outside the cell or stored until a signal for release is received.

Phosphorylation is a major mechanism by which the activity of proteins is controlled. A number of different enzyme systems regulate phosphorylation, which normally occurs on serine and threonine residues of the proteins. However, there is a small amount of phosphorylation on the hydroxy group of tyrosine as well.

When a cell interacts with a protein hormone such as glucagon or the catecholamines, the hormone interacts with a specific receptor protein on the outside of the cell membrane. Inside the membrane is a complex of proteins that bind and hydrolyze GTP. These are called **G-proteins.** Activation of the G-proteins by the hormone receptor can produce second messengers, which in turn produce phosphorylation of cellular proteins.

The production of **cAMP** by the enzyme **adenylate cyclase** is one of the cellular responses to activation of the G-proteins (Fig. 8-7). When cAMP is produced from ATP, it is released into the cytoplasm, where it interacts with a tetrameric cAMP binding protein. Binding of cAMP causes the protein to dissociate into two catalytic and two regulatory subunits. The regulatory subunits each carry two cAMP residues bound to them. The catalytic subunits are protein kinases capable of phosphorylating a number of cellular enzymes. We will encounter regulation of enzymatic activity by cAMP at many times during our study of metabolism. cAMP also binds to another binding protein, cyclic AMP receptor protein (CRP), which then is translocated to the nucleus and serves as a transcription factor by binding to a specific DNA sequence. cAMP is hydrolyzed to 5'-AMP by a **phosphodiesterase.** The activity of the phosphodiesterase is inhibited by caffeine and other methylated xanthines.   ·

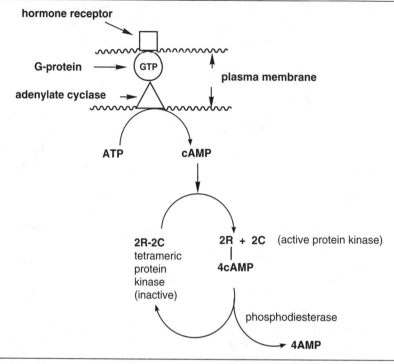

Fig. 8-7. Production of cAMP after hormonal binding.

Another response of the cell to activation of the G-proteins is a cascade begun by the hydrolysis of phospholipids, as shown in Figure 8-8. **Inositol triphosphate** is released from the plasma membrane. This compound, which will be discussed in more detail later, causes the release of calcium. Free calcium then interacts with a specific binding protein called **calmodulin**, which binds four calcium ions. Such binding alters the conformation of the calmodulin molecule, allowing it to interact with other proteins. This activates a number of different protein kinases, which in turn activate or inactivate other enzymes. Another result of the hydrolysis of phospholipids is the production of diacylglycerol in plasma membrane, which, in turn causes the activation and release of **protein kinase C** from the membrane. After interaction with calcium, protein kinase C can phosphorylate a number of different proteins.

Summary

Prokaryotes regulate gene expression primarily at the level of transcription. Genes are turned on by enzyme induction, which involves the binding of an inducer to a repressor protein. This binding releases the repressor from the operator site and allows RNA polymerase to transcribe the genes coding for proteins responsible for catabolizing the substrate. cAMP, acting as a catabolite repressor, binds to a protein that stabilizes RNA polymerase binding to the promoter site. In enzyme repression, the structural genes normally are turned on. When the aporepressor molecule binds to the product of the enzyme pathway (co-repressor), it is activated to bind to the operator, thereby preventing further transcription of the structural genes.

Control of gene expression in eukaryotes also is primarily at the level of transcription. In eukaryotes,

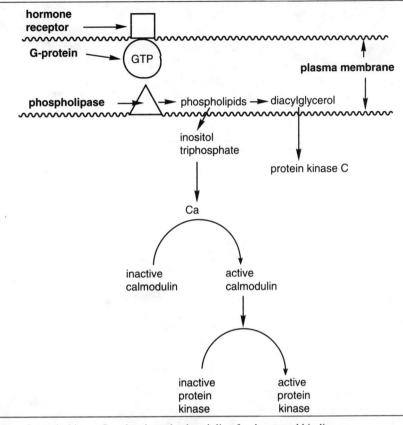

Fig. 8-8. Production of protein kinase C and activated calmodulin after hormonal binding.

transcription factors bind to specific DNA sequences called response elements. This binding enhances or inhibits tight binding of the RNA polymerase to the promoter region. Three major structural motifs are found in DNA binding transcription factors: helix-turn-helix, leucine zipper, and zinc finger. Proteins of this type frequently are dimers that bind in the major groove of DNA. Gene expression in eukaryotes can also be controlled by choice of a poly-A addition site, alternative splicing, gene amplification, gene rearrangement, and mRNA stability.

After synthesis, proteins are modified by adding small groups of carbohydrate or lipid or small prosthetic groups to the amino acids. Disulfide bonds can be formed and segments of the protein removed. Phosphorylation of proteins is another mechanism by which activity is controlled. Phosphorylation takes place by several different enzyme systems including cAMP-dependent protein kinases and calcium-dependent protein kinases. These kinases are activated through the G-proteins after the cell binds a hormone to a specific receptor on the cell membrane.

9 Tools of Molecular Biology

DNA can be fragmented at specific nucleotide sequences, producing a pattern that is characteristic of the individual from whom the DNA has been isolated. This DNA can then be denatured and hydrogen-bonded not only to its original complementary DNA strand but also to other DNAs and RNAs that have identical or nearly identical nucleotide sequences. The ability to form hybrid molecules is the basis for constructing clones of specific genes and for the detection, diagnosis, and treatment of disease. Formation of DNA hybrids also is used in mapping genes to particular chromosome locations, a technique employed in the genetic analysis of particular diseases.

Objectives

After completing this chapter, you should be able to

List the major characteristics of restriction endonucleases.

Outline the steps required for the construction of a recombinant DNA library.

List the requirements for amplifying DNA by the polymerase chain reaction.

Describe the procedure for sequencing DNA by the dideoxynucleotide method.

Describe the process of Southern, Northern, and Western blotting in terms of what type of molecule is being subjected to electrophoresis and what type of molecule is being used as the probe.

Describe the principles of an ELISA.

Restriction Endonucleases

Bacteria are capable of identifying their own DNA and distinguishing it from foreign DNA. The basis of the self-identification is methylation of single bases in a specific DNA sequence. The sequence usually contains four to seven nucleotides that form a **palindrome,** a sequence that is identical in the 5′ to 3′ direction on both strands. DNA is identified as self if it is methylated at the appropriate base in this DNA sequence. DNA such as that of an infecting virus, which is not methylated on at least one of the DNA strands, is recognized by a specific nuclease that cleaves both strands of the DNA. This cleavage may occur at the palindromic identification site or

some distance away. In either case, the double-stranded cleavage usually is lethal.

A virus that eludes this self-identification system and successfully infects the bacteria has its viral DNA modified by the bacterial host's methylation enzymes. Because of this, the viral DNA is not recognized as foreign by other bacteria containing the same DNA modification system. However, the viral DNA will be recognized as foreign by bacteria that carry different DNA modification systems. The enzymes responsible for recognizing the specific DNA sequence and cleaving any DNA that is not properly modified are called **restriction endonucleases.** These enzymes restrict the strains of bacteria that

Table 9-1. Restriction endonucleases

Name	Restriction and recognition site	Source
Eco RI	↓ CAATTC GTTCCG ↑	*E. coli* T 13
Eco RII	↓ CCAGG GGTCC ↑	*E. coli* R 245
Bam I	↓ GGATCC CCTAGG ↑	*B. amyloliquefacians*
Bgl II	↓ AGATCT TCTAGA ↑	*B. globiggi*
Hind III	↓ AAGCTT TTCGAA ↑	*H. influenzae*

the virus can infect successfully to those that carry the same DNA methylation pattern as the virus.

Restriction endonucleases cleave any unmodified DNA at the particular DNA sequence recognized by that restriction endonuclease. The frequency of the cleavage of any given DNA depends on how often the recognition sequence occurs in that particular DNA. Many of the restriction enzymes are commercially available. The recognition sequence and source of some of these enzymes are given in Table 9-1. As you can see, the restriction endonucleases do not necessarily cleave the two strands of DNA at sites directly across from each other. Instead, many of them leave a stagger of four to six bases. These single-stranded or sticky ends can hydrogen-bond with any other DNA cleaved by the same restriction endonuclease. At low temperatures, four to six base pairs are sufficient to hold together two different DNA molecules. The ability of restriction endonucleases to cleave DNA such that sticky ends remain which can then hydrogen-bond to other DNA cleaved by the same enzyme allows for the construction of recombinant DNA libraries.

Construction of Recombinant DNA Libraries

Genomic Library

The construction of recombinant DNA libraries from the genomic DNA of particular organisms has become fairly routine. Libraries from many species are available commercially. However, the identification of the desired recombinant from that library is more difficult.

The general scheme for the construction of a genomic recombinant DNA library is shown in Figure 9-1. First, the DNA is digested with a particular restriction endonuclease. This DNA then is combined with a vector that has been cleaved by the same endonuclease. The vector can be a plasmid, a circular self-replicating segment of DNA, a bacteriophage, a small virus that infects and replicates in bacteria, or a cosmid, a plasmid related to bacteriophage lambda from which most of the viral DNA has been deleted. Recombinant molecules are produced that contain both the vector and a segment of genomic DNA from the organism. The recombinant molecules can be ligated, forming covalently closed circular DNA that can be separated from the original vectors by size. When the recombinant vector is inserted into a suitable host, it will replicate and divide as the bacterial host replicates and divides. This eventually will produce many copies of the recombinant molecule.

Most vectors contain antibiotic resistance genes. In the presence of the antibiotic, only bacteria containing a recombinant vector will be able to grow and divide. The procedure described here produces a **genomic library** in which each segment of the original DNA is represented in at least one of the host bacteria. The progeny of each bacterium originally infected with the recombinant DNA represent a clone of that DNA.

Once identified, the desired DNA can be separated from host DNA and removed from the vector using the same restriction nuclease that was used to construct the recombinant molecule. Then the DNA can be sequenced or made radioactive for use as a probe in chromosome mapping or blotting experiments.

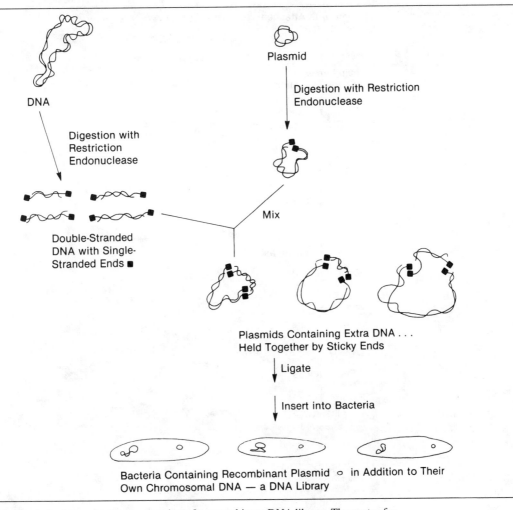

Fig. 9-1. General scheme for the construction of a recombinant DNA library. The vector for the recombinant can be a plasmid (as shown here) or a phage or cosmid.

cDNA Library

It is often difficult to identify the particular clone that contains the DNA of interest because bacteria neither transcribe nor translate eukaryotic genomic DNA. To solve this problem, most investigators have cloned DNA copies (cDNA) of all the mRNAs in a particular cell type or cloned the cDNA for a specific mRNA. If the bacteria can be induced to transcribe and translate the recombinant DNA, then the correct clone can be identified by its reaction with an antibody to the protein.

It is theoretically possible to produce or identify a recombinant cDNA for any mRNA for which the protein has been purified and an antibody prepared. The procedure for producing a recombinant clone to a specific mRNA is outlined in Figure 9-2. First, polysomes synthesizing the protein are precipitated with an antibody to the protein. The mRNA coding for the protein can then be isolated from the polysomes. Because antibodies usually are prepared against a mature protein, they may not recognize a protein in the process of synthesis if it is not in its

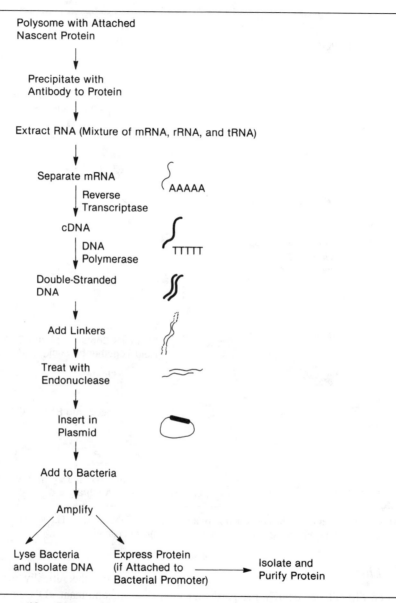

Polysome with Attached
Nascent Protein

↓

Precipitate with
Antibody to Protein

↓

Extract RNA (Mixture of mRNA, rRNA, and tRNA)

↓

Separate mRNA   ⎧ AAAAA
│ Reverse
↓ Transcriptase

cDNA   ⎧ TTTTT
│ DNA
↓ Polymerase

Double-Stranded
DNA

↓

Add Linkers

↓

Treat with
Endonuclease

↓

Insert in
Plasmid

↓

Add to Bacteria

↓

Amplify

↙   ↘

Lyse Bacteria   Express Protein
and Isolate DNA   (if Attached to ──→ Isolate and
                     Bacterial Promoter)   Purify Protein

Fig. 9-2. Procedure for cloning a specific mRNA.

mature conformation. If so, this method will not work. If an mRNA can be isolated, a cDNA is prepared using **reverse transcriptase,** a viral enzyme that synthesizes DNA using an RNA template. A double-stranded DNA then is synthesized, and it is attached to linker DNA sequences, which form the recognition sequence for a restriction endonuclease. The DNA is cleaved by the endonuclease, inserted into a vector cleaved by the same enzyme, and amplified as is done in preparation for a genomic DNA library. This very time-consuming process produces only the clone complementary to a single mRNA.

Alternatively, cDNAs can be prepared from all the mRNAs in a cell or tissue, following the procedure outlined in Figure 9-3. The cDNAs can be attached to linkers, inserted into vectors, and ampli-

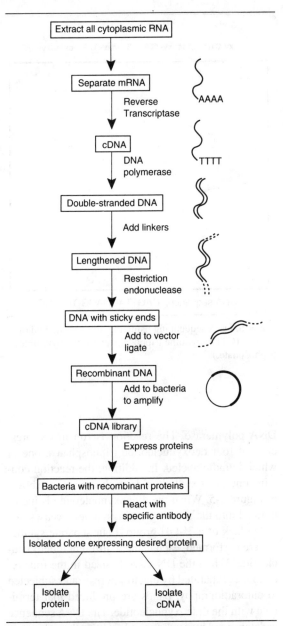

Extract all cytoplasmic RNA

↓

Separate mRNA

↓ Reverse Transcriptase

AAAA

cDNA

↓ DNA polymerase

TTTT

Double-stranded DNA

↓ Add linkers

Lengthened DNA

↓ Restriction endonuclease

DNA with sticky ends

↓ Add to vector ligate

Recombinant DNA

↓ Add to bacteria to amplify

cDNA library

↓ Express proteins

Bacteria with recombinant proteins

↓ React with specific antibody

Isolated clone expressing desired protein

↓   ↓

Isolate protein   Isolate cDNA

Fig. 9-3. Procedure for preparation of a cDNA library and identification of the desired clone.

fied, thus producing a **cDNA library.** Using appropriate vectors, bacterial cells can be induced to express the protein. Clones synthesizing the protein of interest can be identified by reaction with an antibody to the protein. If the protein synthesized by the bacteria does not react with the antibody, it still is

possible to identify the clone of interest. Using the genetic code, oligonucleotides can be synthesized that correspond to a portion of the amino acid sequence of the protein. Once it is made radioactive or otherwise tagged, the oligonucleotides can be used as a probe to identify the clone containing the cDNA of interest. Cloned DNAs can be used for sequencing or as probes for Southern and Northern blotting. Once a cDNA library has been constructed, it can be used repeatedly to isolate cDNAs coding for different proteins.

Production of Recombinant Proteins

Recombinant DNA techniques have been used to manufacture a number of human proteins that are medically useful and difficult to produce by other means. These include insulin, growth hormone, and several of the blood-clotting factors. In addition, proteins used for the preparation of the vaccine for hepatitis B are made using recombinant DNA technology. To achieve both the transcription and translation required to produce human proteins in bacteria, the cDNA must be attached to a bacterial promoter. Activation of this promoter causes transcription and translation of the inserted cDNA. The desired protein can be isolated from the bacterial cells or from the media if the protein is secreted.

Polymerase Chain Reaction

Another method for amplifying specific DNAs is the **polymerase chain reaction** (PCR), which allows for the rapid amplification of a particular DNA even in the presence of a large number of unrelated DNAs. Unlike cloning in a vector in bacteria, PCR requires knowledge of part of the sequence of the DNA to be amplified.

The reactions of automated PCR are outlined in Figure 9-4. First, DNA is denatured and mixed with primers that are complementary to DNA sequences at each end of the DNA to be amplified. DNA synthesis is carried out using a heat-stable DNA polymerase. Then the DNA is denatured and renatured with the primers, and another cycle of synthesis occurs. Each cycle takes 5 minutes or less. Because the reaction contains primers complementary to se-

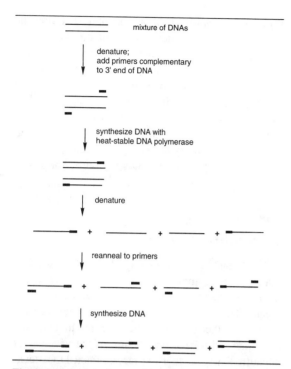

Fig. 9-4. Procedure for amplifying DNA by the polymerase chain reaction.

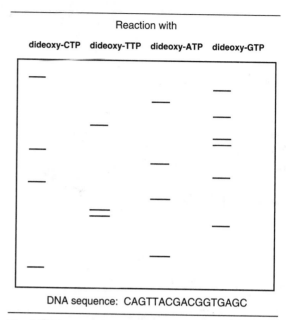

DNA sequence: CAGTTACGACGGTGAGC

Fig. 9-5. DNA sequencing using the dideoxy method of Sanger. (CTP = cytidine triphosphate; TTP = thymidine triphosphate.)

quences present in only one specific DNA, only that DNA will be amplified. A millionfold amplification of a DNA can occur by PCR in as few as 20 cycles. This DNA then is ready for sequencing or other analysis. By appropriate choice of primers, it is possible to determine the existence of a specific mutation in DNA amplified by PCR. This information can be used in diagnosis of a number of inherited genetic diseases.

DNA Sequencing

Determining the nucleotide sequence of DNA has produced much information about the enhancer and promoter regions of genes and the structure of introns and exons. It has also allowed the analysis of the molecular defect in a number of inherited diseases, including cystic fibrosis, thalassemia, and two types of muscular dystrophy. A single-stranded DNA to be sequenced by the Sanger method is hybridized to a primer and then the primer extended by

DNA polymerase. The reaction is run in the presence of four deoxynucleotide triphosphates, one of which is radiolabeled. In addition, the reaction contains one dideoxynucleotide triphosphate, as shown in Figure 9-5. When the dideoxynucleotide is incorporated into the growing chain, synthesis ceases due to the lack of a 3'-OH group. Four separate syntheses are performed, each using a different dideoxynucleotide. When the DNA synthesized in the four reactions is separated by electrophoresis and subjected to autoradiography, bands are produced that terminate with the dideoxynucleotide. The DNA sequence is determined by reading the bands on the gel.

Chromosome Mapping

Hybrids can be created by fusing human cells with mouse cells. These hybrid cells tend to lose human chromosomes in a random fashion. By analyzing a large number of these hybrids containing different combinations of human chromosomes, it is possible to correlate the presence of a particular human chromosome with the presence of a particular human protein product. This will indicate which chromosome carries the gene coding for that particular protein.

It is also possible to determine the approximate location of that gene on the chromosome by hybridizing the chromosome to a cloned DNA for the gene. This is *in situ hybridization* and often is the first step in mapping genes to particular chromosomes. DNA for which the chromosomal location is known can be used to probe a genomic library to identify neighboring DNA, if the two have overlapping sequences. The neighboring DNA, in turn, can be sequenced and used to identify more DNA. This method is known as *chromosome walking*. Eventually, all the genes on all the human chromosomes will be mapped and sequenced using this and other methods, which is the goal of the Human Genome Project.

Blotting

If DNA that has been fragmented by digestion by a particular restriction endonuclease is separated by size using electrophoresis, it will form a specific and reproducible pattern. Denaturation and transfer of the separated DNA fragments to nitrocellulose allows the fragments to be hybridized to a radioactive DNA probe. This probe can be a cDNA, an RNA, or a copy of a DNA sequence that is repeated in human DNA. The probe will react with the immobilized DNA fragments that contain complementary nucleotide sequences. The hybrid can be detected by exposing the nitrocellulose to x-ray film, which is the basis of the **Southern blot.**

Southern blots are used in **prenatal diagnosis** to determine whether the fetus has inherited a defective gene. This method is used for the diagnosis of sickle-cell anemia, cystic fibrosis, Tay-Sachs disease, Huntington's chorea, and one type of neurofibromatosis. In the analysis of sickle-cell anemia (Fig. 9-6), the DNA is treated with a specific restriction nuclease and is Southern-blotted using as a probe a radioactive cDNA complementary to the mRNA coding for β-globin. With the use of this particular restriction nuclease, the gene for normal β-globin is found in a DNA segment 1150 nucleotides long. The mutation producing sickle-cell anemia results in the loss of one of the recognition-restriction sites for this enzyme. Restriction then produces a DNA fragment 1350 nucleotides long. If the autoradiogram of the Southern blot shows a single band of between 1100 and 1200 nucleotides in size, the gene coding for β-globin is normal. If two bands occur, the individual has one normal β-globin gene and one with the sickle-cell mutation. If a single band occurs with a size between 1300 and 1400 nucleotides, the individual from whom the DNA was obtained has two mutated β-globin genes. In Figure 9-6, the fetus in question has inherited one normal β-globin gene and one carrying the mutation causing sickle-cell anemia.

The fragmentation pattern produced by a specific restriction endonuclease varies between individuals. Combining the results of fragmentation by a number of different endonucleases that have been probed with DNAs repeated in the human genome, a pattern of bands characteristic of a single individual, or a **DNA fingerprint,** can be produced. These DNA fingerprints can be used for positive identification of an individual, for identification of suspects in criminal investigations, and for determination of paternity.

If the material separated by electrophoresis is RNA, then the product of probing the RNA with a radioactive cDNA is a **Northern blot.** Northern blots are used to determine the presence or absence of an mRNA coding for a specific protein. In addition, Northern blots can determine whether the mRNA is altered in size.

Fig. 9-6. Prenatal diagnosis of the gene coding for sickle-cell anemia by a Southern blot.

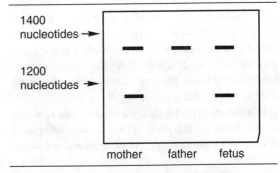

1400 nucleotides →

1200 nucleotides →

mother   father   fetus

The final blotting technique is **Western blotting.** Here, proteins are separated by size by electrophoresis. After transfer and immobilization, they are probed by specific antibodies. The definitive test for human immunodeficiency virus (HIV) infection is a Western blot in which purified viral proteins are separated by electrophoresis. If the virusspecific proteins are recognized by antibodies in patient serum, then the patient has been exposed to HIV and has produced antibodies to HIV proteins.

Enzyme-Linked Immunosorbent Assay

The principles of Western blotting are used as well in the less complicated enzyme-linked immunosorbent assay or **ELISA** system to detect either the presence of a specific protein or an antibody to a specific protein, as shown in Figure 9-7. In a **sandwich ELISA,** antibody is coated in wells on a plastic plate. Then samples that may contain a protein specific for the antibody are added. Finally, a second antibody to the protein is added. The second antibody will be bound to the plate only if the protein has bound to the antibody coated onto the wells of the plate. The second antibody contains a covalently bound enzyme such as horseradish peroxidase. Given the proper substrate, the catalytic activity of the enzyme will produce a colored product. The more enzyme that is bound, the more color will be produced. This method quantitates the amount of protein in the sample. ELISA is the first screening test used for individuals with possible HIV infection.

There are different variations of ELISA depending on what is bound to the wells of the plastic plate and in what order reagents are added. In a **direct ELISA,** the wells of the plate are coated with samples containing the protein at different concentrations. Then the wells are reacted with an antibody linked to an enzyme. Color is produced in the same way as in a sandwich ELISA: The more protein bound to the plate, the more antibody that is bound and the more color produced.

In a **competitive ELISA,** purified protein is bound to the wells of the plate. Then the test sample is mixed with a limiting amount of antibody and added

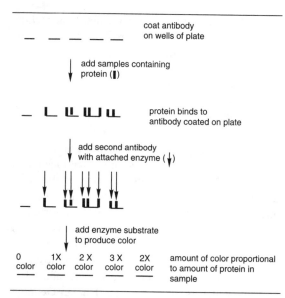

Fig. 9-7. Procedure for a sandwich enzyme-linked immunosorbent assay.

to the wells. Antibody reacting with protein in the test sample will not be available to react with the protein coated on the plate. In this case, the more protein in the test sample, the less color will be produced.

Summary

The ability of DNA to be fragmented by specific restriction endonucleases forms the basis of DNA cloning and Southern blotting. In cloning, DNA is fragmented by a restriction enzyme that produces staggered double-stranded breaks in the DNA at a specific DNA sequence, which is usually a palindrome. The fragmented DNA then is combined with a vector cleaved by the same restriction enzyme. The recombinant molecules are ligated and inserted into bacteria, where they are amplified. The DNA that is cloned can be either genomic DNA or DNA complementary to mRNA (cDNA). Transcription and translation of cDNA in a bacteria has allowed the production of human proteins unavailable by other means. DNA can also be amplified by PCR, which cyclically replicates only the segment of DNA for which specific primers have been provided in the reaction.

Cloned DNA can be sequenced using the dideoxynucleotide method of Sanger. DNA synthesis is terminated when the dideoxynucleotide is added to the growing chain. Separation of the DNA by electrophoresis will produce a pattern of chain termination characteristic of the nucleotide sequence. Cloned DNA can also be used as a probe in Southern and Northern blotting. In Southern blotting, used for prenatal diagnosis, DNA is fragmented and separated by size. It is probed with a radioactive DNA that is a complement of a specific mRNA or DNA that has been repeated in the human genome. If a number of different restriction nucleases are used and several different probes, a DNA fingerprint characteristic of the individual will be produced. In Northern blotting, RNA is separated and then probed by a radioactive cDNA.

In Western blotting and ELISA tests, the reaction is between a protein and an antibody to that protein. In a Western blot, proteins are separated by size and probed with an antibody. In ELISA, the reaction between protein and antibody takes place in the wells of a plastic plate.

Part II Questions: Molecular Biology

1. Which of the following statements best describes the B form of DNA?
   A. The bases are stacked at a 45-degree angle to the axis of the helix.
   B. The number of purine residues must be 50% of the total on each strand.
   C. Covalent bonding of nucleotide base pairs is the major stabilizing force holding the structure together.
   D. It is a right-handed helix with 10 base pairs per helical turn.
   E. The three strands of DNA run parallel.

2. The structure depicted here is

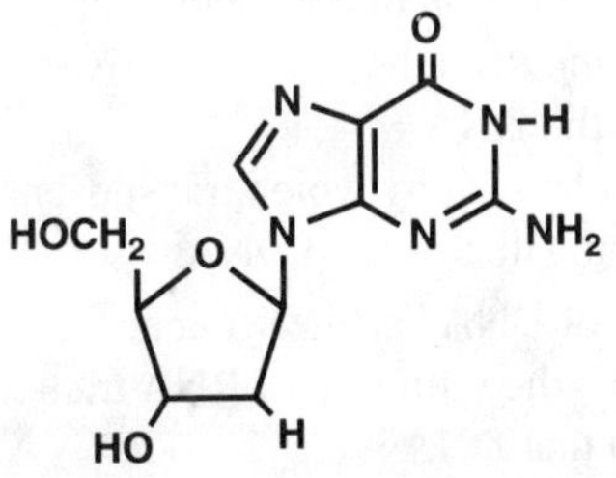

   A. a pyrimidine nucleoside.
   B. guanosine.
   C. deoxyguanosine.
   D. a purine nucleotide.
   E. thymidine.

3. Nucleosomes
   A. contain covalently linked RNA and DNA.
   B. are the secondary structure of mitochondrial chromosomes.
   C. include a core of two each of histones H1, H2, and H5.
   D. form the tertiary structure of DNA in the eukaryotic cell nucleus.
   E. require a specific base sequence for formation.

4. If one strand of double-stranded DNA contains the following number of nucleotides—A = 1500, T = 1100, C = 1700, and G = 1600—the total numbers of each nucleotide in the entire structure of DNA would be

   A. A = 3100, T = 2600, C = 2600, G = 3100.
   B. A = 3000, T = 2200, C = 3400, G = 3200.
   C. A = 2600, T = 2600, C = 3300, G = 3300.
   D. A = 3300, T = 3300, C = 2600, G = 2600.
   E. A = 2200, T = 2700, C = 2200, G = 2700.

5. When DNA is denatured by heat
   A. the Tm is linearly related to the G + C content of the DNA.
   B. the primary structure is destroyed.
   C. the absorption at 260 nm decreases.
   D. the pH must be less than 4 or greater than 10.
   E. the rate is greater for Z DNA than for B DNA.

6. When DNA is replicated in the eukaryotic cell nucleus
   A. replication begins at a single unique site on each chromosome.
   B. old histones associate primarily with the lagging strand.
   C. the primer can use either RNA or DNA as a template.
   D. synthesis occurs on both strands in small segments called *Okazaki fragments.*
   E. synthesis occurs in the 5′ to 3′ direction.

7. Place the following in the temporal order in which they occur during DNA replication:
   A. helicase, primase, DNA polymerase, 5′-nuclease, ligase.
   B. helicase, DNA polymerase, 5′-nuclease, ligase, primase.
   C. primase, DNA polymerase, ligase, 5′-nuclease, helicase.
   D. DNA polymerase, primase, 5′-nuclease, helicase, ligase.
   E. helicase, DNA polymerase, 5′-nuclease, primase, ligase.

8. Okazaki fragments
   A. occur only in the leading strand.
   B. are produced when histone biosynthesis is inhibited.

99

C. demonstrate the existence of an RNA primer for DNA synthesis.

D. are needed for finishing the synthesis of circular chromosomes.

E. occur when mitochondrial DNA is replicated.

9. All the following reactions require the hydrolysis of high-energy phosphate bonds *except*

A. DNA synthesis by DNA polymerase.

B. helix unwinding by helicase.

C. supercoil removal by topoisomerase II.

D. primer removal by nuclease.

E. phosphodiester bond formation by DNA ligase.

10. Which of the following is characteristic of all DNA polymerases?

A. They synthesize in the 3′ to 5′ direction.

B. They require a primer of RNA or DNA.

C. They read a template in the 5′ to 3′ direction.

D. They contain a 5′ to 3′ nuclease activity.

E. They require monovalent ions.

11. The product of irradiation of DNA by ultraviolet light

A. is breakage of the phosphodiester backbone.

B. is repaired by long-patch excision repair.

C. requires the action of a uracil glycosylase for repair.

D. produces dimers between thymidine residues located on opposite strands of the DNA helix.

E. is repaired in humans by photoreactivation.

12. Postreplication repair

A. occurs primarily during the M phase of the cell cycle.

B. is the major mechanism for removal of x-ray damage.

C. uses a crossover mechanism to complete replication of a damaged template strand.

D. prevents ligation of Okazaki fragments.

E. produces DNA with many improperly paired bases.

13. X-irradiation of DNA can produce all the following *except*

A. pyrimidine dimers.

B. apurinic sites.

C. repair by short-patch excision repair.

D. repair by the action of DNA ligase alone.

E. breakage of the phosphodiester backbone.

14. O^6-methylguanosine

A. is produced by depurination of DNA.

B. spontaneously occurs at elevated temperatures.

C. is repaired by long-patch excision repair.

D. repair is defective in persons with xeroderma pigmentosum.

E. is repaired by enzymatically removing the methyl group without breaking the DNA backbone.

15. General recombination

A. is the mechanism for integration of viruses into chromosomal DNA.

B. occurs by crossing over at meiosis.

C. requires the action of a 5′-nuclease.

D. involves the resynthesis of extensive segments of the DNA.

E. is the mechanism by which transposons move from site to site in DNA.

16. If a segment of DNA has the sequence ACGTTGAC, the sequence of RNA transcribed from that DNA is

A. TGCAACTG.

B. UGCAACUG.

C. GTCAACGT.

D. CAGUUGCA.

E. GUCAACGU.

17. Major characteristics of tRNA include

A. an amino acid binding site at the 5′ end.

B. an anticodon site on the D arm.

C. inclusion of deoxynucleotides.

D. short stretches of double-stranded structure in the A form.

E. the nucleotides GGC at the 3′ end.

18. The major features of mRNA in eukaryotic cells include all the following *except*

A. a GTP-containing cap at the 5′ end.

B. methylation of the 2′-OH of the first nucleotide.

C. a large number of minor nucleosides such as ribothymidine, inosine, and dimethyladenosine.

D. an untranslated region at the 5′ end before the sequence AUG.

E. a tail of 200 AMP residues on the 3′ end.

19. The small nuclear RNAs
    A. are necessary for the processing of ribosomal RNA.
    B. function to position tRNA during protein synthesis.
    C. serve as primers for DNA replication.
    D. bind to splice junctions during the maturation of mRNA.
    E. guide proteins into the mitochondria.

20. Properties common to both RNA and DNA include
    A. hydrolysis by base.
    B. a linear arrangement of nucleotides exhibiting polarity of structure.
    C. a large number of modified nucleotides.
    D. a double helix in the B form.
    E. G:U base pairs.

21. tRNA processing in eukaryotic cells can include all the following steps *except*
    A. removal of an intron.
    B. export to mitochondria.
    C. addition of the sequence CCA to the 3′ end.
    D. modification of nucleotides by methylation.
    E. removal of nucleotides from the 5′ end.

22. Ribosomal RNA
    A. is synthesized by RNA polymerase I.
    B. has a primary transcript the same size as the sum of the sizes of the mature products.
    C. contains information for the synthesis of the ribosomal proteins.
    D. is spliced using the small nuclear RNAs.
    E. is inhibited in its maturation process by methylation of the 2′-OH groups.

23. At sufficiently high concentrations, both actinomycin D and α-amanitin directly inhibit
    A. rRNA synthesis by RNA polymerase I.
    B. DNA replication in mitochondria.
    C. protein synthesis in bacteria.
    D. tRNA synthesis by RNA polymerase III.
    E. protein transport into the mitochondria.

24. During transcription
    A. only one strand of DNA is copied.

B. in bacteria, recognition requires the function of a protein factor called *rho*.

C. the first nucleotide usually is a pyrimidine.

D. the RNA polymerase binds to the DNA at the operator site.

E. the DNA is read in the 5′ to 3′ direction.

25. Introns
    A. are sequences included in both the primary transcript and the mature product.
    B. direct proteins to the nucleus.
    C. contain sequences coding for protein.
    D. are found exclusively at the ends of chromosomes.
    E. are removed from pre-mRNA in spliceosomes.

26. Tetracycline
    A. inhibits translocation in bacterial ribosomes.
    B. inhibits peptidyl transferase in eukaryotic cytoplasmic ribosomes.
    C. inhibits the binding of amino acyl tRNA to the A site on the ribosome.
    D. structurally resembles tyrosyl tRNA.
    E. prevents the nonribosomal synthesis of glutathione.

27. A change from the codon UGG to UAG is an example of
    A. a transversional mutation.
    B. a frameshift mutation.
    C. a silent mutation.
    D. a nonsense mutation.
    E. an acceptable missense mutation.

28. During the activation phase of protein synthesis,
    A. energy from the hydrolysis of GTP is used to decarboxylate amino acids.
    B. amino acids are attached to the 2′- or 3′-hydroxyl group of an adenosine residue on the tRNA.
    C. mRNA is combined with the large and small ribosomal subunits.
    D. the initiator tRNA binds to the P site of the ribosome.
    E. amino acids are attached to a phosphate at the 5′ end of the tRNA.

29. When the codon AUG is in the P site of a ribosome,

A. the initiator tRNA binds to the P site.

B. a silent mutation occurs.

C. ATP is hydrolyzed as the growing peptide is moved from the A site to the P site.

D. protein synthesis stops.

E. a frameshift mutation occurs.

30. The codons CUU, CUC, CUA, and CUG all code for leucine. This indicates that the genetic code

A. is ambiguous.

B. has wobble in the third base of the anti-codon.

C. is degenerate.

D. is universal.

E. is not determined by base pairing.

31. In enzyme induction,

A. the repressor requires an inducer before it can bind to the promoter site.

B. the RNA polymerase binds to the operator site.

C. the inducer is structurally related to the product of the reaction pathway.

D. the structural genes are transcribed only in the presence of the inducer.

E. the mechanism occurs most frequently in the control of synthetic pathways.

32. Enzyme repression

A. occurs primarily in synthetic pathways.

B. in bacteria is controlled at the level of mRNA transport from the nucleus to the cytoplasm.

C. occurs when the repressor molecule binds to the promoter site.

D. is facilitated by high levels of cAMP binding to the operator site.

E. occurs when the co-repressor, which structurally resembles the substrate, binds to the promoter.

33. A signal sequence in a protein

A. is often at the N-terminal end.

B. directs the protein to ribosomes for synthesis.

C. is an indication of a missense mutation.

D. directs the protein to the lysosomes for degradation.

E. demonstrates wobble in the genetic code.

34. All the following are possible mechanisms of control of gene expression in eukaryotes *except*

A. binding leucine zipper proteins to specific DNA sequences.

B. gene rearrangement.

C. attenuation.

D. binding of a zinc finger protein to a response element.

E. an increase in the stability of a specific mRNA.

35. Reactions in the cascade that control protein activity by phosphorylation can include

A. binding of cAMP to a G-protein.

B. release of membrane inositol triphosphate and diacylglycerol by the action of a phospholipase.

C. binding of Zn ions to calmodulin.

D. activation of the soluble form of protein kinase C.

E. binding of calcium to the regulatory subunits of protein kinase A.

36. A Southern blot

A. requires that RNA is first digested with a restriction nuclease.

B. uses an antibody as a probe.

C. separates protein according to size as the first step.

D. transfers full-size DNA to nitrocellulose before probing it with RNA and separating the hybrid according to size.

E. can be used for prenatal diagnosis of inherited gene defects.

37. A genomic library is

A. a bacterial cell containing DNA from a foreign source.

B. a set of bacteria that together contain all the DNA sequences found in the nucleus of an organism.

C. the entire set of cDNA clones from one cell type.

D. produced by the action of a restriction nuclease on cDNA.

E. a small extrachromosomal DNA capable of self-replication.

38. When DNA is sequenced,

A. both strands of the DNA are sequenced at the same time.

B. the 5′ end of the DNA must be radiolabeled.

C. the strand of DNA is cleaved by enzymes
specific for single bases.

D. an RNA copy of the DNA is separated by
size.

E. the DNA is amplified by use of specific
primers and a heat-stable DNA polymerase.

39. Restriction nucleases are characterized by

A. cleaving DNA at random sites.

B. cleaving DNA at a palindromic sequence.

C. being used in the analysis of RNA in Western blots.

D. splicing DNA at specific sites.

E. being necessary for DNA synthesis.

40. A characteristic of both ELISA and Western
blots is that both

A. involve a reaction between DNA and an antibody.

B. involve separation of a protein by electrophoresis.

C. require reaction between an antibody and a
protein.

D. allow prenatal diagnosis of genetic defects.

E. involve a reaction between RNA and a
cDNA.

Part III Proteins

10 Protein Structure

The human genome contains enough DNA to code for at least 50,000 different proteins. These may have a variety of functions including catalysts, transporters, regulators, recognition sites, structural elements, or some combination of these functions. Although proteins are composed of only 20 different amino acids, their diverse functions require different structures. In human cells, the observed protein structures vary from large aggregates containing many separate protein chains with a combined molecular weight exceeding 1 million to small, single-chain peptides with molecular weights of only a few thousand. By combining a number of different secondary structures in segments of a single protein molecule or adding coenzymes or other proteins, the variety of structures necessary for human existence is produced.

Objectives

After completing this chapter, you should be able to

Draw the structures of the 20 amino acids that occur naturally in proteins.

Define the terms *isoelectric point* and *peptide bond*.

Discuss the primary structure of proteins and describe the molecular interactions that stabilize the higher levels of structure.

Describe the structure and the bonding that occur in the α-helix and in the β-structure or pleated sheet.

List the major characteristics of globulin, α- and β-keratin, collagen, and tropocollagen.

Describe the principles for separating proteins by gel filtration, ultracentrifugation, electrophoresis, isoelectric focusing, ion exchange, and affinity chromatography.

Amino Acids

As we saw in Chapter 7, amino acids attach to other amino acids to form proteins. These serve as the biological catalysts essential for most life processes. Amino acids can be combined with carbohydrate, lipids, or nucleic acids to form other essential compounds. The naturally occurring amino acids have a carboxyl group and an amino group attached to the same carbon atom: that is, they are α-amino acids. In their ionized form, they can be represented as follows:

$$H_3N^+-\underset{\underset{R}{|}}{\overset{\overset{H}{|}}{C}}-\overset{\overset{O}{||}}{C}-O^-$$

With the exception of glycine, there are four different groups attached to the α-carbon: NH_3, COO^-, H, and R. Because of this, the structure of an amino acid cannot be superimposed on its mirror image. This property makes amino acids optically active, meaning that a solution of a pure amino acid will rotate a beam of polarized light to either the right or the left. The 20 amino acids that occur naturally in proteins have the same orientation around the α-carbon and are designated as L. The structures of the common amino acids are given in Figure 10-1. Amino acids with the opposite orientation, the D forms, are produced by some plants and occasionally occur in antibiotics.

The amino acids in Figure 10-1 are drawn in the ionized form in which they would be found in cells. The pK of the α-carboxyl group (an acid) is approximately 2, whereas the α-amino group (a base) dissociates at approximately pH 9. Thus, at pH 7.4,

Fig. 10-1. Structures of the 20 amino acids found naturally in proteins. The structures are drawn in the ionized form in which they would be found in cells.

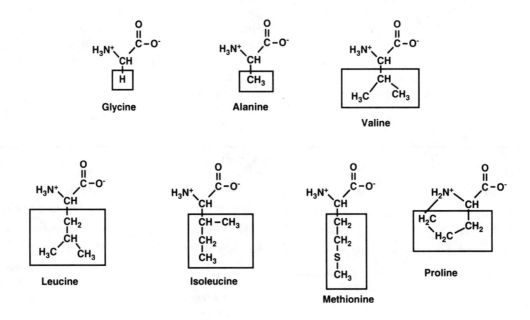

POLAR AMINO ACIDS

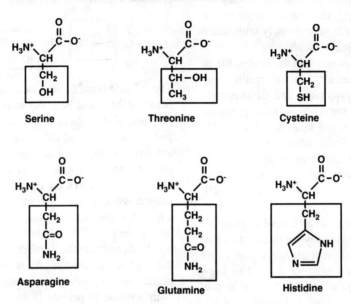

Serine   Threonine   Cysteine

Asparagine   Glutamine   Histidine

NEGATIVELY CHARGED AMINO ACIDS

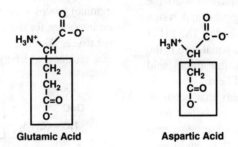

Glutamic Acid   Aspartic Acid

POSITIVELY CHARGED AMINO ACIDS

Lycine   Arginine

Fig.10-1. (continued)

the α part of these amino acids contains one positive charge and one negative charge. Because of the positive and negative charges, this part of the molecule is hydrophilic and interacts very strongly with water and other polar or charged molecules.

The other part of the amino acid molecules, the R group, of alanine, isoleucine, leucine, methionine, phenylalanine, proline, tryptophan, and valine is nonpolar. This part of the molecule is not capable of forming hydrogen bonds and tends to cluster with other nonpolar groups. Serine, threonine, and tyrosine contain OH groups and cysteine an SH group, which are polar and can form hydrogen bonds. Asparagine, glutamine, and histidine contain NH groups that can form hydrogen bonds. However, none of these amino acids has a charged R group at cellular pH. They are electrically neutral; that is, the sum of the positive and negative charges equals zero.

In contrast, aspartic and glutamic acids have negatively charged R groups at pH 7.4. The pK of the terminal carboxyl group is near 4. Arginine and lysine contain basic R groups, which have a positive charge at cellular pH. In an electrical field at pH 7.4, aspartic and glutamic acids will move toward the positive pole (anode), whereas arginine and lysine will move to the negative pole (cathode). The pH at which the charges on an amino acid equal zero is the **isoelectric point** (pI). At this pH, the amino acid will no longer move in an electrical field.

In addition to the 20 amino acids that occur in proteins, cells contain other amino acids that have important metabolic functions. These serve as precursors for hormones, neurotransmitters, and metabolic intermediates. The structures and functions of these compounds will be discussed as we come across them in our discussion of metabolism.

Proteins

Primary Structure

Amino acids can be combined with the loss of water to produce polypeptides or proteins. The primary structure of a protein is the sequence of amino acids. The **peptide bond** that joins amino acids in proteins combines the α-carboxyl of one amino acid with the

α-amino group of another. It produces long unbranched chains of amino acids.

The peptide bond has partial double-bond character between the C and N atoms so it is planar and rigid. Because of this, there is no free rotation around it, and the side chains of the amino acids (the R groups) are fixed in space relative to the peptide bond. In proteins, the R groups are in a *trans*-configuration (on opposite sides of the peptide bond). This conformation reduces steric interaction, especially if the adjacent amino acids contain bulky R groups. In addition, the partial double-bond character of the peptide bond restricts the number of configurations that the peptide can assume. As we will see, this is important in determining the secondary structure.

In addition to peptide bonds, amino acids can be joined by amide bonds and disulfide linkages. Amide bonds occur between R groups containing carboxyl (glutamic and aspartic acids) and amino (lysine and arginine) groups. Later we will see how amide linkages are important for the structure of collagen in connective tissue. Disulfide linkages occur between two cysteine residues, which then form the amino acid cystine.

glutamic acid   lysine

cysteine   cysteine

This type of linkage can join two sections of a single polypeptide chain or link two separate polypeptide chains. For example, the two chains of insulin are joined only by disulfide bonds.

In addition to amino acids, proteins can contain phosphate, acetate, and sulfate as esters of hydroxyl groups. Carbohydrates can be attached to the hydroxyl groups of serine or threonine as O-glycosides or to the amides of glutamine or asparagine as N-glycosides. This type of modification occurs almost exclusively on proteins destined for incorporation into membranes or export from the cell. Small organic molecules or coenzymes that are derived from vitamins can also be integral parts of proteins.

Secondary Structure

The secondary structure of a protein is determined by the primary structure. It is the lowest energy state for a particular order of amino acids in a particular environment. Most of the secondary structure is formed at the time the protein is synthesized and remains the same even if modifications occur to the amino acids or if segments of the protein are removed. Proteins without a regular or repeating structure are considered random coils, an unfortunate term as the nature of hydrophobic and hydrophilic amino acids will make the arrangement definitely nonrandom.

α-Helix

There are multiple possible secondary structures, but two occur frequently enough that you should know their characteristics. The first of these is the α-**helix** (Fig. 10-2). Like the α-helix of DNA, the protein α-helix is right-handed. Left-handed helices are possible, but the right-handed one happens to be more stable for the naturally occurring L-amino acids.

The protein α-helix has 3.6 amino acids per turn, with the R groups of the amino acids extended outward perpendicular to the axis of the helix. The structure is held together by hydrogen bonds between the carboxyl of one peptide bond and the NH of the amino acid four residues above it in the helix. Because every peptide bond, except those at the ends, is involved in two hydrogen bonds, the α-helix is a very stable structure.

The α-helix is most stable when the primary structure contains amino acids with uncharged side chains: alanine, leucine, phenylalanine, tyrosine, tryptophan,

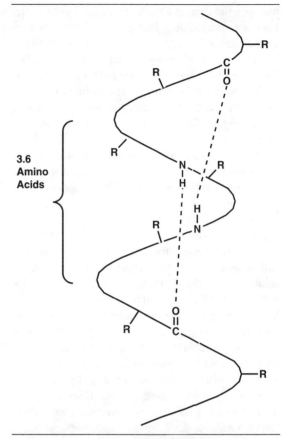

3.6 Amino Acids

Fig. 10-2. Structure of the protein α-helix. The protein α-helix is a right-handed helix with 3.6 amino acids per turn. Solid line represents the backbone. The R groups of the amino acids extend outward perpendicular to the helix axis. Hydrogen bonds (dotted lines) are formed between the carboxyl group in one peptide bond and the NH of the peptide bond located four amino acids above it in the helix.

cysteine, methionine, histidine, asparagine, glutamine, and valine. The α-helix is destabilized by amino acids with bulky or charged side chains such as serine, isoleucine, threonine, glutamic and aspartic acids, lysine, and arginine. The only amino acid that cannot be accommodated in an α-helix is proline, due to the rigidity and bond angles of its structure. If an α-helix exists in a protein, it will stop when a proline residue is encountered in the primary structure.

α-**Keratin,** the major protein of skin, hair, nails, and some of the intracellular filaments, is predominantly in the form of an α-helix. Several of these helices are twisted together to form long fibers or ropes. The tertiary structure of α-keratin is stabilized by disulfide bonds between adjacent helices. Because the individual helices are held together by hydrogen bonds, α-keratin can be denatured by detergents, heat, and other agents that disrupt hydrogen bonds. When this happens, the molecule can be stretched out to considerably longer than its original length. To see this change for yourself, stretch out a wet strand of hair. Permanent waves disrupt the disulfide bonds between helices. When the bonds reform at new locations, the shape of permanently waved hair is changed and curls result.

Like α-keratin, the **globulins** contain a high content of α-helix. Because these proteins are compact and spherical rather than ropelike, they are water-soluble. The globulins include hemoglobin, myoglobin, the blood plasma proteins, the blood-clotting factors, immunoglobulins, and the proteins necessary for the transport of iron and copper. In the globulins, the hydrophobic amino acids are clustered on the inside and the hydrophilic amino acids on the outside. Globulins often contain carbohydrates and, occasionally, lipids. These proteins frequently have more than one possible conformation under cellular conditions.

β-Structure

The second of the important secondary structures is the **β-structure** or **pleated sheet** (Fig. 10-3). In this structure, the protein chains are extended, with the R groups alternating above and below the backbone. Hydrogen bonding occurs between peptide bonds on chains running parallel or antiparallel. These bonds can be between different chains or the same chain folded back on itself. The hydrogen bonds in the β-structure are oriented perpendicular to the direction of the protein backbone. As with the α-helix, all peptide bonds can be involved in hydrogen bonding. Because of steric interactions between R groups, the β-structure can form only when the primary structure contains amino acids with small side chains.

β-Keratin, found in silk and the beaks of birds, is arranged almost entirely in a pleated-sheet structure. It does not change conformation when wet or when heated at low temperature. Because β-keratin con-

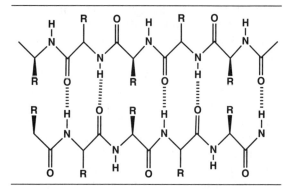

Fig. 10-3. Structure of the β-structure or pleated sheet. The β-structure is stabilized by hydrogen bonds between peptide bonds on protein chains running parallel or antiparallel. Hydrogen bonds (stacked dashes) are perpendicular to the direction of the backbone of the protein chains.

tains no cysteine, it has no disulfide bonds and is not altered by treatment with disulfide-reducing agents. The preponderance of amino acids that have small R groups allows β-keratin to assume the β-structure.

Tertiary Structure

The tertiary structure is the configuration that the protein takes in space. Like the secondary structure, the tertiary structure is determined primarily by the original amino acid sequence. Ionic interactions between charged R groups, hydrophobic interactions, and disulfide bonds are all important for the stability of the tertiary structure. In an aqueous environment, nonpolar amino acids in proteins generally are found on peptide segments folded to the inside away from the water. The polar amino acids usually are located on the outside where they can interact with water and other polar compounds. In addition to these interactions, the tertiary structure includes any conformation induced by the covalent addition of prosthetic groups or other posttranslational modifications.

Proteins often are divided into fibrous and globular types, based on their tertiary structure. Fibrous proteins, such as α-keratin, are elongated fibers, rods, or ropes that are strong and insoluble in water. Globular proteins, on the other hand, are more spherical and are readily soluble in water or a dilute salt solution.

Collagen, the most abundant protein of the human body, is an insoluble fiber found in tendons, bone, cartilage, skin, and cornea. Collagen is not a single protein, as at least five different collagens and eight different genes coding for collagen precursors have been identified. Nevertheless, all the collagens have a similar structure.

Collagen is an unusual protein in that it contains 30% glycine, 20% proline and hydroxyproline, and the modified amino acid hydroxylysine.

4-hydroxyproline

5-hydroxylysine

Its secondary structure is a helix, but not an α-helix, because the high content of proline makes an α-helix impossible. Three of the collagen helices are twisted into a larger left-handed helix, forming **tropocollagen.** Tropocollagen helices, in turn, are lined up parallel, staggered by one-fourth to one-fifth of the length of the molecule (400 nm), and crosslinked together to form the final collagen fiber. The major crosslinks in collagen are between hydroxylysine residues in different protein chains. These can be located in the same tropocollagen helix or between different helices. Failure to form these crosslinks is one of the causes of **Ehlers-Danlos syndrome.** There are several varieties of this syndrome, but all are characterized by stretchable skin and hypermobile joints. Because vitamin C is required for the hydroxylation reactions, defects in collagen formation also occur with vitamin C deficiency.

Quaternary Structure

The quaternary structure of a protein consists of the arrangement of one protein chain with others and with noncovalently bound coenzymes. Individual protein chains can be combined with other chains (identical or different) as subunits in a larger structure. Hemoglobin, which is a combination of two α-chains, two β-chains, and four heme groups, is an example of subunits combining to make up a final functional protein. We will discuss the structure and function of the hemoglobin molecule in more detail in Chapter 12.

Another example of quaternary structure is the combination of several individual enzymes into a complex entity in which the enzymes act sequentially on a given substrate. Such quaternary complexes are held together by noncovalent interactions, and, occasionally, disulfide bonds.

Denaturation

Because their secondary structures are held together by hydrogen bonds, proteins can be denatured by the same agents that denature nucleic acids, by extremes of pH or temperature, and by hydrogen bond–disrupting agents such as urea and detergents. Substances that reduce disulfide bonds can also cause denaturation of proteins. As with nucleic acids, all degrees of structure higher than the primary one are disrupted when a protein is denatured. High temperatures combined with extremes of pH can also hydrolyze peptide bonds and destroy the primary structure. In base hydrolysis, the amino acids are racemized (lose their optical activity), and serine and threonine are destroyed. In acid hydrolysis, tryptophan is destroyed, whereas glutamine and asparagine are deaminated to their respective acids.

Denaturation causes the biological activity of a protein to be lost. The protein generally decreases markedly in solubility as well. You are all familiar with the solubility change that occurs when ovalbumin (the major protein in egg white) is denatured by heat as an egg is hard-boiled or by mechanical agitation in the preparation of meringue. The same effect, although not the same taste, can be produced by adding concentrated HCl or NaOH.

Once a protein is denatured, it is difficult (but not impossible) for it to resume its original conformation even after the denaturing agent has been removed. This is usually because the protein has been altered after synthesis by modification of amino acid residues, formation of disulfide bonds, removal of protein segments, or other posttranslational modifications. The altered protein cannot assume the structure established originally by the unaltered protein even if placed in a similar environment.

Separation and Detection of Proteins

You should be familiar with the current methods of separating proteins from one another and of determining size and purity. One of the major methods is **gel filtration.** In this process, proteins are separated according to size as they pass through a porous gel. Very large proteins are excluded from the gel and emerge from the column quickly, in what is called the *void volume.* Smaller proteins enter the gel to varying degrees, depending on size. The larger proteins exit the column first. Smaller ones take a longer time to work their way through the pores and are eluted later.

Proteins that bind to a specific compound, such as a substrate or coenzyme, can be separated from other proteins by virtue of this binding. In this process, called **affinity chromatography,** a solution of the protein is passed through a column on which the substrate or coenzyme is immobilized. Any protein with an affinity for this compound will bind to the column; all others will pass through. The protein then is removed by creating nonbinding conditions. Affinity chromatography can provide major purification of a protein in a single step.

Because proteins are made up of amino acids, their electrical charge is the sum of the charges of the R groups of the individual amino acids and any added functional group that has an electrical charge. The net electrical charge can be used to separate proteins on **ion exchange** columns. Proteins with a net negative charge bind to columns that have a positive charge, whereas proteins with a positive charge or no charge will not bind. Similarly, proteins with a net positive charge will bind to columns that have a negative charge. Changing the pH or the salt concentration of the buffer running through the column will elute the bound proteins.

Like amino acids, proteins will move in an electrical field at all pHs except for their isoelectric point—that is, when the net charge is zero. This property can be used to determine the purity of proteins by **electrophoresis** and by **isoelectric focusing.** Electrophoresis in the presence of detergents not only separates individual protein subunits but also provides an estimate of molecular weight. This is because the detergent (usually sodium dodecyl sulfate) surrounds the protein, giving all proteins approximately the same electrical charge and destroying the secondary and tertiary structures. The molecules then separate by size during their passage through the pores of the electrophoresis gel.

In isoelectric focusing, proteins move in an electrical field in a gel containing a pH gradient. The proteins move until they encounter the portion of the gel in which the pH is the isoelectric point of the protein. Isoelectric focusing is a method for determining whether a protein is pure or contaminated with other proteins with very similar physical properties.

Another method of estimating protein size is by **ultracentrifugation.** Here proteins are separated under the force of increased gravity produced by high-speed centrifugation. Larger proteins move faster; smaller ones or ones with nonspherical shapes move more slowly.

Proteins separated by these methods can usually be detected by their absorbance of ultraviolet light at 280 nm. This absorbance is due to the presence of amino acids with aromatic side chains, such as phenylalanine, tyrosine, and tryptophan. Proteins that do not contain any aromatic amino acids cannot be detected by this method. However, they can be detected by their reaction with various dyes such as Coomassie blue and certain silver stains. The N-terminal end of a protein and individual amino acids can be detected by their reaction with triketohydrindene hydrate (Ninhydrin), fluorescein, or dansyl chloride.

Summary

There are 20 α-amino acids that occur naturally in proteins. The nonpolar amino acids include alanine, glycine, isoleucine, leucine, methionine, phenylalanine, tryptophan, and valine. The polar amino acids with uncharged R groups are serine, threonine, tyrosine, cysteine, asparagine, glutamine, and histidine. Aspartic and glutamic acids have a negative charge at cellular pH. Arginine and lysine have a positive charge. Proteins are composed of a linear array of amino acids connected by peptide bonds. Individual amino acids can also be linked by amide and disulfide bonds.

The primary structure of a protein is determined by the amino acid sequence. This, in turn, determines the secondary, tertiary, and quaternary structures. The secondary structure is held together by hydrogen bonds between the carboxyl and amino groups of the peptide bond. In the α-helix, the bonding occurs between one carboxyl and the NH of the amino acid four residues further along in the chain. α-Keratin, found in hair and nails, is predominantly in an α-helix. The separate helices are held together by numerous disulfide bonds. The globulins, including hemoglobin, myoglobin, and the soluble proteins of the blood, contain a large amount of α-helical structure, with the hydrophobic amino acids predominantly on the inside of the molecule and the hydrophilic ones predominantly on the outside.

In the β-structure or pleated sheet, hydrogen bonding occurs between peptide bonds in chains running parallel or antiparallel. The secondary structure of β-keratin, found in silk, is a β-structure. β-Keratin contains no disulfide bonds, but the structure is stabilized by hydrogen bonds between separate chains. The structures of both the α-helix and β-structure are most stable with amino acids with small, uncharged side chains.

Collagen is an unusual protein that contains 30% glycine and 20% proline and hydroxyproline. Each protein strand forms a helix that combines with two other strands into a triple helix called *tropocollagen*. Collagen is held together by crosslinks between hydroxylysine residues. Abnormal collagen occurs in Ehlers-Danlos syndrome and vitamin C deficiency.

The tertiary structure of proteins is stabilized by interactions of the R groups of the amino acids: positive and negative charges, hydrophobic interactions, disulfide bonds, and a variety of posttranslational modifications. The quaternary structure consists of the noncovalent interaction of protein chains with other proteins and coenzymes.

All these levels of structure except for the primary one can be destroyed by denaturation by heat, extremes of pH, urea, and detergents. A combination of heat and extremes of pH can hydrolyze the primary structure as well. Unlike nucleic acids, renaturation of proteins does not readily occur. Proteins can be separated by gel filtration, ultracentrifugation, affinity and ion exchange chromatography, isoelectric focusing, and electrophoresis.

11 Proteins as Catalysts

Many of the proteins synthesized in cells function as enzymes or biological catalysts. To understand metabolism, it is necessary to understand the factors that control enzymatic reactions. Be sure to remember that the reaction conditions found to be optimal in a laboratory setting, although yielding much useful information, often bear little resemblance to the conditions that occur in cells. This is especially true for rate-controlling reactions. For the study of human biochemistry, we are concerned with the conditions and reaction rates that take place in living cells.

Objectives

After completing this chapter, you should be able to

Classify an enzyme by the type of reaction catalyzed.

Define ΔG and be able to recognize reactions that are energetically favorable.

Define E_a.

Write the Michaelis-Menten and Lineweaver-Burk equations.

Define the terms *equilibrium constant, initial velocity, maximal velocity, first order,* and *zero order.*

Recognize competitive, noncompetitive, and uncompetitive inhibition by their Lineweaver-Burk plots.

Describe single- and double-displacement reactions in terms of the order of adding and leaving substrates and products.

Describe first-order and second-order reactions.

Define the terms *allosteric effector, apoenzyme, holoenzyme, isozyme, specific activity,* and *turnover number.*

Discuss the effect of temperature, pH, and phosphorylation on enzymatic activity.

List the amino acids that occur frequently in the active sites of enzymes.

Classification of Enzymes

Enzymes are classified into six categories depending on the type of reaction they catalyze. These categories are summarized in Table 11-1. Many enzymes also have common names that do not include the category names from the table. You will encounter a number of them as you proceed through this text. Try to classify these enzymes by category as you read about them.

The first class of enzymes is the **oxidoreductases.** In these reactions, one substrate is oxidized while, at the same time, the other is reduced.

$$\text{oxidized A} + \text{reduced B} \rightleftarrows \text{reduced A} + \text{oxidized B}$$

NAD and dihydronicotinamide adenine dinucleotide (NADH) frequently are substrates or products in oxidoreductase reactions.

Table 11-1. Classification of enzymes

Type	Reaction	Example
Oxidoreductase	Oxidized A + reduced B ↑↓ reduced A + oxidized B	Alcohol dehydrogenase: RH-OH + NAD$^+$ ↑↓ R=O + NADH + H$^+$
Transferase	A-X + B $\rightleftarrows$ A + B-X	Hexokinase: glucose + ATP ↑↓ glucose 6-phosphate + ADP
Hydrolase	A-B + H$_2$O $\rightleftarrows$ A + B	Lactase: lactose + H$_2$O ↑↓ glucose + galactose
Lyase	$-\overset{\mid}{\underset{A}{C}}-\overset{\mid}{\underset{B}{C}}- \rightleftarrows -\overset{\mid}{C}=\overset{\mid}{C}- \;+\; A\text{-}B$	Aldolase: fructose 1,6 biphosphate ↑↓ dihydroxyacetonephosphate + glyceraldehyde 3-phosphate
Isomerase	$-\overset{A}{\underset{B}{C}}=C- \rightleftarrows -\overset{\mid}{C}=\overset{A}{\underset{B}{C}}-$	UDP galactose 4-epimerase: UDP-galactose ↑↓ UDP-glucose
Ligase	ATP + A + B ↑↓ A-B + ADP + P$_i$ or ATP + A + B ↑↓ A-B + AMP + PP$_i$	DNA ligase

The second group of enzymes is the **transferases,** by which a functional group is transferred between two substrates.

$$A\text{-group} + B \rightleftarrows A + B\text{-group}$$

Among the groups that can be transferred are one-carbon fragments, carbohydrate moieties, phosphate, sulfate, alkyl, and amino and acyl groups. An enzyme that transfers a phosphate from ATP to another compound is called a *kinase.*

The third category of reactions is catalyzed by **hydrolases.** In these reactions, water splits an ester, ether, peptide, glycosyl linkage, acid anhydride, carbon-carbon bond, carbon-halide bond, or phosphorus-nitrogen bond.

$$A\text{-B} + H_2O \rightleftarrows A + B$$

Most of the digestive enzymes are hydrolases. Enzymes that remove phosphate are called *phosphatases.*

Lyases are the fourth class of enzymes. In lyase reactions, two groups are removed from a substrate, leaving a double bond.

$$-\overset{|}{\underset{A}{C}}-\overset{|}{\underset{B}{C}}- \;\underset{\longleftarrow}{\longrightarrow}\; -\overset{|}{C}=\overset{|}{C}- \; + \; A\text{-}B$$

Lyase reactions are common in carbohydrate metabolism, especially in the pentose phosphate pathway.

Isomerases are enzymes that interconvert isomers—optical, geometrical, and positional. There is no net oxidation or loss of atoms in these one-substrate reactions. These enzymes are also called *mutases.*

$$-\overset{\overset{A}{|}}{C}=\overset{|}{\underset{B}{C}} \;\underset{\longleftarrow}{\longrightarrow}\; -\overset{|}{C}=\overset{\overset{A}{|}}{\underset{B}{C}}-$$

The final group of enzymes is **ligases.** These catalyze the formation of covalent bonds, utilizing the hydrolysis of ATP or some other high-energy compound releasing inorganic phosphate (P_i) or inorganic pyrophosphate (PP_i).

$$ATP + A + B \rightleftarrows A\text{-}B + ADP + P_i$$

We encountered ligases in our discussion of DNA synthesis.

Energetics

Enzymatic activity usually is studied at one of two points during the reaction. First, at the beginning of the reaction, there is a large amount of substrate compared to very little product. The second point is at the steady state where there is no net change in the amount of substrate or product with time. For a reaction at steady state or equilibrium, the substrate is converted to product at rate constant k_1 and the product to substrate at rate consant k_{-1}.

$$\text{substrate} \underset{k_{-1}}{\overset{k_1}{\rightleftarrows}} \text{product}$$

The **equilibrium constant** (K_{eq}) is the ratio of the two reaction constants, k_1 and k_{-1}.

$$K_{eq} = \frac{k_1}{k_{-1}}$$

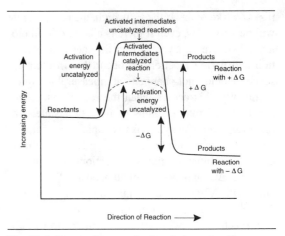

Fig. 11-1. Energy diagram of a chemical reaction. For a reaction to take place, the products must have a lower energy than the substrates: That is, the energy change (ΔG) must be negative. In most cases, additional energy (the activation energy) must be added to produce activated intermediates or a transition state. Only then will the reaction actually take place.

At equilibrium, the rate of the forward reaction that converts substrate to product is the same as the backward rate that converts product to substrate. This does not mean that there is no reaction taking place but only that the overall concentrations of substrate and product do not change with time.

Chemical reactions occur only when the total energy of the substrates is greater than the total energy of the products, as shown in the energy diagram in Figure 11-1. The difference in energy between the substrates and products, **ΔG,** must be negative if the reaction is to take place spontaneously in the direction of substrate to product. If ΔG is positive, the reaction will tend to go in the opposite direction, product to substrate. At equilibrium, there is no net change in the concentration of substrate or product and ΔG equals zero. For biological reactions, ΔG is measured at pH 7 with all reactants at 1-M concentration and is called **$\Delta G°'$.** However, these are not the concentrations at which reactants are found inside cells. ΔG is related to the equilibrium constant at pH 7 by the following equation:

$$\Delta G°' = -2.3 \; RT \log K'_{eq}$$

where R is the gas constant and T is the absolute temperature.

Not all reactions with a negative ΔG occur to any extent in a measurable amount of time. The **energy of activation** (E_a) must be supplied before the reaction can take place. If the E_a is high, the reaction rate will be very slow, even if there is a large ΔG. In biological systems, enzymes lower the energy of activation and permit reactions to occur that in their absence would not happen at measurable rates. Frequently, this occurs when the enzyme stabilizes the activated intermediates or transition state. Another function of enzymes is to couple two reactions, one that is energetically favorable and another that is not. In combination, the two reactions may have a negative ΔG and occur spontaneously. Such coupling with ATP or another high-energy compound is the function of ligases. Enzymes can change the rate of a reaction but will not change the ratio of substrate to product present after equilibrium is reached.

Kinetics

An enzymatic reaction can be diagrammed as follows:

substrate + enzyme $\rightleftarrows$ enzyme-substrate complex $\rightleftarrows$ enzyme + product

or

$$S + E \rightleftarrows ES \rightleftarrows E + P$$

At the very beginning of the reaction, there is no product, and the rate of the forward reaction (substrate $\rightarrow$ product) can be measured as the **initial velocity** (V_i). Initial velocity can be measured as either the rate of disappearance of substrate or the rate of appearance of product. With a constant amount of enzyme, as the concentration of the substrate is increased from zero, the V_i of the reaction increases (Fig. 11-2). The increase in rate continues until the enzyme is saturated with substrate. Then it reaches a **maximum velocity** (V_{max}). Above this concentration of substrate, the V_i will no longer increase even in the presence of additional substrate. Most one-substrate enzymes show this kind of relationship between initial reaction rate and substrate concentration. The initial velocity can also be measured later in the reaction as long as the substrate concentration is much greater than the product.

The **Michaelis-Menten equation** relates the V_i to the substrate concentration (S) and the V_{max}:

$$V_i = \frac{V_{max}(S)}{K_m + (S)}$$

where K_m, **the Michaelis constant**, is defined as the concentration of substrate that will produce one-half maximum velocity.

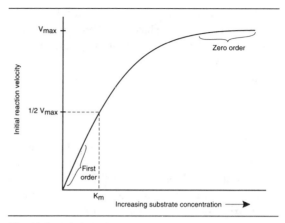

Fig. 11-2. Relationship between initial reaction velocity and substrate concentration. At a constant concentration of enzyme, the initial rate of conversion of substrate to products increases as the amount of substrate is increased. With substrate concentration that fully saturates the enzyme, the initial velocity will no longer increase with increasing substrate. This is the V_{max} of the reaction.

When the substrate concentration is small compared to the binding capacity of the enzyme, there is not enough substrate to saturate the enzyme. Under these conditions, not all the enzyme will be in the enzyme-substrate complex. The V_i of the reaction will be proportional to the amount of enzyme-substrate complex, which in turn is proportional to the amount of substrate available. When this occurs, the reaction is described as **first order** with respect to the substrate.

When the substrate concentration is very large compared to the binding capacity of the enzyme, all the enzyme will be in the enzyme-substrate complex. Then the rate of the reaction will not change with the addition of more substrate. Under these conditions, the initial rate of the reaction will be close to the maximal rate. Such reactions are described as **zero order** with respect to the substrate.

Graphically, K_m and V_{max} can be determined by using a linear version of the Michaelis-Menten equation, the **Lineweaver-Burk equation:**

$$\frac{1}{V_i} = \frac{K_m}{V_{max}(S)} + \frac{1}{V_{max}}$$

A plot of $1/V_i$ versus $1/(S)$ produces a straight line where the y-intercept = $1/V_{max}$, the x-intercept = $-1/K_m$, and the slope = K_m/V_{max}.

Inhibition

If an inhibitor of an enzymatic reaction competes for the same binding site on the enzyme as does the substrate, the inhibition is called **competitive.** This usually happens when the substrate and the inhibitor are structurally similar. Because the substrate and the inhibitor compete for the same site on the enzyme, in the presence of the inhibitor a higher concentration of substrate will be needed before the same rate of reaction will occur. If sufficient substrate is present, all the enzyme will be in a productive enzyme-substrate complex, and the reaction will reach its uninhibited V_{max} (Fig. 11-3). Competitive inhibition can be recognized by a Lineweaver-Burk plot in which the K_m increases in the presence of increasing inhibitor, but the V_{max} is unchanged.

In **noncompetitive inhibition,** V_{max} is reduced but K_m is unchanged, as shown in Figure 11-4. A noncompetitive inhibitor is able to combine with both the free enzyme and the enzyme-substrate complex. This reduces the amount of enzyme available to catalyze the reaction. No matter how much substrate is added, the reaction never will achieve the maximum velocity attainable in the absence of the inhibitor. A noncompetitive inhibitor usually does not resemble the substrate in structure and does not bind to the same site on the enzyme as does the substrate. In practice, it is difficult to distinguish reversible noncompetitive inhibition from inhibition

Fig. 11-3. Reaction curve (*A*) and Lineweaver-Burk plot (*B*) of competitive inhibition. In competitive inhibition, the inhibitor and the substrate have similar structures and compete for the same site on the enzyme. Competitive inhibition is characterized by a K_m that is increased in the presence of the inhibitor, while the V_{max} remains unchanged.

Fig. 11-4. Reaction curve (*A*) and Lineweaver-Burk plot (*B*) of noncompetitive inhibition. In noncompetitive inhibition, the inhibitor combines with both the free enzyme and the enzyme-substrate complex. This results in a decreased V_{max} but an unchanged K_m in the presence of the inhibitor.

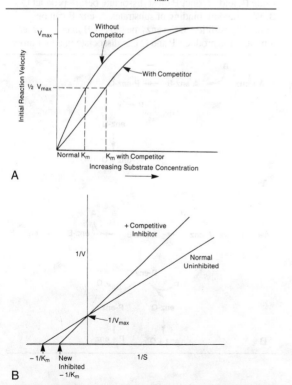

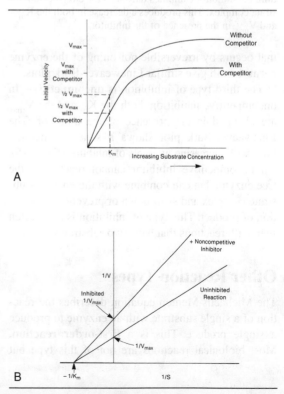

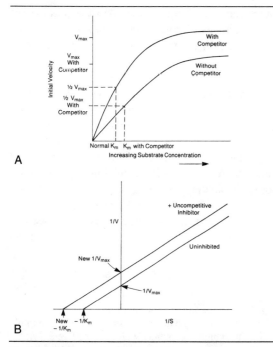

A

B

Fig. 11-5. Reaction curve (*A*) and Lineweaver-Burk plot (*B*) of uncompetitive inhibition. In uncompetitive inhibition, the inhibitor combines only with the enzyme-substrate complex. This produces a decrease in both the K_m and V_{max} in the presence of the inhibitor.

that occurs by irreversible poisoning of the enzyme because both give similar Lineweaver-Burk plots.

The third type of inhibition is **uncompetitive.** In uncompetitive inhibition, both the K_m and the V_{max} are changed in the presence of the inhibitor. The Lineweaver-Burk plot shows a series of parallel lines with increasing amounts of inhibitor (Fig. 11-5). An uncompetitive inhibitor cannot react with the free enzyme but can combine with the enzyme-substrate complex and slow down or prevent the formation of product. This type of inhibition is most often seen with reactions that have two substrates.

Other Reaction Types

The Michaelis-Menten equation describes the reaction of a single substrate with an enzyme to produce a single product. This is a **first-order reaction.** Most biological reactions are not of this type but have more complicated kinetics. However, at a high concentration of one of the substrates, reactions can become zero order for this substrate. Then the reaction will show typical Michaelis-Menten kinetics for the other substrate.

Reactions such as those involving oxidoreductases, transferases, and ligases require at least two substrates. The initial velocity depends on the concentration of both substrates and is a **second-order reaction** under most biological conditions. These reactions can occur as either single- or double-displacement reactions. In a **single-displacement reaction,** both substrates must bind to the enzyme before any products are released:

$$A + B + enzyme \rightleftarrows EAB \rightleftarrows + P + Q$$

The addition of substrates A and B to the enzyme may be ordered or random, as shown in Figure 11-6. Because the chance of all three molecules (A, B, and

Fig. 11-6. Two-substrate single-displacement reaction. A. With ordered binding, substrate A must bind before substrate B, and product P must dissociate before product Q. B. With random binding of substrates, A-enz-B can be formed by addition to the enzyme of A or B in either order. Similarly the products P and Q can dissociate in either order.

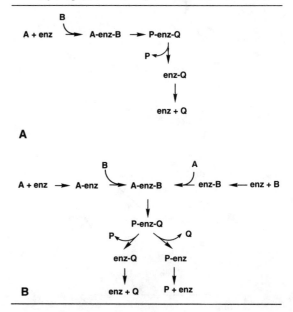

A

B

enzyme) colliding at the same time is remote, there must be some order of addition of the substrates to the enzyme. However, the order does not always have to be the same. Similarly, the release of the products P and Q can be either ordered or random.

In **double-displacement** (or *ping-pong*) **reactions,** one substrate binds to the enzyme and one product is released before the second substrate binds. Transferase reactions are of this type, with the functional group to be transferred attached temporarily to the enzyme.

A + enzyme ⟶ enzyme-A ⟶ enzyme' + P
                                        ↳ B
                            enzyme'-B
                                 ↓
                            enzyme + Q

Other Modulators

In addition to the concentration of substrate, enzyme, and any inhibitors, the rate of an enzymatic reaction can be altered by temperature, pH, and compounds that are structurally unrelated to the substrate. In general, an increase in temperature of 10°C tends to increase the rate of a reaction twofold. At higher temperatures, the reactants collide more frequently and do so with greater energy. More molecules have sufficient energy to surmount the activation energy, and the reaction rate increases. For enzymatic reactions, an increase in temperature also tends to denature the enzyme protein, destroying its ability to catalyze the reaction. Because of this, most enzymes are effective catalysts only over a very small temperature range.

Enzymatic activity also can be affected by pH. Even relatively small variations in pH can alter the protonation state of the substrate, product, or specific amino acids of the enzyme. This can prevent the substrate from binding or reacting, can prevent the product from dissociating, or can even denature the enzyme. Because of these effects, enzymes usually show a sharp pH optimum for activity.

Enzymatic activity can be modulated by other small molecules as well. If these molecules are not structurally related to the substrate or product and bind at a site different from the one where enzymatic activity takes place, they are called **allosteric effectors.** These can act negatively, reducing the rate of reaction, or positively, increasing it (Fig. 11-7). These effects can occur by altering the K_m, the V_{max}, or both. Allosteric effectors are very important for biological systems, and we will encounter them frequently in our study of metabolism. The final product of a reaction pathway often acts as an allosteric effector, inhibiting the first enzyme of that pathway. This is called **feedback inhibition.** There are also allosteric effectors that are protein subunits of a large multiprotein complex. In the presence of the regulatory subunit, the enzyme may be more or less active than in its absence.

Phosphorylation is another mechanism for the modulation of enzymatic activity. For enzymes regulated by phosphorylation, only one form of the enzyme, phosphorylated or dephosphorylated, is catalytically active. The other is essentially inactive. By using phosphorylation, catalyzed by a protein kinase, or dephosphorylation, catalyzed by a protein phosphatase, as an on-off switch, a cell can control a reaction without having to synthesize and degrade the protein that catalyzes it. This allows a faster response to changing conditions and saves the metabolic energy used in the synthesis of the protein. As you saw in Chapter 3, the proteins that regulate the cell cycle are controlled primarily by phosphorylation and dephosphorylation.

Fig. 11-7. Reaction curve showing the effect of positive and negative allosteric effectors.

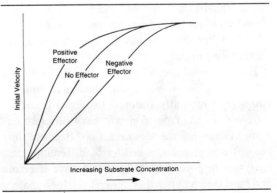

Active Site

Enzymes are specific for only a small group of related compounds. The specificity applies to optical as well as structural isomers. Some substrates appear to fit into the enzyme's active site much as a key fits into a lock. Substrates that are too large are not able to enter the active site, and those that are too small do not bind in the correct configuration to allow the reaction to take place. In other enzyme systems, the substrate appears to alter the structure of the enzyme when it binds. This produces an induced fit between enzyme and substrate. The conformational change in the enzyme facilitates the enzymatic reaction. Compounds that do not induce the same change in the structure of the enzyme do not cause the catalytic reaction to occur.

The amino acids most frequently found at the site where a substrate binds are those that are charged at cellular pH (glu, asp, lys, arg), those that can form hydrogen bonds, those that can form covalent bonds (cys, lys, ser, asp, glu), and histidine, the only amino acid whose level of protonation is readily changed at cellular pH. Enzymes with hydrophobic substrates tend to have hydrophobic amino acids at the substrate binding site. Hydrophobic amino acids are important structural elements near the active site in enzymes that must exclude water from the active site.

In addition to amino acids, coenzymes, prosthetic groups, and metal ions frequently are found at the active site. We will discuss the structure and function of coenzymes and prosthetic groups as we encounter them in our study of metabolism. Enzymes from which these groups have been removed are called **apoenzymes** and are inactive. The apoenzyme combined with the coenzyme or metal ion is called a **holoenzyme** and is the catalytically active form of the protein.

Enzyme proteins can be structurally different and still catalyze the same reaction. These are known as **isozymes.** Frequently, different tissues or tissues at different stages of development have different enzymes catalyzing the same reaction. Because they are different proteins, coded for by different genes, isozymes respond differently to allosteric effectors and have different reaction rates, K_m values, molecular weights, and pH and temperature optima. Lactate dehydrogenase (LDH), which has different forms in different tissues (e.g., heart, muscle, and liver) is an example of a well-studied isozyme system. The amount of specific LDH isozymes in the blood is used as a diagnostic tool to determine the extent of cardiac damage in myocardial infarct and the existence and extent of some types of liver disease.

Measurement of Enzymatic Activity

To measure the activity of an enzyme, it is necessary first to determine its kinetic parameters. The activity, as measured by the initial rate of reaction, is determined with increasing substrate concentration. These data, graphed in a Lineweaver-Burk plot, will give the K_m of the reaction. At high substrate concentrations ($S \gg K_m$), most of the enzyme will be in an enzyme-substrate complex, and the rate of the reaction will be proportional to the amount of enzyme. From this information, the **specific activity** (activity per milligram of protein) can be calculated.

Specific activity is the parameter usually followed during the purification of a protein. It generally increases with each purification step as unrelated and nonreactive proteins are removed. However, the specific activity will decrease if the protein is denatured during the purification process. If the molecular weight of the enzyme is known, the **turnover number** (number of substrate molecules reacted per second per enzyme molecule) can also be determined. To measure an accurate specific activity or turnover number, it is necessary to control all factors that can affect enzymatic activity, including temperature, pH, concentration of substrates, inhibitors, disulfide reagents, cofactors, and metal ions. In cells, enzymes that control the overall rate of conversion of the initial substrate to the final product in a reaction pathway frequently operate under suboptimal conditions. This allows the enzymatic activity to be modulated by altering the concentration of the substrate, product, or allosteric effectors without the necessity of synthesizing or degrading the enzyme protein.

Summary

Enzymes are classified into six categories. Oxidoreductases catalyze the oxidation of one substrate while reducing the other. Transferases transfer functional groups from one substrate to another. Hydrolases split covalent bonds by utilizing water. Lyases remove functional groups, leaving behind a double bond. Isomerases rearrange single substrates. Ligases form covalent bonds utilizing the energy from the hydrolysis of ATP or another high-energy compound.

Reactions take place spontaneously only when the ΔG is negative. Enzymes do not change the ΔG nor the ratio of substrate to product at equilibrium. They do reduce the activation energy, E_a, and allow equilibrium to be achieved sooner. Enzymes also couple reactions so that the overall ΔG is negative. ΔG for biological reactions is measured at pH 7, with all the reactants at 1-M concentration. The resulting term $\Delta G°'$ is related to the equilibrium constant by the following equation:

$$\Delta G°' = -2.3 \text{ RT log } K'_{eq}$$

In the absence of product, the initial rate of a reaction is proportional to the amount of enzyme-substrate complex. If there is a limiting amount of substrate, not all the enzyme will be in the enzyme-substrate complex. Then the reaction rate will be proportional to the concentration of substrate, or first order with respect to substrate. If the enzyme is the limiting factor, all the enzyme will be in the enzyme-substrate complex and the rate will be independent of substrate concentration, or zero order with respect to substrate. The Michaelis-Menten equation relates initial reaction velocity, substrate concentration, and maximum velocity.

A linear version of the Michaelis-Menton equation, the Lineweaver-Burk equation, is used graphically to determine the values of K_m and V_{max}:

$$\frac{1}{V_i} = \frac{K_m}{V_{max}(S)} + \frac{1}{V_{max}}$$

It can also be used to classify inhibitors of enzyme reactions. Competitive inhibitors resemble the substrate and bind to the same site on the enzyme. These compounds change the K_m but not the V_{max}. Noncompetitive inhibitors do not resemble the substrate and bind to a different site on both the enzyme and the enzyme-substrate complex. They reduce the V_{max} but not the K_m. Uncompetitive inhibitors bind only to the enzyme-substrate complex and change both the K_m and the V_{max}.

Michaelis-Menten kinetics describe a first- or zero-order reaction in which the rate of production of the enzyme-substrate complex is much higher than its conversion to enzyme plus product. Second-order reactions are the single-displacement type when both substrates bind before any product is released. In double-displacement reactions, one substrate binds and one product is released before the second substrate binds. Enzyme activity also can be affected by pH, temperature, and allosteric effectors. The latter compounds bind at different sites from the substrate and either increase or decrease the activity of the enzyme. Often the final product of a reaction pathway acts as an allosteric inhibitor of the first enzyme in that pathway. Subunits of a complex enzyme also can act as allosteric effectors. Phosphorylation and dephosphorylation are still another mechanism for the control of enzymatic activity.

The active sites of enzymes contain amino acids that can form covalent bonds, salt linkages, or hydrogen bonds, or undergo reversible protonation. Metal ions are found in the active site as well. Enzymes without their metal ions or prosthetic groups are called *apoenzymes*. When the group is present, the enzyme is called a *holoenzyme*. Two different enzymes that catalyze the same reaction are *isozymes*. Specific activity is the activity of an enzyme per milligram of protein. It increases as the enzyme is purified. The turnover number is the number of substrate molecules reacted per unit time per enzyme molecule. It can be calculated from the specific activity if the enzyme is pure and the molecular weight is known.

12 Specialized Proteins

In this chapter, we discuss three groups of proteins that demonstrate the relationship between protein structure and protein function. The first group contains the soluble oxygen transport proteins hemoglobin and myoglobin. Hemoglobin is the major protein in red blood cells. Myoglobin is located in muscle. The difference in the proteins surrounding the heme group determines the difference in the ability of these two proteins to take up and release oxygen reversibly.

The second group of proteins includes the blood-clotting proteins, which function in an enzyme cascade. Each factor circulates as a soluble zymogen. On activation, it becomes an enzyme that activates another enzymatic factor that activates still another enzymatic factor. Because each factor acts catalytically, a small amount of the first factor leads to more of the second and still more of the third. This results in a very large reaction at the end of the pathway. In the final reaction, a soluble protein, fibrinogen, becomes the insoluble protein, fibrin, which forms the basis for the stable blood clot.

The third group contains the proteins involved in muscle contraction—actin, myosin, tropomyosin, and troponin. Acting together, these proteins convert the energy from the hydrolysis of ATP into the mechanical energy needed for physical motion. The function of these proteins is tightly controlled so that muscle can contract and relax as needed.

Objectives

After completing this chapter, you should be able to

Describe the structures of hemoglobin and myoglobin and their oxygen saturation curves and relate these to the physiological functions of the two proteins.

List the allosteric effectors of oxygen binding and recognize how they act.

Define the terms *Heinz body, Hill plot, Bohr effect,* and *chloride shift.*

Discuss the alterations in hemoglobin that occur in sickle-cell anemia, diabetes mellitus, and exposure to carbon monoxide and cyanide.

Outline the extrinsic and intrinsic pathways for blood clotting, using both the numbers and the common names for the factors.

List the conditions and agents that inhibit clotting both in drawn blood and in circulation.

Report how clots are dissolved.

Relate the inherited diseases that affect clotting to the site of the defect.

Define *prothrombin time, partial thromboplastin time,* and *thrombin time.*

Describe the arrangement of proteins in muscle.

Discuss the reactions that take place in muscular contraction and relaxation and how these reactions are controlled.

Hemoglobin and Myoglobin

Structure

Hemoglobin and myoglobin are evolutionarily conserved proteins. When isolated from a variety of species, they are similar in secondary structure. In addition, changes in the primary structure tend to be conservative—one aliphatic amino acid exchanged for another, one positively charged amino acid for another, or one negatively charged amino acid for another. Hemoglobin and myoglobin are compact globular proteins whose secondary structure consists of a number of α-helical segments. Similar to other globular proteins, the interior region of myoglobin and hemoglobin contains primarily hydrophobic amino acids, whereas the outside contains the hydrophilic amino acids. Adult hemoglobin, designated (Hb A), contains two α-chains, two β-chains, and four heme groups, one associated with each of the four protein chains. Myoglobin contains only one protein chain and a single heme group. The heme, which is contained in a pocket of hydrophobic amino acids, is capable of binding a single oxygen molecule.

The structure of **heme** is given in Figure 12-1. Note that heme is asymmetrical. It has four methyl groups, two vinyl groups, and two propionic acid groups. Only the molecule with the correct arrangement of these groups will bind to the protein to create a molecule capable of reversibly carrying oxygen.

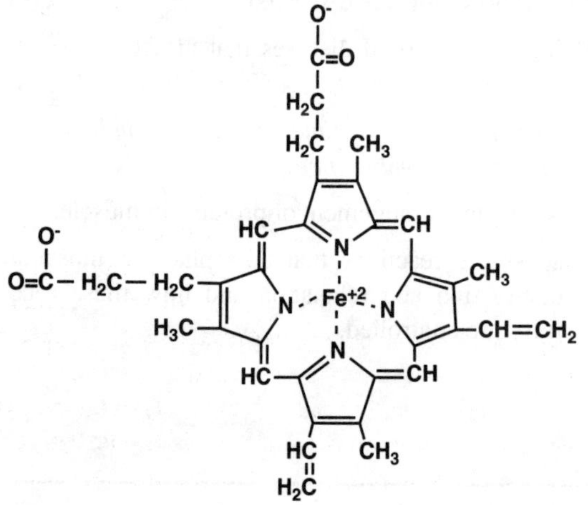

Heme contains iron in the ferrous or +2 oxidation state (Fe^{2+}). The iron is coordinated to four nitrogen atoms in the heme. The fifth coordination position is occupied by a histidine residue of the globin protein. When oxygen binds to the heme, it occupies the sixth coordination position. If the iron in heme is oxidized to ferric iron (Fe^{3+}), then the heme is called **hemin.** In this form, it will not bind oxygen reversibly. When the heme is sheltered from amino acids that could oxidize it, from water, and especially from other heme groups, it can be kept in the ferrous state and bind oxygen reversibly. Hemoglobin and myoglobin in which the iron has been oxidized to the ferric form are called **methemoglobin** and **metmyoglobin,** respectively. Hemin can be reduced to heme by methemoglobin reductase:

$$metHb + NADH + H^+ \rightarrow Hb + NAD^+$$

This reduction is important for maintaining the oxygen transport ability of blood. Lack of sufficient NADH causes an increase in the amount of methemoglobin. In reticulocytes, reducing equivalents come from glycolysis or the pentose phosphate pathway. An inherited deficiency of glucose 6-phosphate dehydrogenase, an enzyme in the pentose phosphate pathway, causes a decrease in overall reducing equivalents in the cell. This, in turn, decreases the conversion of methemoglobin to hemoglobin. Methemoglobin accumulates and precipitates, forming visible clumps called **Heinz bodies.** Cells containing Heinz bodies are lysed, causing one form of hemolytic anemia.

Oxygen Binding

In myoglobin, oxygen binding follows Michaelis-Menten kinetics and produces the percent saturation–versus–oxygen concentration curve shown in Figure 12-2. In hemoglobin, binding of one molecule of oxygen facilitates the binding of other oxygen molecules. When oxygen is released, the loss of one molecule of oxygen facilitates the loss of additional oxygen molecules, a process called **positive cooperativity.** It produces the sigmoidal curve seen in Figure 12-3 when percent saturation is plotted against the partial pressure of oxygen.

Fig. 12-1. Structure of heme. Note the asymmetrical arrangement of the methyl, vinyl, and propionyl groups. In addition to coordination with four nitrogens, the heme is coordinated to a histidine residue on the globin protein and to oxygen.

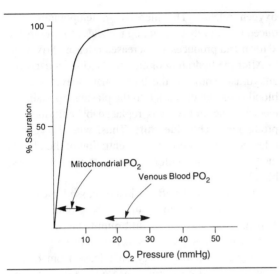

Fig. 12-2. Oxygen-binding curve for myoglobin.

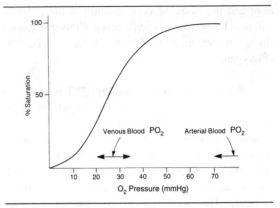

Fig. 12-3. Oxygen-binding curve for hemoglobin.

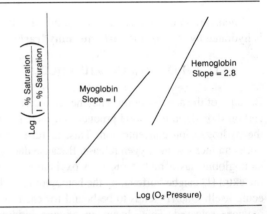

Fig. 12-4. Hill plot for oxygen binding of myoglobin and hemoglobin.

The degree of cooperativity between the binding of substrate molecules can be measured by a **Hill plot** in which log (Y/l − Y) is plotted against log PO_2, where Y = the percent saturation (Fig. 12-4). The **Hill coefficient** is the slope of the line at one-half oxygen saturation. For myoglobin, the Hill coefficient is 1, indicating no cooperativity. For hemoglobin, it is 2.8, indicating positive cooperativity.

The difference in oxygen binding between hemoglobin and myoglobin fits their different physiological functions. Myoglobin has a higher affinity for oxygen than does hemoglobin. Because of this, myoglobin is capable of taking up oxygen from hemoglobin and releasing the oxygen at the low oxygen tensions found inside cells. This binding and release takes place over a very small range of oxygen pressure values. Hemoglobin, on the other hand, has a lower oxygen affinity. It can bind oxygen at the higher pressures found in the lungs and will release it only at the much lower pressures found in the tissues. In addition, because of positive cooperativity, when one molecule of oxygen is lost in the tissues, all four tend to be lost. Similarly, when one molecule of oxygen is bound in the lungs, four tend to be bound, fully saturating the hemoglobin.

Fetal hemoglobin has an oxygen affinity intermediate between those of hemoglobin and myoglobin. The oxygen-binding curve shown in Figure 12-3 is shifted to the left. An oxygen affinity greater than

adult hemoglobin is necessary in order for fetal hemoglobin to pick up oxygen from the maternal circulation. The affinity must be low enough to be able to release oxygen into fetal tissues. Fetal hemoglobin, like adult hemoglobin, shows a sigmoidal oxygen-binding curve with positive cooperativity.

Allosteric Effectors

Hemoglobin is a dynamic molecule having two major quaternary structures. These structures interconvert when one of the pairs of α/β protein chains is moved relative to the other pair. The **R** or **relaxed form** binds oxygen several hundred times more readily than does the **T** or **tense form** of hemoglobin. In the T form, the ends of the β-chains are in salt bridges. This stabilizes the T form and reduces its affinity for oxygen. Any change that alters the equilibrium and increases the concentration of the R form will tend to increase oxygen binding. Conversely, any change that stabilizes the T form will promote the loss of oxygen.

$$\text{R form (high oxygen binding)} \rightleftarrows \text{T form}$$
$$\text{(lower oxygen binding)}$$

Several allosteric effectors increase the concentration of the T form and promote the release of oxygen in the tissues. These effectors include H^+, dissolved carbon dioxide, temperature, and 2,3-diphosphoglycerate. They cause the oxygen saturation curve to be shifted down and to the right, as shown in Figure 12-5.

When carbon dioxide enters the red blood cell, it is hydrated by the enzyme **carbonic anhydrase:**

$$CO_2 + H_2O \rightarrow H_2CO_3 \rightarrow H^+ + HCO_3^-$$

Because of the activity of this enzyme, an increase in carbon dioxide in the blood produces an increase in the hydrogen ion concentration. This, in turn, produces an increase in oxygen release. Because deoxyhemoglobin has a higher pK than oxyhemoglobin, the extra H^+ can be buffered by the hemoglobin molecule itself. One H^+ tends to be bound for each two oxygens released. This change in oxygen binding with a change in pH is called the **Bohr effect.** The release of lactic acid from rapidly metabolizing tissues also produces an increase in the hydrogen ion concentration, which has the desired effect of increasing

Fig. 12-5. Effect of the allosteric effectors H^+, dissolved carbon dioxide, increased temperature, and 2,3-diphosphoglycerate (DPG) on the oxygen-binding curve of hemoglobin.

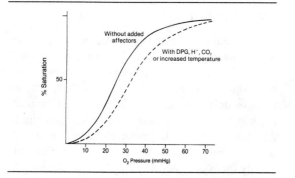

oxygen release. The increase in temperature produced by actively contracting muscles also favors the T form and produces an increased release of oxygen.

After the hydration of carbon dioxide by carbonic anhydrase, some of the bicarbonate leaves the red blood cells and diffuses into the plasma. To maintain electrical neutrality, it is replaced by chloride ion, producing a **chloride shift.** Thus, when oxygen is released, the bicarbonate concentration of the plasma increases, and the chloride concentration of the red blood cell also increases.

In addition to the effects just discussed, bicarbonate can bind to amino groups in the globin protein. This binding shifts the equilibrium from the R form to the T form and promotes oxygen release.

Another effector of oxygen release from hemoglobin is **2,3-diphosphoglycerate** (DPG). DPG, the structure of which is provided here, is produced by the glycolysis pathway:

$$\begin{array}{ccc} & O^- & \\ & | & \\ O^- & O=C & \\ | & | & \\ O=P-O-HC & & O^- \\ | & | & | \\ O^- & CH_2-O-P=O \\ & & | \\ & & O^- \end{array}$$

One DPG molecule binds to each hemoglobin and stabilizes the T form, promoting oxygen release. Fetal hemoglobin does not bind DPG. This is one of the major differences between adult and fetal hemoglobin and accounts for much of the observed difference in oxygen binding.

Other Effectors of Oxygen Binding in Hemoglobin

The ability of hemoglobin to carry oxygen can be altered by binding of small molecules such as carbon monoxide and cyanide and by mutations of the globin protein. **Carbon monoxide** (CO) has an affinity for hemoglobin that is 200 times that of oxygen. Thus, once bound, the CO does not readily dissociate. The pocket arrangement of the heme in hemoglobin does not provide sufficient space for a linear alignment of the iron and the C and O atoms of carbon monoxide. This arrangement decreases the ability of CO to bind tightly to hemoglobin and allows for human survival. Usually, less than 5% of hemoglobin molecules contain bound CO. If the percentage is increased by exposure to CO, oxygen transport is adversely affected.

Unlike CO, **cyanide** binds to methemoglobin and metmyoglobin but not to the reduced forms. Once cyanide is bound, the methemoglobin cannot be reduced to the active form. After even a short exposure to cyanide, the ability of the blood to transport oxygen is impaired. Prompt medical attention is required to prevent death. In addition, cyanide will bind to other enzymes that contain heme groups. We will encounter the effects of cyanide again when we discuss the cytochromes of the electron transport chain.

Sickle-cell anemia, a common disease in the black population of the United States, is due to a mutation in the gene coding for the β-chain of hemoglobin, which changes a codon for glutamic acid to a codon for valine. The loss of the positive charge from glutamic acid causes the solubility of the deoxygenated form of hemoglobin to be decreased. Under conditions of low oxygen pressure, the deoxygenated hemoglobin polymerizes and precipitates, distorting the shape of the red blood cell. The cell assumes a sickled shape that interferes with its passage through capillaries and blocks the vessel, producing extreme pain. Persons with only one gene for sickle cell have **sickle-cell trait** and usually no clinical symptoms. Those with two sickle-cell genes have no normal β-chains and no adult hemoglobin. They have **sickle-cell disease.** Although the clinical symptoms of the disease are variable, they can be severe.

A number of other globin mutations have been isolated that alter the secondary and tertiary structures of hemoglobin. The ability of these hemoglobin molecules to carry oxygen reversibly also is altered.

Glucose can bind covalently to the N-terminal end of the β-chains of hemoglobin, producing hemoglobin A_c. This modification does not change the ability of the hemoglobin to carry oxygen. Whereas the amount of hemoglobin A_c is less than 5% in normal adults, in individuals with poorly controlled diabetes mellitus, it can constitute more than 10% of the total hemoglobin. Measurement of glucose in blood and urine indicates the glucose level at a specific point in time; average long-term levels can be estimated by the amount of hemoglobin A_c.

Blood Clotting

Overview

Blood clotting occurs via two pathways, intrinsic and extrinsic. The intrinsic pathway is composed of enzymes that are activated when injured tissue, especially collagen, is exposed directly to blood. The extrinsic pathway requires at least one noncirculating component and is activated by a factor released by injured tissues. After a number of reactions unique to each pathway occur, the two pathways join to produce the final clotting process. The factors are all proteins or glycoproteins and are identified by Roman numerals as well as common names. The inactive factors are identified by Roman numerals and the activated forms by a Roman numeral and a lowercase *a*. For example, Factor II is the inactive form, whereas IIa is the active factor. Table 12-1 lists these factors.

Table 12-1. Names and functions of the clotting factors

Zymogen	Active factor	Common name	Pathway	Function
Fibrinogen	Ia	Fibrin	Common	Forms soft clot
Prothrombin	IIa	Thrombin	Common	Catalyzes I to Ia
V	Va		Common	Accelerates conversion of prethrombin to thrombin
VII	VIIa	Proconvertin	Extrinsic	Converts X to Xa
VIII	VIIIa	Antihemolytic factor	Intrinsic	Needed for conversion of X to Xa
IX	IXa	Christmas factor	Intrinsic	Converts X to Xa
X	Xa	Stuart factor	Common	Converts prothrombin to prethrombin
XI	XIa	Plasma thromboplastin	Intrinsic	Binds to exposed tissue, activates IX
XII	XIIa	Hageman factor	Intrinsic	Binds to collagen, activates XI
XIII	XIIIa	Transaminase	Common	Crosslinks fibrin

The active factors for blood clotting circulate as inactive proteins. On activation, most function as serine proteases that contain serine as an essential amino acid in the active site. Each factor acts catalytically to activate the next factor, thereby producing a cascade in which a small amount of the first factor can produce a very large reaction at the end of the pathway. Because the factors are in the circulation, a rapid response to blood vessel breakage can occur, limiting the extent of blood loss.

The first response to a broken blood vessel is the aggregation of platelets to plug the leak partially. When platelets adhere to exposed collagen, they release adenosine diphosphate (ADP). In response, the platelets swell and release more ADP. Their membranes become sticky, and they aggregate to form the initial clot. Platelets also release a prostaglandin, thromboxane, which is a vasoconstrictor and further reduces blood loss.

Intrinsic Pathway

The reactions of the intrinsic pathway are outlined in Figure 12-6. Follow the diagram as you read this section. All the factors in the intrinsic pathway are found circulating in the plasma as inactive zymogens. In the first reaction, Factor XII (Hageman factor) binds to exposed tissue, especially collagen. This binding changes the conformation of the protein and allows it to convert prekallikrein to kallikrein. Kallikrein, in turn, activates the bound Factor XII, producing Factor XIIa. Activation of Factor XII by kallikrein also requires high-molecular-weight kininogen (HMW kininogen) as a cofactor.

In the next step, Factor XIIa reacts with Factor XI (plasma thromboplastin antecedent) to produce Factor XIa. This reaction also requires HMW kininogen as a cofactor. In its activated form, Factor XIa binds to the exposed tissue surface. Here, it converts Factor IX (Christmas factor) to Factor IXa. This latter reaction requires calcium. The final reaction unique to the intrinsic pathway is the conversion of Factor X (Stuart factor) to Factor Xa, a reaction catalyzed by Factor IXa. In addition, the reaction requires platelet factor, found on the surface of aggregated platelets, Factor VIII (antihemolytic factor), and calcium. Factor Xa is the final active product of both the intrinsic and extrinsic pathways.

Extrinsic Pathway

The extrinsic pathway is outlined in Figure 12-7. It derives its name from the fact that it requires a nonplasma protein to initiate the cascade. The initiating reaction in this pathway is the conversion of Factor VII (proconvertin) to Factor VIIa. This reaction requires calcium, thromboplastin (a tissue factor available only after blood vessel wall injury), and phospholipids from injured tissue or the surface of aggregated platelets. The reaction is accelerated by the presence of Factor XIIa, kallikrein, and Factor IXa from the intrinsic pathway. Factor VIIa then converts Factor X to Xa.

Fig. 12-6. Intrinsic pathway for blood clotting. (*HMW* = high-molecular-weight.)

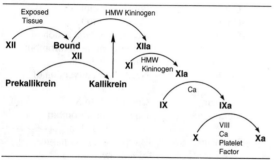

Fig. 12-7. Extrinsic pathway for blood clotting.

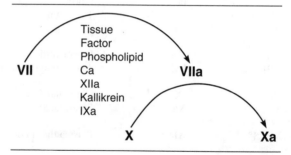

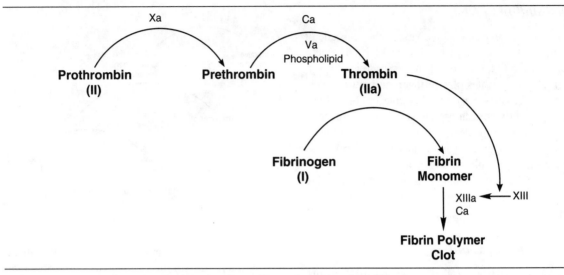

Fig. 12-8. Common pathway for blood clotting. Factor Xa can be produced by either the intrinsic or extrinsic pathway.

Common Pathway

The production of Factor Xa occurs by both the intrinsic and extrinsic pathways. It is the starting reaction for the common pathway outlined in Figure 12-8. Despite the fact that both pathways produce Factor Xa, deficiencies in either pathway produce prolonged bleeding times. Factor Xa converts prothrombin (Factor II) to prethrombin. Prethrombin can slowly be converted to thrombin (Factor IIa) by autocatalysis. The reaction is vastly accelerated by calcium, phospholipids, and Factor Va. All these reactions take place on the surface of platelets.

Thrombin then reacts with fibrinogen (Factor I) to produce fibrin (Factor Ia). Fibrinogen is a complex protein composed of three different protein pairs held together by disulfide bonds. The reaction with thrombin cleaves a portion of the amino terminal end of two of the chains (α and β). This allows the fibrin molecules to aggregate, forming a soft clot that traps platelets and red blood cells.

Firm clots occur by crosslinking the fibrin monomer. Factor XIIIa, a transaminase activated by thrombin, crosslinks the terminal carboxyl group on glutamine with the terminal amino group of lysine, as shown in Figure 12-9. This links different chains and solidifies the clot.

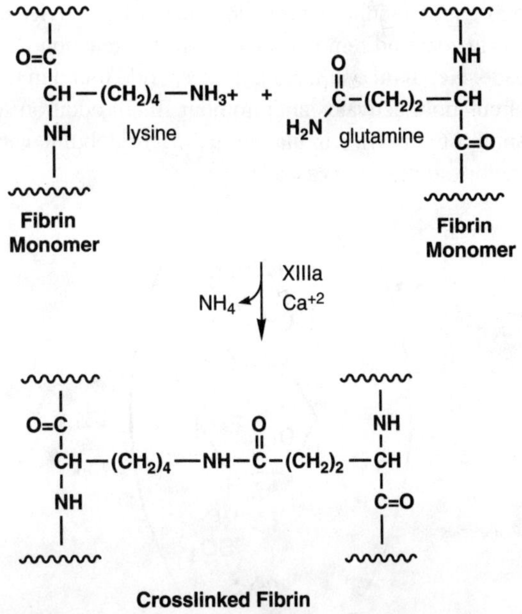

Fig. 12-9. Crosslinking of fibrin, the final step in clot formation.

Dissolution of Clots

The dissolution of clots involves another cascade of serine proteases, outlined in Figure 12-10. The activator of the first reaction of this cascade is fibrin itself. Fibrin activates **plasminogen activator** which, in turn, activates plasmin. This digests fibrin. Plasminogen activator can be activated in the absence of fibrin in cases of cancer, shock, or prolonged bleeding. Plasminogen activators such as **streptokinase** and **tissue-type plasminogen activator** (t-PA) are used clinically to dissolve blood clots quickly and to reduce the tissue damage produced by myocardial infarction.

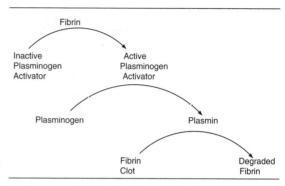

Fig. 12-10. The mechanism of dissolution of clots.

Prevention of Clotting

Blood clotting is controlled by several mechanisms. First, activated factors are removed by the liver as the blood passes through it. This is a slow response. More immediate inhibition of clotting is produced by **antithrombin III.** This natural clotting inhibitor accounts for 50% of the anticlotting activity of blood. Antithrombin III binds all the serine proteases of the cascade, especially thrombin, in a one-to-one complex. Such binding inhibits the enzymatic activity of the factors and removes them from the reaction cascade. **Heparin,** a sulfated polysaccharide found in the circulation, activates antithrombin III. In addition to antithrombin III, circulating α_2-macroglobulin also inhibits the clotting cascade.

heparin

Whereas heparin and antithrombin III immediately interfere with the clotting cascade, derivatives of the drug **dicoumarol,** a **vitamin K** analog, interfere more slowly.

dicoumarol

Dicoumarol inhibits the γ-carboxylation of glutamic acid residues in clotting Factors II, VII, IX, and X. This reaction, which occurs in the liver, is necessary for the maturation of these factors. The γ-carboxyl groups are required for these factors to bind calcium (Fig. 12-11). Continued exposure to high doses of dicoumarol eventually will produce a lack of mature clotting factors, which prevents all clotting. Low doses of oral dicoumarol can be used to slow but not entirely prevent clotting. Because dicoumarol is a competitive inhibitor, its effect can be reversed by vitamin K.

Aspirin also inhibits blood clotting, by preventing the aggregation of platelets and the production of thromboxane. Thromboxane is one of the eicosanoids, which functions as a vasoconstrictor. The structure and synthesis of the eicosanoids, including the inhibition by aspirin, will be discussed in Chapter 20. If the platelets do not aggregate at the site of the injury, the initial plug

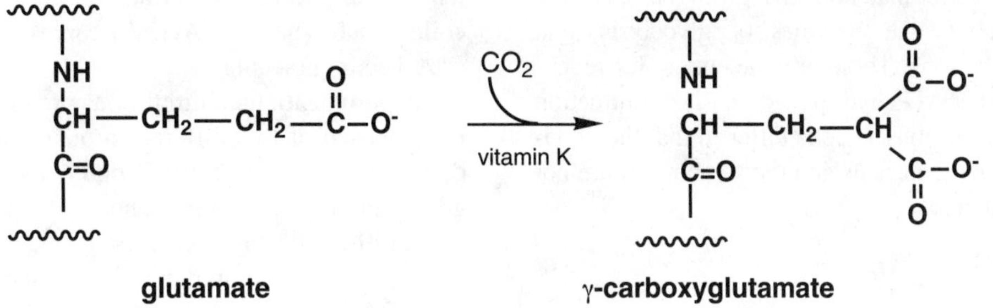

Fig. 12-11. Gamma-carboxylation of glutamate residues in clotting Factors II, VII, IX, and X. This reaction, which takes places in the liver, requires the participation of vitamin K.

will not form and, without thromboxane, the vasoconstriction will not occur. In addition, the platelet surface required for several of the reactions will not be available. This ultimately prevents the clot from forming. Because of its effect on the clotting process, aspirin is not given to patients within a week before surgery. Similarly, aspirin should not be taken by pregnant women during the final month before delivery. Acetaminophen, another over-the-counter analgesic, does not have a similar effect on the clotting process.

Measurement of Clotting

When blood clots, the liquid remaining is **serum.** It contains all the proteins of blood except for the aggregated clotting factors. If clotting is prevented by chelating calcium or the addition of heparin and the red blood cells are removed, the liquid remaining is **plasma.** Plasma is capable of clotting if calcium is restored or the heparin is removed.

Prothrombin time measures the activity of the extrinsic pathway. Blood is collected in the presence of citrate to chelate calcium ions. Then tissue factor (thromboplastin) is added, and the time until the formation of a fibrin clot is measured.

The activity of thrombin can be measured by adding thrombin to plasma and waiting for a clot to form. This is **thrombin time. Partial thromboplastin time** measures the intrinsic pathway (Factors I, II, VIII, IX, X, XI, and XII). To measure it, plasma is exposed to a surface to provide activation of Factor XII. Then calcium and phospholipids are added, and the time to clot formation is measured.

Inherited Defects in the Clotting Process

The most studied of the bleeding diseases is **hemophilia A.** It is due to a deficiency of Factor VIII. Because hemophilia A is inherited as X-linked recessive, it is more prevalent in males than females. The royal line of hemophilia A appears to have been passed down from Queen Victoria. By intermarriage among families, it eventually affected most of the royal houses of Europe.

Hemophilia B or *Christmas disease* is due to a deficiency of Factor IX. It is inherited as an autosomal recessive trait. All the other clotting factors are carried as autosomal dominant traits. **Von Willebrand's disease** is due to a deficiency in platelet adherence and a defect in Factor VIII. Unlike hemophilia A and B, there is clotting with von Willebrand's disease, but the time is prolonged.

Muscle Contraction

Overview

Muscle tissue makes up approximately 40% of the body mass of a healthy adult man. This tissue can be divided into three types: smooth, skeletal, and cardiac muscle. The latter two appear striated when examined under polarized light. Striated muscle is composed of long nucleated cells surrounded by an excitable plasma membrane, the **sarcolemma.** The repeating units of muscle, the **sarcomeres,** are aligned in striated muscle, producing its characteristic appearance. Surrounding the sarcomere is the **sarcoplasm,** an intracellular fluid that contains glycogen granules, phosphocreatine, the enzymes of glycolysis, and many mitochondria. These provide energy for regeneration of the ATP hydrolyzed in the contraction process. Smooth-muscle cells differ in that they contain only a single nucleus and the sarcomeres are not linearly arranged.

Proteins of Muscle

Striated muscle contains four major proteins: myosin, actin, tropomyosin, and troponin. Smooth muscle lacks troponin but contains the other three. **Myosin** is the major protein of all muscle. The myosin monomer contains three different proteins—one heavy chain and two different light chains. The heavy chain is 1800 amino acids long, one of the longest single-chain proteins known. The secondary structure of myosin heavy chain is an α-helix with a globular head end. Proline residues are grouped together, providing a hinge between the helix and the head. The globular head has an associated ATPase activity necessary for muscle contraction. It also has binding sites for the light chains and for the protein actin. The arrangement of the myosin chains is shown in Figure 12-12. It is drawn as a dimer as that is its normal physiological state.

Actin is a smaller protein with a molecular weight of only 45,000. When first synthesized, actin has a globular structure and is called **g-actin.** After binding ATP and one molecule of calcium, it is converted to a form that can polymerize into a fiber. This form is called **f-actin.** The polymer subunits are noncova- lently bound to each other and arranged in a super- coiled double helix. α-**Actinin** connects the polym- erized actin molecules.

Tropomyosin, the third fibrous protein of muscle, is composed of two different protein chains, α and β, arranged in an α-helix. Tropomyosin binds to a groove in the f-actin helix, where it can interact with seven different actin molecules.

Troponin, found only in striated muscle, contains three different subunits. The T subunit binds the mol- ecule to tropomyosin. The I subunit inhibits the inter- action of myosin and actin and prevents muscle con- traction. The C subunit is capable of binding four calcium ions. This binding is necessary for the initia- tion of muscle contraction. The arrangement of these proteins is shown schematically in Figure 12-13.

Mechanism of Contraction

When a nerve stimulates a muscle cell, the sar- coplasmic reticulum, a membranous vesicle located in the sarcomere, releases calcium so that the cal- cium concentration rises to approximately 10^{-5} M. At this concentration, the C subunit of troponin can bind calcium. This binding alters the protein struc- tures so that the myosin binding sites on actin are opened. Seven binding sites open for each troponin C subunit that binds calcium.

In the resting state, the ATPase that is located on the globular head of the myosin contains bound ADP and phosphate. On contraction, the head swivels and binds to one of the myosin binding sites on actin. This causes the bound ADP and phosphate to be re-

Fig. 12-12. Arrangement of proteins in the myosin dimer.

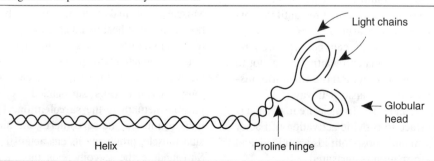

Light chains

Globular head

Helix   Proline hinge

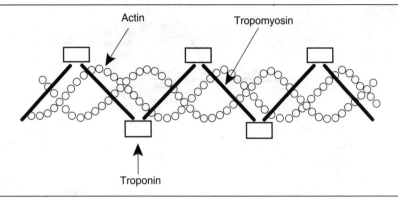

Fig. 12-13. Arrangement of the actin, tropomyosin, and troponin subunits.

leased. The binding of myosin to actin positions the myosin head group at an energetically unfavorable angle to the rest of the myosin molecule. The head group returns to a more normal position by pulling the actin molecule toward the middle of the sarcomere. The myosin then is released from the actin, binds and hydrolyzes another ATP, moves to another binding site, and pulls the actin still further toward the middle. Each of these contractions decreases the size of the sarcomere by approximately 1%.

After the initial release of calcium, the sarcoplasmic reticulum begins to pump calcium back inside, using the energy provided by the hydrolysis of ATP. Eventually, this reduces the calcium concentration to less than 10^{-7} M. At this concentration, the C subunit no longer binds calcium, and the conformation change of troponin is reversed. The inhibitory (I) subunit of troponin prevents further myosin-actin interaction by masking the myosin-actin binding sites. The myosin is trapped with its hydrolyzed but unreleased ADP and phosphate until another stimulus for contraction occurs. Because there is no force holding the actin in its contracted position, it returns to the relaxed state. The contraction-relaxation mechanism is summarized in Figure 12-14.

In cardiac tissue, the calcium required for muscle contraction does not come from the sarcoplasmic reticulum but from the extracellular space. In the absence of extracellular calcium, cardiac cells quickly will stop contracting. Because there is intracellular calcium in the sarcoplasmic reticulum, skeletal muscle cells will continue contracting for a considerable time in the absence of added calcium.

There is no troponin in smooth muscle. Contraction is controlled by phosphorylation and dephosphorylation of one of the myosin light chains. In the presence of elevated levels of calcium, a myosin light-chain kinase phosphorylates one of the light chains. In the phosphorylated state, the head can bind to actin. As long as the calcium concentration remains high, contraction will continue in the same manner as in striated muscle. At low calcium concentrations, a myosin light-chain phosphatase removes the phosphate. The dephosphorylated light chain can no longer bind to actin and contraction ceases.

Energy

The energy for muscle contraction comes from the hydrolysis of ATP, which is needed for the movement of the actin fiber relative to the myosin and also for the transport of calcium back into the sarcoplasmic reticulum. If there is no source of ATP, the calcium will remain on the C subunit and the muscle will remain contracted—that is, in rigor.

The ATP used in contraction is regenerated at the expense of phosphocreatine by the following reaction:

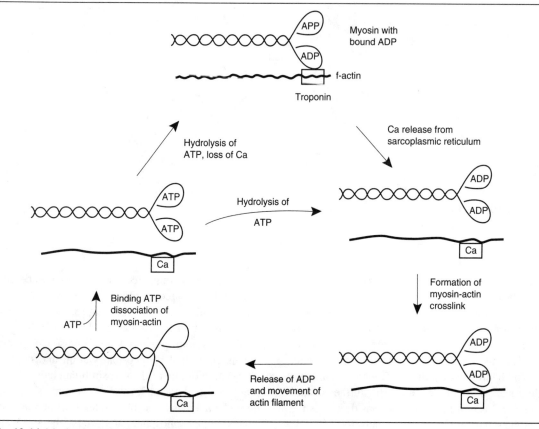

Fig. 12-14. Mechanism of actin and myosin interaction during muscle contraction.

phosphocreatine   creatine

The enzyme responsible for the phosphorylation of creatine, **creatine phosphokinase** (CPK) is released from cardiac tissue after cells are damaged by events such as myocardial infarction. The level of this enzyme serves as an indicator of the extent of cardiac damage. ATP also is regenerated by **myokinase,** which catalyzes the following reaction:

$$2ADP \rightleftarrows AMP + ATP$$

These are all rapid reactions for production of ATP from other high-energy compounds. In the absence of oxygen, ATP can be generated from glucose. This is inefficient because the net reaction produces only two ATPs and two lactic acids from each glucose molecule. Nevertheless, this is the primary energy source in fast-twitch muscle.

In slow-twitch muscle, there is an abundance of myoglobin to provide oxygen. Glucose can be completely oxidized to water and carbon dioxide, producing 38 ATP. If these reactions start from glycogen, one additional ATP is produced for each glucose oxidized. The mechanism by which ATP is generated from glucose will be covered in Part IV.

Summary

Myoglobin is a monomeric globular protein containing a single heme group. It exhibits saturation in its binding and release of oxygen. Hemoglobin contains four subunits and four heme groups, but each subunit has a secondary and tertiary structure similar to myoglobin. The oxygen-binding curve for hemoglobin is sigmoidal with positive cooperativity. The binding of one oxygen facilitates the binding of the next. Likewise, the release of one oxygen promotes the release of others. The amount of cooperativity is assessed by a Hill plot.

In both myoglobin and hemoglobin, the iron of the heme group is in the +2 form (Fe^{2+}). When the iron is in the +3 state (Fe^{3+}), it is called *hemin,* and the proteins are metmyoglobin and methemoglobin. They can be reduced by an NADH-requiring enzyme. If not reduced, they cannot carry oxygen reversibly and tend to precipitate in Heinz bodies.

Oxygen release from hemoglobin is modulated by H^+, dissolved carbon dioxide, increased temperature, and DPG. These tend to stabilize the T form of hemoglobin and to promote oxygen release. The effect of pH on the oxygen binding of hemoglobin is the Bohr effect. Entry of chloride ions into the red blood cell when bicarbonate diffuses into the plasma is known as the *chloride shift.*

Sickle-cell anemia is due to a change in one amino acid in the β-chain of hemoglobin and causes a decrease in the solubility of deoxyhemoglobin. This produces sickling of the red blood cell and severe clinical symptoms. In individuals with poorly controlled diabetes mellitus, glucose becomes bound to the β-chain of hemoglobin, forming HbA_c. Measurement of HbA_c provides an estimate of the control of blood glucose over long time periods. Carbon monoxide binds to hemoglobin 200 times more strongly than oxygen, but cyanide binds irreversibly to methemoglobin and metmyoglobin.

Blood clotting occurs by two separate pathways that converge for the last three steps. All the factors circulate as zymogens that are activated in a cascade pathway. In the intrinsic pathway, Factor XII is activated by contact with exposed injured tissue and catalyzes the conversion of prekallikrein to kallikrein. This allows the complete activation of Factor XII to Factor XIIa, which then activates XI, which activates IXa. In the final reaction, IXa converts X to Xa. The last two reactions require calcium. In the extrinsic pathway, a tissue factor in the presence of calcium, XIIa, kallikrein, and IXa, activates VII to VIIa, which converts X to Xa. In the common pathway, Xa in the presence of calcium, phospholipid, and Factor V converts prethrombin to thrombin. This, in turn, converts fibrinogen to fibrin. The final fibrin clot is crosslinked by XIIIa.

Clots can be prevented by antithrombin III, heparin, dicoumarol compounds, calcium chelation, and aspirin. Clots are dissolved by the reaction of fibrin with plasminogen activator, producing plasmin, which digests the fibrin. Lack of Factor VIII produces hemophilia A; lack of Factor IX, hemophilia B or Christmas disease; and a defect in Factor VIII along with decreased platelet aggregation, von Willebrand's disease.

Striated muscle contains four types of contractile proteins: myosin, actin, tropomyosin, and troponin. Smooth muscle contains the first three. Myosin is the most abundant and is located in the thick filaments. The heavy myosin chains are α-helices with globular heads. The head regions contain the light chains, which bind to actin and have ATPase activity. In the presence of calcium and ATP, g-actin polymerizes to f-actin, which makes up the thin fibers of the muscle. Tropomyosin is found in a groove of the actin helix. Troponin contains three different subunits. The C subunit binds calcium. The I subunit prevents actin-myosin interaction. The T subunit binds troponin to tropomyosin.

When muscle receives a stimulus to contract, calcium is released from the sarcoplasmic reticulum and binds to the C subunit of troponin. This causes the myosin binding sites on actin to be opened. Myosin, which has hydrolyzed ATP but not released the products, binds to actin. The ADP and phosphate are released. The myosin head returns to a more energetically favorable position by pulling the actin toward the center of the sarcomere. Another ATP binds, the myosin is released from actin, and the process is repeated. When the calcium concentration decreases to 10^{-7} M by removal by the sarcoplasmic reticulum calcium pump, the C subunit no longer can bind calcium. The I subunit masks the myosin binding sites, and the muscle returns to a relaxed state.

In smooth muscle, contraction is controlled by the phosphorylation and dephosphorylation of one of the myosin light chains rather than by binding of calcium to troponin.

Energy for contraction comes from the hydrolysis of ATP. The ATP can be regenerated by the reaction of ADP with phosphocreatine or with another ADP residue, or by the oxidation of glucose or glycogen.

Part III Questions: Proteins

1. The primary structure of a protein chain
    A. determines the secondary structure.
    B. is disrupted by urea and detergents.
    C. involves the association between a protein and prosthetic groups.
    D. can be characterized by the pattern of hydrogen bonding between peptide bonds.
    E. is an α-helix in β-keratin.

2. All the following are true about the protein α-helix *except* it
    A. is a right-handed helix with 3.6 amino acids per turn.
    B. is found in high amounts in globular proteins.
    C. is not a possible structure for proteins containing large amounts of proline.
    D. is held together by disulfide bonds.
    E. can be formed only if the R groups of the amino acids are polar.

3. Globular proteins are characterized by
    A. glycine at every third residue.
    B. large amounts of hydroxyproline.
    C. high concentrations of hydrophobic amino acids on the outside of the molecule.
    D. α-helix as a major structural feature.
    E. many polar and charged amino acids in the interior of the molecule.

4. Which one of the following amino acids would be considered *most* polar?
    A. Serine
    B. Alanine
    C. Proline
    D. Valine
    E. Phenylalanine

5. All the following statements are correct *except*
    A. enzymes are proteins and most can be inactivated by heating to 100°C.
    B. the loss of enzymatic activity on heating is due to denaturation of the protein.
    C. The denaturation of protein involves breaking of peptide bonds.

    D. Hydrogen bonds play an important role in the secondary structure of proteins.
    E. The primary structure of a protein is a major influence on the secondary, tertiary, and quaternary structure of the protein.

6. Which of the following statements is true?
    A. Oxidoreductases frequently utilize NAD as a reactant or product.
    B. Lyases can form covalent bonds utilizing the energy from the hydrolysis of ATP.
    C. Digestive enzymes that hydrolyze covalent bonds using water are classified as ligases.
    D. The following reaction would be an example of an isomerase:

    $$-A-B- + -D-E \rightarrow -A-E- + -B-D-$$

    E. Transferases remove two groups from a substrate, leaving a double bond.

7. Which of the following statements is true about energetics?
    A. All reactions with a negative ΔG take place spontaneously.
    B. Enzymes reduce the activation energy for a reaction.
    C. Enzymes change the ratio of substrate to product at equilibrium.
    D. Under standard conditions, $\Delta G^{\circ\prime} = 2.3$ RT log K_{eq}.
    E. At equilibrium, the ratio of substrate to product is 1.

8. Which of the following are commonly found at the active site of enzymes?
    A. Amide and ester
    B. Thiol and hydroxyl
    C. Ether and methyl
    D. Potassium and sodium
    E. Purine and pyrimidine

9. Which of the following statements is true about kinetics?
    A. The K_m of an enzyme is measured when half of the substrate has been converted into product.

B. A plot of 1/v versus 1/S will allow the determination of K_m but not V_{max}.
C. In a ping-pong reaction mechanism, the substrates bind in a random fashion but the products leave in a specific order.
D. The rate of an enzymatic reaction is first order if the concentration of the substrate is significantly less than the K_m.
E. The Lineweaver-Burk equation provides information about the reaction mechanism.

10. An enzyme
A. is called an *isozyme* when more than one protein is required to catalyze the reaction.
B. must have a known molecular weight before the K_m can be calculated.
C. without its prosthetic group is called a *zymogen*.
D. increases in specific activity during purification if denaturation is prevented.
E. increases the activation energy of a reaction.

11. All the following statements are true of both hemoglobin and myoglobin *except* they both
A. bind one oxygen per heme group.
B. contain iron in the ferrous form.
C. have a high concentration of α-helix.
D. show a Bohr effect.
E. in the ferric form are inhibited by cyanide.

12. All the following could make the indicated change in the oxygen-binding curve of hemoglobin *except*

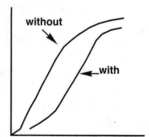

A. lactic acid.
B. carbon dioxide.
C. 2,3-diphosphoglycerate.
D. chloride.
E. increased temperature.

13. All the following will prevent clotting of blood *except*
A. citrate.
B. EDTA.
C. heparin.
D. aspirin.
E. vitamin K.

14. Which of the following substances is *directly* required for the conversion of fibrinogen to fibrin?
A. Calcium
B. Vitamin K
C. Thrombin
D. Prothrombin
E. Christmas factor

15. In the absence of ATP,
A. g-actin will dissociate into f-actin.
B. muscle will remain in the contracted state.
C. calcium is released from the sarcolemma.
D. myosin heavy chain is dephosphorylated.
E. tropomyosin dissociates into subunits.

Part IV Metabolism of Carbohydrates: The Central Pathways

13 Glycolysis

The oxidation of carbohydrates is a major source of energy for most cells. The first step in the metabolism of carbohydrate is glycolysis or the Embden-Meyerhof pathway. This pathway converts glucose and other sugars to pyruvate. The enzymes of the pathway are soluble proteins found in the cytoplasm of all cells. The reactions of glycolysis take place both in the presence and in the absence of oxygen. In the presence of oxygen, pyruvate is further oxidized in the mitochondria to carbon dioxide and water. In the absence of oxygen, pyruvate is converted to lactate. Although the energy yield of glycolysis is low, the advantage of the pathway is its ability to take place under conditions of low oxygen tension such as during extreme exercise. Glycolysis is also a major pathway for ATP production in cells that lack mitochondria.

Objectives

After completing this chapter, you should be able to

Identify the structures of glucose, fructose, galactose, ribose, sucrose, lactose, and glyceraldehyde.

Trace the carbons from glucose, galactose, and fructose to pyruvate.

List the rate-controlling reactions of glycolysis and the factors that control these reactions.

Calculate the energy yield of the glycolysis pathway starting from glucose, glycogen, fructose, or galactose.

Identify the points at which arsenate, iodoacetate, fluoride, and ethanol inhibit glycolysis.

Match specific inherited defects with the enzymes involved.

Structure of Carbohydrates

Carbohydrates are compounds that contain multiple hydroxyl groups and either a ketone or an aldehyde group. Although carbohydrates are the most abundant type of biological compound on earth, in humans they constitute only 1% of the body. They are ingested as part of the diet, stored as glycogen, and used as a primary cellular fuel. Carbohydrates are also important for the structure of cartilage and bone.

The nomenclature of carbohydrates is complex. They can be classified by their ability to be hydrolyzed to smaller compounds, by the number of carbon atoms, by the direction of rotation of polarized light, and by the structural relationship to the three-carbon sugar glyceraldehyde. Glucose, ribose, galactose, and fructose are classified as **monosaccharides** as they cannot be converted to smaller molecules by hydrolysis. Lactose is a disaccharide and glycogen a polysaccharide, as they can be converted to smaller carbohydrates by hydrolysis.

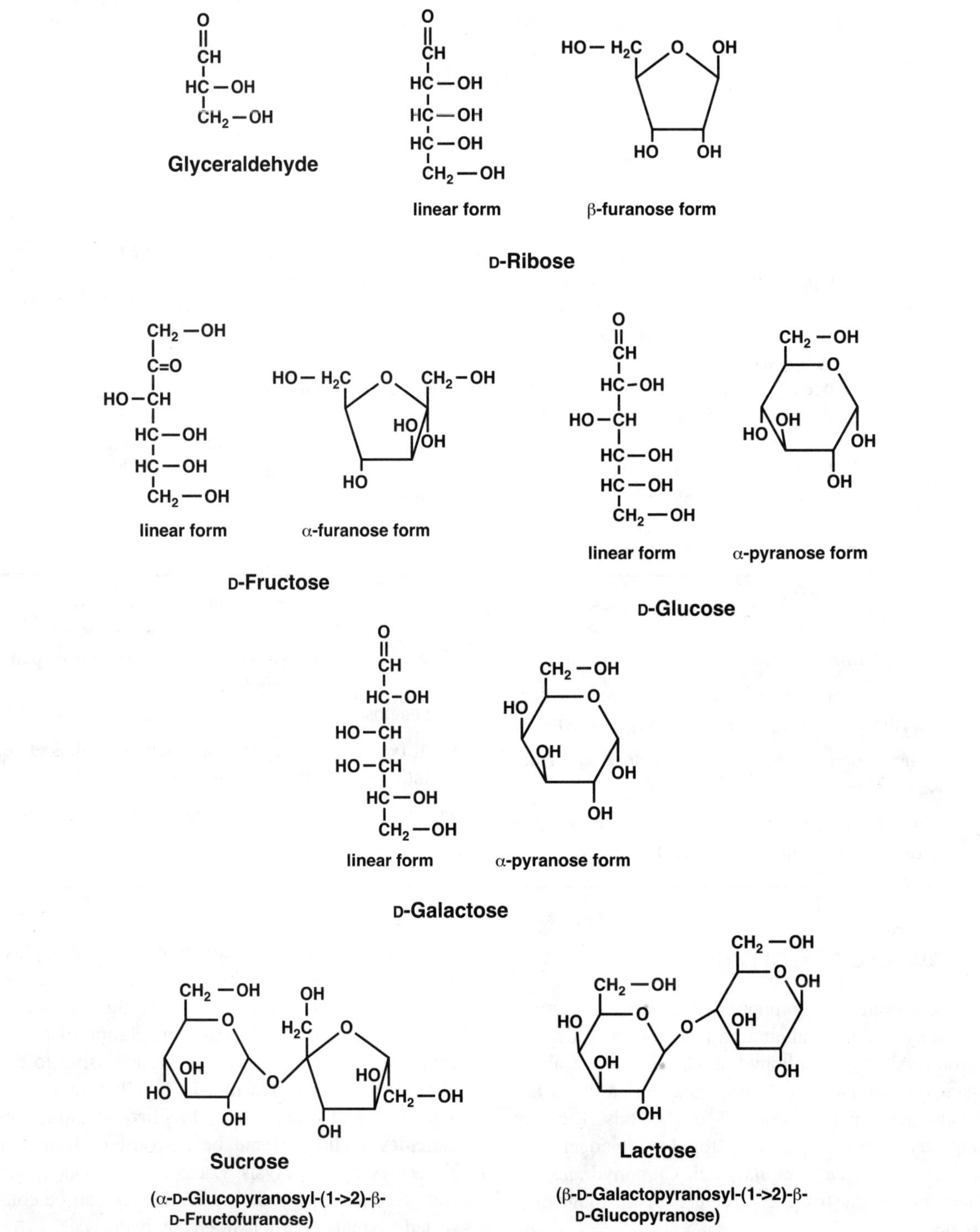

Glyceraldehyde

linear form   β-furanose form

D-Ribose

linear form   α-furanose form

D-Fructose

linear form   α-pyranose form

D-Glucose

linear form   α-pyranose form

D-Galactose

Sucrose

(α-D-Glucopyranosyl-(1->2)-β-
D-Fructofuranose)

Lactose

(β-D-Galactopyranosyl-(1->2)-β-
D-Glucopyranose)

Fig. 13-1. Structures of some of the important cellular carbohydrates. Those sugars that naturally form cyclic structures are shown in both the linear and cyclic forms.

Sugars with three to seven carbons are the most important carbohydrates for human cellular metabolism. Glyceraldehyde is a triose (three-carbon sugar), and glucose is a hexose (six-carbon sugar). If a sugar contains an aldehyde group, it is called an **aldose.** If it has a ketone group, it is a **ketose.** Figure 13-1 depicts the structures of some of the more important cellular carbohydrates.

Sugars, like amino acids, have asymmetrical carbon atoms. A molecule with four asymmetrical carbons has 2^4 or 16 possible isomers. Rotation of a beam of polarized light to the right gives the compound a plus (+) designation, rotation to the left a minus (–).

The + and – notation differs from the D and L designation. The latter comes from the structural relationship of the sugar to glyceraldehyde. Compounds that have the last asymmetrical carbon in the same orientation as D-glyceraldehyde are called *D sugars*. If they are related to L-glyceraldehyde, they are *L sugars.* Most sugars important for cellular metabolism are D sugars.

Monosaccharides such as glucose, fructose, and ribose do not usually exist in a linear form. Instead they condense into rings of five or six atoms. These are called **furanose** or **pyranose,** respectively, after the related organic ring structures. Almost all the glucose in cells is in the pyranose form. Depending on the orientation of the hydroxyl group at carbon 1, the ring structures are called α or β. Because only the addition and subtraction of water is needed to interconvert the α and β forms, they exist together in solution. The exchange from the α to the β form is called **mutarotation.**

Sugars frequently are found attached covalently to other entities—methyl or amino groups, other carbohydrates, a nitrogenous base, or proteins. The linkage to the sugar is through one of the hydroxyl groups. Because there are several hydroxyl groups on each sugar, these structures can be either linear or branched. The functions of these complex carbohydrates will be discussed as they are encountered in our study of metabolism.

Glycolysis Pathway

The 10 reactions of the glycolysis pathway are given in Figure 13-2. Follow this diagram as you read about the individual reactions. In the figure, glucose and fructose are drawn in their linear forms so that you can see more easily the reactions that occur at each step.

The first reaction (1) is the phosphorylation of glucose by ATP catalyzed by **hexokinase.** This enzyme is found in all cells and has a high affinity (low K_m) for glucose. Its affinity for other hexoses is significantly lower. Because the phosphate comes from ATP, this reaction is irreversible under cellular conditions. The liver contains an additional enzyme, **glucokinase,** which has a high K_m for glucose. This enzyme is capable of phosphorylating glucose only under conditions of high glucose concentration, such as after a large meal.

Phosphorylation of glucose is important both as the first reaction in the glycolysis pathway and for preventing glucose from being transported out of the cell. Very little free glucose is found inside the cell because of the high activity of the phosphorylating enzymes and the irreversibility of the reaction. Because the transport system for glucose has a low affinity for glucose 6-phosphate, phosphorylation prevents glucose release from cells by converting glucose to a form that is not readily transported.

The next step (2) in the pathway is the isomerization of glucose 6-phosphate to fructose 6-phosphate. This is a readily reversible reaction that changes an aldopyranose (glucose) to a ketofuranose (fructose). The enzyme catalyzing this reaction is **phosphoglucose isomerase.**

The third step (3) is the phosphorylation of fructose 6-phosphate to fructose 1,6-diphosphate by **phosphofructokinase.** This reaction uses ATP as the source of the phosphate. Phosphofructokinase is an important regulatory enzyme that will be discussed in detail later in this chapter. It catalyzes the first committed step toward the oxidation of glucose by the glycolysis pathway.

In the next step (4), fructose 1,6-diphosphate is cleaved by **aldolase** to two triose phosphates: glyceraldehyde 3-phosphate and dihydroxyacetone phosphate. The two trioses are interconvertible in a reaction catalyzed by **triosephosphate isomerase** (5). The equilibrium of the reaction favors the formation of dihydroxyacetone phosphate. Because glyceraldehyde 3-phosphate is continually being removed by the next step in the pathway (reaction 6), the reaction proceeds in cells in the direction in which it is written in Figure 13-2.

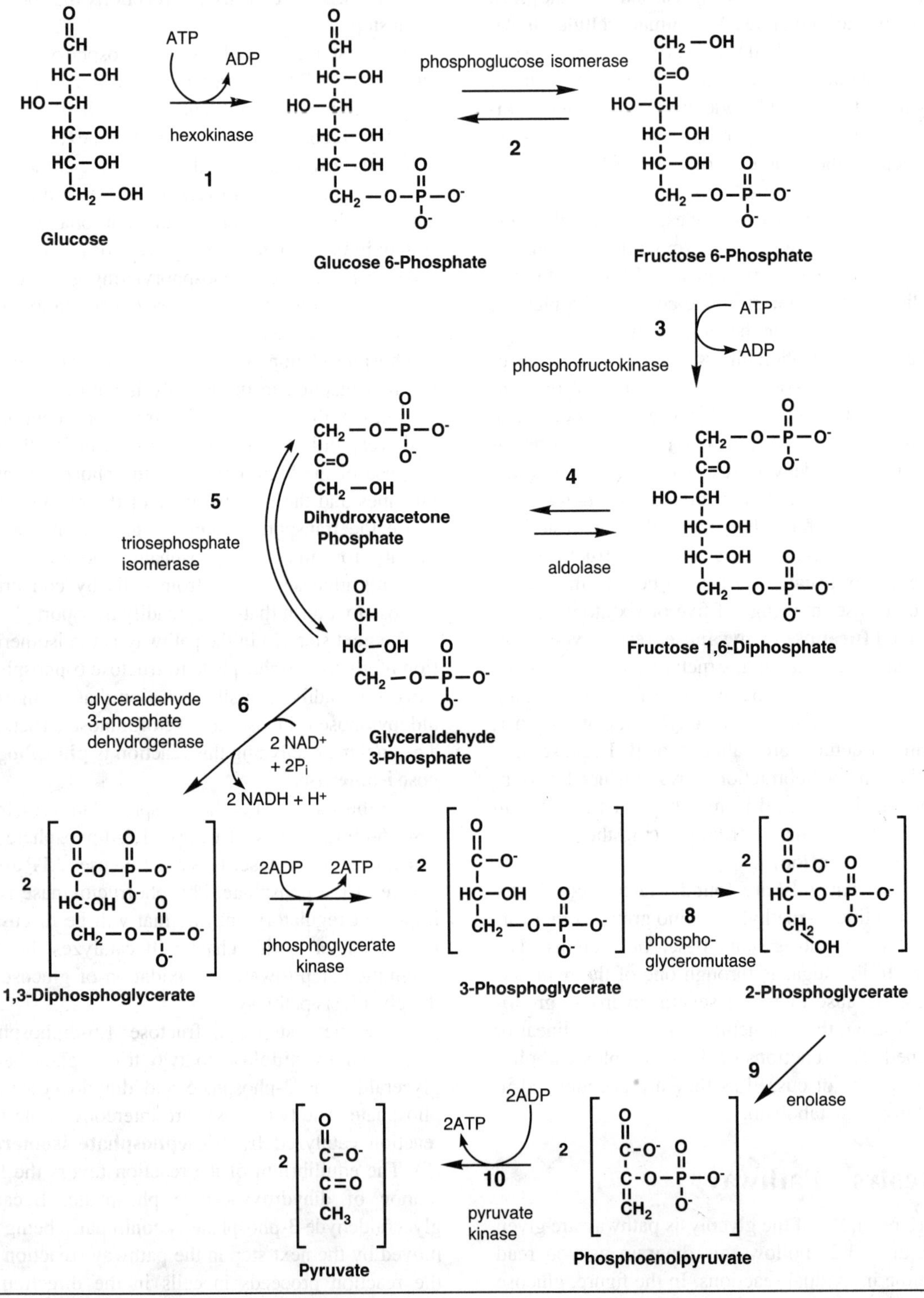

Fig. 13-2. Reactions of the glycolysis pathway.

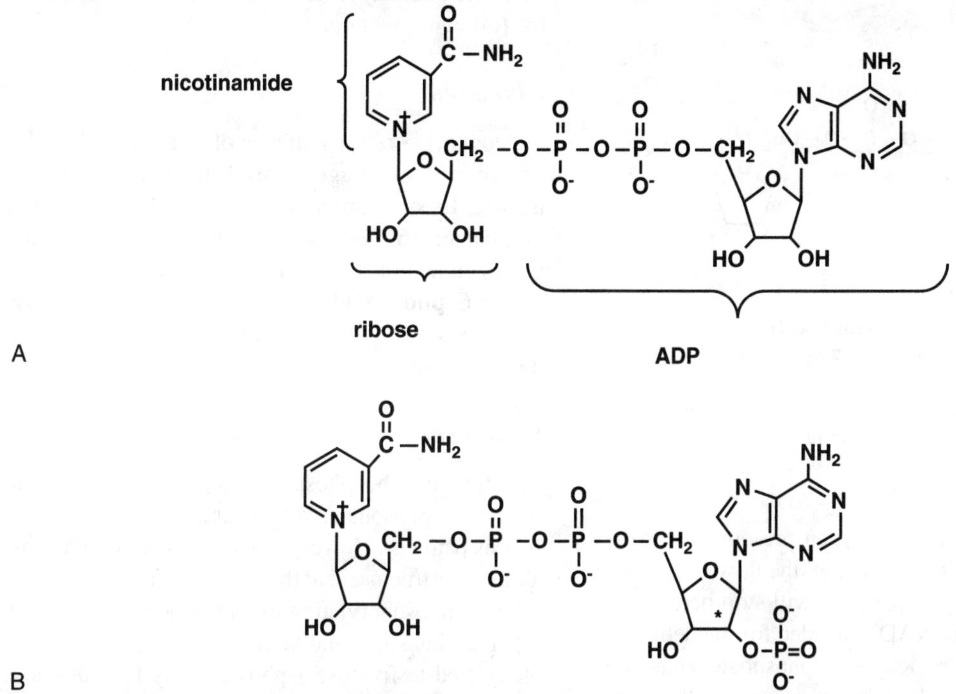

Fig. 13-3. A. Structure of NAD. B. The addition of a second phosphate at the carbon (marked with an *) creates NADP.

The next reaction (6) is the oxidation and phosphorylation of glyceraldehyde 3-phosphate by NAD to produce 1,3-diphosphoglycerate. **Glyceraldehyde 3-phosphate dehydrogenase** catalyzes this reaction, which uses inorganic phosphate, not ATP, as the source of phosphate. The product is a mixed anhydride that has a large negative $\Delta G^{\circ\prime}$ for hydrolysis. Therefore, 1,3-diphosphoglycerate is a high-energy compound.

This is your first encounter with a reaction that uses **nicotinamide adenine dinucleotide** or **NAD.** Let us digress a moment to discuss this important compound. NAD is made up of the base nicotinamide, ribose, and ADP (Fig. 13-3A). A second phosphate may be attached to the hydroxyl group at the * carbon, creating NADP (Fig. 13-3B). NAD and NADP serve as mobile electron acceptors for a number of oxidation-reduction reactions. They usually are not covalently bound to the enzymes with which they react. In their reduced forms, NADH and NADPH (dihydronicotinamide adenine dinucleotide phosphate) have the structures shown in Figure 13-4.

In reaction 7, **phosphoglycerate kinase** transfers the high-energy phosphate from 1,3-diphosphoglycerate to ADP, producing 3-phosphoglycerate and ATP. This reaction is a **substrate-level phosphorylation.** It is an unusual reaction in metabolism because most reactions producing ATP utilize the electron transport chain in the mitochondria rather than the transfer of phosphate from a high-energy compound that is not a nucleotide.

Next (8), **phosphoglyceromutase** transfers the phosphate from position 3 to position 2, forming 2-phosphoglycerate. This sets up the production of another high-energy phosphate and another ATP two reactions later.

Enolase catalyzes the dehydration of 2-phosphoglycerate (9) to phosphoenolpyruvate, another high-energy compound. Phosphoenolpyruvate then transfers phosphate to ADP in reaction 10, producing pyruvate and ATP. The enzyme catalyzing this reaction is **pyruvate kinase,** which is named for the reverse reaction.

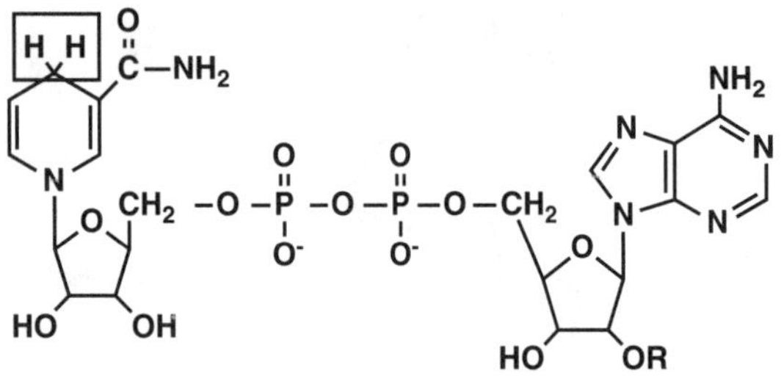

R = H for NADH

R = PO$_3^{-2}$ for NADPH

Fig. 13-4. Structures of NADH and NADPH. The two hydrogens that are added to NAD and NADP to produce NADH and NADPH are indicated by a box.

If there is insufficient oxygen present or impairment of mitochondrial oxidation, the flow of glucose through the glycolysis pathway will stop because of a lack of NAD. The NAD is needed for the dehydrogenation of glyceraldehyde 3-phosphate (reaction 6). NADH produced by this reaction normally is oxidized back to NAD in the mitochondria. If this does not occur, NAD can be regenerated in the cytoplasm from NADH by reducing pyruvate to lactate in a reaction catalyzed by **lactate dehydrogenase** (LDH) (Fig. 13-5). By utilizing this reaction, glycolysis can continue even in the absence of oxygen. In the presence of oxygen, pyruvate is decarboxylated and the two-carbon fragment is oxidized further in the mitochondria to carbon dioxide and water. These reactions will be discussed in Chapter 14.

There are two different genes coding for LDH. One codes for the *M* or *muscle form* and the other the *H* or *heart form.* Because LDH has four subunits, there are five possible isozymes of LDH: M_4, M_3H, M_2H_2, MH_3, and H_4. Skeletal muscle contains some of all five isozymes but predominantly M_4. Heart has predominantly H_4. LDH is released to the circulation after tissue injury. The concentration of the various isozymes serves as a marker for the site and extent of such injury.

Metabolism of Other Sugars

Glucose is not the only carbohydrate metabolized by the glycolysis pathway. Galactose, fructose, and glycogen are converted to hexose phosphates and enter the pathway at different levels as described in the following sections.

Glycogen

Glycogen, an α-1,4 polymer of glucose with α-1,6 branches, is a storage form of glucose in liver and muscle. Its structure and synthesis will be discussed in Chapter 16. Glucose residues are released from glycogen as glucose 1-phosphate. As shown in Figure 13-6, **phosphoglucomutase** can convert glucose 1-phosphate to glucose 6-phosphate, at which point it can enter the glycolysis pathway.

Fructose

Fructose can be phosphorylated by hexokinase to fructose 6-phosphate, which directly enters the glycolysis pathway. However, hexokinase has such a low affinity for fructose that this is not a major enzyme for fructose phosphorylation except in adipose tissue. In the liver, intestine, and kidney, fructose can be phosphorylated to fructose 1-phosphate by **fructokinase.** Fructose 1-phosphate then is split by aldolase to glyceraldehyde and dihydroxyacetone phosphate. Dihydroxyacetone phosphate can be converted to glyceraldehyde 3-phosphate by triosephosphate isomerase (reaction 5 of glycolysis). As such, it can enter the glycolysis pathway. The glyceraldehyde can be oxidized by aldehyde dehydrogenase to glycerate, which can be phosphorylated by ATP to 2-phosphoglycerate and enter the glycolysis pathway. The direct phosphorylation of glyceraldehyde to 2-phosphoglycerate does not produce 1,3-diphosphoglycerate as an intermediate. Therefore, the substrate-level phosphorylation of ADP that normally occurs is lost. This pathway is summarized in Figure 13-7.

In red blood cells, **2,3-diphosphoglycerate,** a modulator of oxygen release from hemoglobin, is produced from 1,3-diphosphoglycerate (Fig. 13-8). This pathway also eliminates the substrate-level ATP production step.

Galactose

The pathway for **galactose** oxidation is outlined in Figure 13-9. Galactose is derived primarily from dietary lactose by hydrolysis in the intestine by the enzyme **lactase.** It then is phosphorylated by galactose

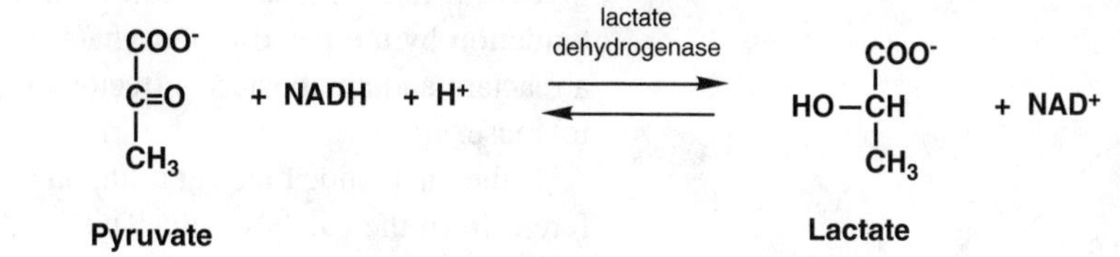

Fig. 13-5. Formation of lactate from pyruvate by lactate dehydrogenase. This reaction regenerates NAD and allows the glycolysis pathway to continue to function in the absence of oxygen.

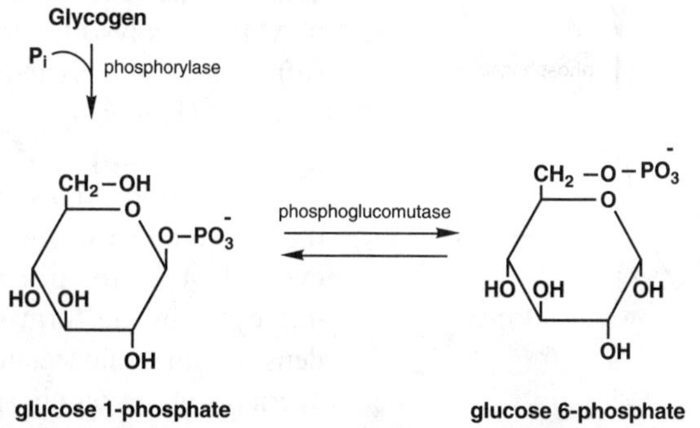

Fig. 13-6. Pathway for the conversion of glucose 1-phosphate to glucose 6-phosphate by phosphoglucomutase. This reaction allows glucose released from glycogen as glucose 1-phosphate to be converted to glucose 6-phosphate, an intermediate of the glycolysis pathway.

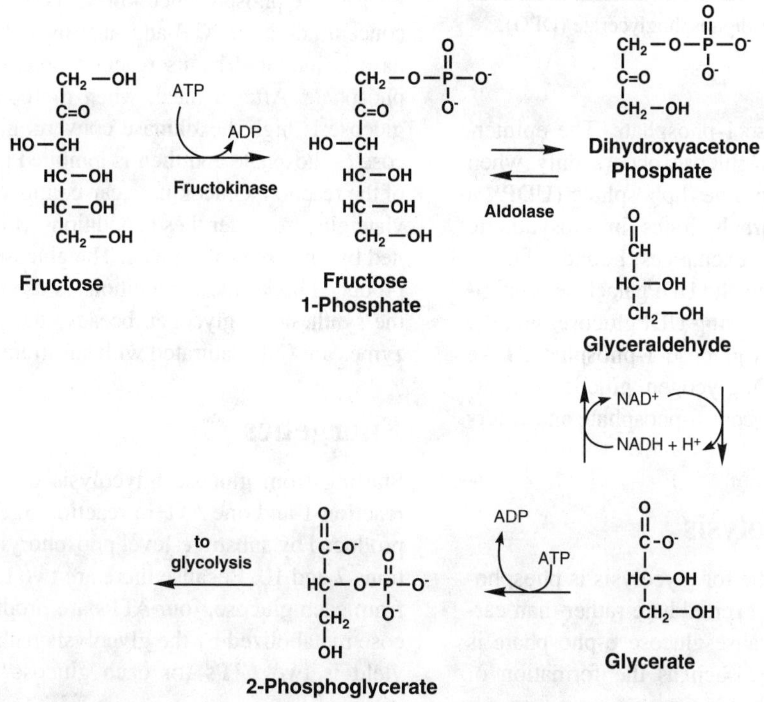

Fig. 13-7. Pathway for the entry of fructose into glycolysis.

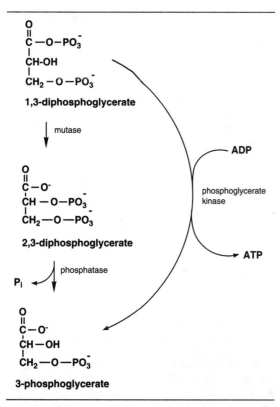

Fig. 13-8. Synthesis of 2,3-diphosphoglycerate (DPG).

kinase to yield galactose 1-phosphate. The epimerization of galactose to glucose occurs only when galactose is linked to uridine diphosphate (UDP), a common carrier of carbohydrates in biosynthetic reactions. UDPglucose exchanges bound glucose for bound galactose. Then, the UDPgalactose is epimerized at position 4, producing UDPglucose. Finally, the sugar is released as glucose 1-phosphate. Like the glucose derived from glycogen, glucose 1-phosphate is converted to glucose 6-phosphate and enters the glycolysis pathway.

Control of Glycolysis

The main control enzyme for glycolysis is phosphofructokinase. Control is exerted here rather than earlier in the pathway because glucose 6-phosphate is used in other reactions, such as the formation of glycogen, release from the cell as free glucose, and oxidation by the pentose phosphate pathway. ATP is a reactant and an allosteric effector for phosphofructokinase. In the presence of a high concentration of ATP, the nucleotide binds at a site on the enzyme different from the catalytic site. This binding increases the K_m and reduces the affinity of the enzyme for fructose 6-phosphate, thereby slowing the reaction. The effect of excess ATP can be reversed by AMP. When a cell needs energy, as indicated by an increase in AMP, glycolysis proceeds rapidly. When a cell has sufficient energy, as indicated by a high concentration of ATP, glycolysis is inhibited. Any glucose 6-phosphate formed is then diverted to other pathways.

Phosphofructokinase also is inhibited by citrate, the first product of the tricarboxylic acid or Krebs cycle. In the presence of oxygen, pyruvate enters this cycle in the form of acetylcoenzyme A, condenses with oxaloacetate, and forms citrate. Citrate exerts its effect on phosphofructokinase by enhancing the inhibitory action of ATP.

Two other glycolysis enzymes are regulated as well. However, they do not exert the degree of control on the pathway shown by phosphofructokinase. Pyruvate kinase, like phosphofructokinase, is inhibited by high concentrations of ATP and also by NADH. Hexokinase is inhibited by its reaction product, glucose 6-phosphate. After a meal, when the concentration of glucose is high, hexokinase converts glucose to glucose 6-phosphate and then is inhibited by the product of the reaction. Glucokinase can continue to phosphorylate glucose under these conditions as it is not inhibited by glucose 6-phosphate. The glucose 6-phosphate produced under these conditions is used primarily for the synthesis of glycogen because the glycolysis enzymes are fully saturated with substrate.

Energetics

Starting from glucose, glycolysis uses one ATP in reaction 1 and one ATP in reaction 3. Two ATPs are produced by substrate-level phosphorylation in reactions 7 and 10. Because there are two trioses formed from each glucose, four ATPs are produced per glucose metabolized by the glycolysis pathway. The net yield is two ATPs for each glucose converted to

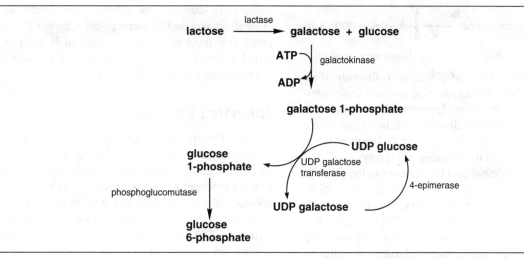

Fig. 13-9. Pathway for the entry of galactose into glycolysis. (*UDP* = uridine diphosphate.)

pyruvate. In addition, two NADHs are produced. In the presence of oxygen, the electrons from NADH can go through mitochondrial oxidative phosphorylation and produce more ATP. In the absence of oxygen, lactate is formed and the net energy produced is two ATPs. The same energy yield occurs when the start point is galactose rather than glucose. If the starting substrate is glycogen, the initial phosphate transfer from ATP is not required because glucose is phosphorylated using inorganic phosphate when it is released from glycogen. Then, only a single ATP is used (reaction 3) and four are produced, for a net yield of three.

Oxidation of fructose by the glycolysis pathway produces a net yield of only one ATP. ATP is used in the phosphorylation of fructose to fructose 1-phosphate, and two ATPs are produced by the oxidation of dihydroxyacetone phosphate to pyruvate. The oxidation of glyceraldehyde uses one ATP and produces one ATP, so the net yield is one ATP for each fructose metabolized by glycolysis.

Inhibitors

There are several inhibitors of glycolysis that have no clinical use but have provided much interesting biochemical data about the mechanism of action of the enzymes of glycolysis. The first of these is **arsenate** (AsO_4), which resembles phosphate (PO_4). Arsenate competes with phosphate in the formation of 1,3-diphosphoglycerate, producing 1-arseno-3-phosphoglycerate.

$$
\begin{array}{ccc}
& O & O \\
& \| & \| \\
& C-O-As-O^- & \\
HO-CH_2 & O^- & O^- \\
& | & \\
& CH_2-O-P{=}O & \\
& | & \\
& O^- &
\end{array}
$$

This compound decomposes spontaneously. Although 3-phosphoglycerate is produced after the decomposition, no phosphate is transferred to ADP. Therefore, arsenate uncouples the chemical oxidation from the production of ATP. We will encounter several other uncouplers when we discuss oxidative phosphorylation in Chapter 15. In the presence of arsenate, two ATPs are used by the glycolytic pathway and two are produced by reaction 10, for a net yield of 0.

Iodoacetate is a reagent that binds preferentially to the SH group of cysteine. In glycolysis, the main enzyme affected by iodoacetate is glyceraldehyde 3-phosphate dehydrogenase. In other pathways, other enzymes with SH groups in their active sites are also inhibited by this compound.

$$\{-SH \ + \ I-CH_2-COO^- \longrightarrow \ \}-S-CH_2-COO^- \ + \ HI$$

enzyme   **inhibited enzyme**

The final inhibitor of glycolysis is **fluoride.** It inhibits enolase, the enzyme that converts 2-phosphoglycerate to phosphoenolpyruvate in reaction 9. Although fluoride in small amounts is beneficial for the formation of bone and teeth, it is lethal at larger doses because of its inhibition of glycolysis.

Ingested **ethanol** also is inhibitory to the glycolysis pathway. Ethanol is first converted to acetaldehyde by **alcohol dehydrogenase,** as outlined in Figure 13-10. The enzyme uses NAD, which then becomes unavailable for glycolysis. Again using NAD, acetaldehyde is converted to acetate. After attachment to coenzyme A, it can be oxidized to carbon dioxide and water in the mitochondria or synthesized into fatty acids. The lack of NAD due to its use in the metabolism of ethanol inhibits all NAD-requiring reactions, not just those of glycolysis.

After ingestion of **methanol,** NAD is used by alcohol dehydrogenase to produce formaldehyde as the first metabolic product. Because formaldehyde reacts readily with many important cellular compounds, production of formaldehyde from methanol usually is a lethal event. However, ethanol is the preferred substrate for alcohol dehydrogenase. In the presence of ethanol, methanol is oxidized very slowly, and only small amounts of formaldehyde are produced.

Similarly, ethanol can prevent the lethal effect of ingestion of **ethylene glycol,** the major constituent of antifreeze. In the absence of ethanol, ethylene glycol is converted in several steps to oxylate, which irreversibly damages the kidney. Where both are present, ethanol and ethylene glycol compete for alcohol dehydrogenase. Because ethanol is the preferred substrate, the ethylene glycol is not oxidized but is excreted unchanged.

Inherited Defects

Because the glycolysis pathway is the source of 2, 3-DPG, defects in glycolysis affect the oxygen affinity of hemoglobin. For example, a deficiency in hexokinase produces a block at the beginning of the glycolysis pathway. All intermediates, including DPG, are at low concentration. Because of the low concentration of DPG, hemoglobin exhibits a very high affinity for oxygen and does not release it completely in the tissues. A deficiency of pyruvate kinase, which functions at the end of the glycolysis pathway, has the opposite effect. The concentration of all intermediates found earlier in the pathway is increased. Because of the increase in DPG concentration, hemoglobin has a low oxygen affinity and does not become fully saturated with oxygen in the lungs.

The absence of fructokinase produces **essential fructosuria,** a relatively benign condition. When fructose is part of the diet, its concentration in the blood rises rapidly and some is found in the urine. Eventually, most of it is metabolized or excreted. **Hereditary fructosuria,** a more serious condition, is due to the absence of aldolase, which cleaves fructose 1,6-diphosphate. Glycolysis is blocked, and fructose 1,6-diphosphate, fructose 6-phosphate, and glucose 6-phosphate accumulate. Hereditary fructosuria is characterized by hypoglycemia and vomiting if fructose is part of the diet.

Fig. 13-10. Metabolism of ethanol to acetate.

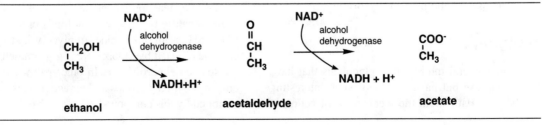

Defects in galactokinase, galactose UDP transferase, or the 4-epimerase produce **galactosemia.** A lack of the first enzyme causes only a mild condition, but a lack of either of the other two results in cataracts, growth retardation, vomiting, and diarrhea if galactose is not excluded from the diet.

Summary

Glucose is oxidized to pyruvate by the glycolysis pathway in ten reactions. In the first reaction, glucose is phosphorylated by ATP to produce glucose 6-phosphate. This is then converted to fructose 6-phosphate. A second phosphorylation by ATP, the major rate-controlling step of glycolysis, forms fructose 1,6-diphosphate. This reaction is inhibited by ATP and citrate. This inhibition can be relieved by AMP. Fructose 1,6-diphosphate is split into dihydroxyacetone phosphate and glyceraldehyde 3-phosphate, which can be interconverted by triosephosphate isomerase. Glyceraldehyde 3-phosphate is oxidized and phosphorylated to 1,3-diphosphoglycerate. The phosphate is transferred to ADP in a substrate-level phosphorylation, producing 3-phosphoglycerate. The phosphate is moved to position 2, and then the compound is dehydrated to produce phosphoenolpyruvate. Finally, another substrate-level phosphorylation occurs when phosphoenolpyruvate is converted to pyruvate and the phosphate transferred to ADP. In the absence of oxygen, pyruvate is converted to lactate to regenerate NAD. The net yield is two ATPs per glucose oxidized by the glycolysis pathway.

The glycolysis pathway also metabolizes glucose derived from glycogen and galactose. The carbons from fructose enter the pathway as dihydroxyacetone phosphate and 2-phosphoglycerate. Defects in fructose and galactose metabolism produce intolerance to these two sugars. Arsenate and iodoacetate inhibit 3-phosphoglyceraldehyde dehydrogenase, whereas fluoride inhibits enolase. Ethanol, methanol, and ethylene glycol are oxidized by alcohol dehydrogenase to their respective aldehydes using NAD. This produces a lack of NAD, which then inhibits glycolysis.

14 Krebs Cycle

The **Krebs cycle** is made up of a group of mitochondrial enzymes. These enzymes oxidize acetyl-CoA from pyruvate to carbon dioxide and water. This pathway also is called the **citric acid** or the **tricarboxylic acid** (TCA) **cycle.** With one exception, all of the Krebs cycle enzymes are located in the matrix of the mitochondria. In contrast, the enzymes of glycolysis are found in the cytoplasm.

The Krebs cycle provides the major site for the oxidation of a large number of important cellular compounds and also provides intermediates for biosynthesis. In addition to carbohydrates, carbons from amino acids and fatty acids are oxidized by the Krebs cycle. Krebs cycle intermediates are used in the biosynthesis of heme and many of the amino acids. Acetyl-CoA serves as the precursor for the synthesis of fatty acids and steroids. Because of the importance of the Krebs cycle in both degradation and biosynthesis, it is central to your understanding of cellular metabolism.

Objectives

After completing this chapter, you should be able to

List the reactions of the Krebs cycle, indicating the reactions where NADH, FADH$_2$, and GTP are formed.

Discuss the regulation of the Krebs cycle.

Calculate the energy yield of the Krebs cycle.

Match fluoroacetate, arsenite, mercury, and malonate with the reactions they inhibit.

Formation of Acetyl-CoA

Because the oxidative enzymes of the Krebs cycle are located in the mitochondria, the first step of the cycle is the transport of pyruvate formed from glucose in the cytoplasm into the mitochondrial matrix. This is accomplished by a specific carrier that binds pyruvate and H$^+$ in the cytoplasm and releases them in the mitochondria. Pyruvate can also be transported into mitochondria in exchange for hydroxyl ions or citrate.

Next, pyruvate is oxidized, decarboxylated, and attached to coenzyme A (CoA) to form **acetyl-CoA.**

This reaction requires a complex of four different enzymatic activities and is controlled by two others. Because the amount of acetyl-CoA directly influences the overall rate of activity of the Krebs cycle, the conversion of pyruvate to acetyl-CoA is an important rate-controlling reaction. The net result of the action of the four-enzyme complex is as follows:

$$\text{pyruvate} + \text{CoA} + \text{NAD}^+ \rightarrow \text{acetyl-CoA} + \text{CO}_2 + \text{NADH} + \text{H}^+$$

Owing to the decarboxylation, this is an irreversible reaction.

In the first step of the reaction, pyruvate is added to one carbon of an enzyme-bound **thiamine pyrophosphate** and then is decarboxylated. This reaction is catalyzed by **pyruvate decarboxylase** (Fig. 14-1). Thiamine pyrophosphate is derived from the vitamin **thiamine** and is a common reactant in decarboxylation reactions.

Next, the two-carbon fragment is transferred to **lipoamine** by a transacetylase. The fragment is oxidized and the lipoamine is reduced in the next reaction. Finally, the acetyl group is transferred to **coenzyme A** by the same transacetylase (Fig. 14-2). CoA is made up of the vitamin **panthothenic acid,** thioethanolamine (from cysteine), ADP, and phosphate. CoA is the carrier of acetyl and acyl groups for many biosynthetic reactions including fatty acid synthesis and degradation.

To regenerate the lipoamine that was reduced in reaction 2, a third enzyme—a dehydrogenase—is required. This enzyme uses a bound FAD to accept the reducing equivalents. The $FADH_2$ produced by this reaction then is regenerated by NAD using enzyme 4 (Fig. 14-3).

Acetyl-CoA can also be generated from sources other than pyruvate oxidation. Fatty acid degradation produces acetyl-CoA, as does the degradation of the amino acids leucine, lysine, phenylalanine, tyrosine, tryptophan, and isoleucine.

Reactions of the Krebs Cycle

The Krebs cycle consists of nine reactions, summarized in Figure 14-4, that oxidize the two carbons of the acetyl group to carbon dioxide. As you did for glycolysis, follow this diagram as you read about the individual reactions.

In the first reaction (1), acetyl-CoA is condensed with oxaloacetate to form **citrate.** The enzyme catalyzing this reaction is **citrate synthetase,** an important control enzyme for the Krebs cycle.

Next (Steps 2 and 3), citrate is isomerized by dehydration and rehydration to **isocitrate.** *Cis-aconitate* is an intermediate in this reaction that is catalyzed by the enzyme **aconitase.** Even though citrate is a symmetrical compound, the two carbons that entered the cycle as acetyl-CoA are distinguishable from the carbons that entered as oxaloacetate. This can happen only if there is a three-point binding of the citrate to aconitase. As indicated in Figure 14-4, the carbons entering the Krebs cycle as acetyl-CoA are the upper two carbons in all the intermediates and are indicated by a bracket.

Fig. 14-1. The first reaction in the production of acetyl-CoA from pyruvate. Pyruvate is added to thiamine pyrophosphate and then decarboxylated by the enzyme pyruvate decarboxylase. Carbons that will form acetyl group at acetyl-CoA are in box.

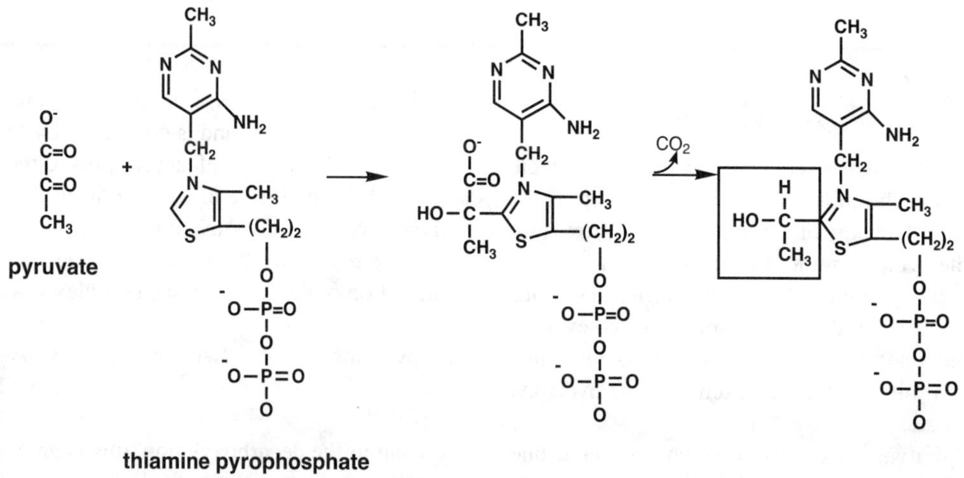

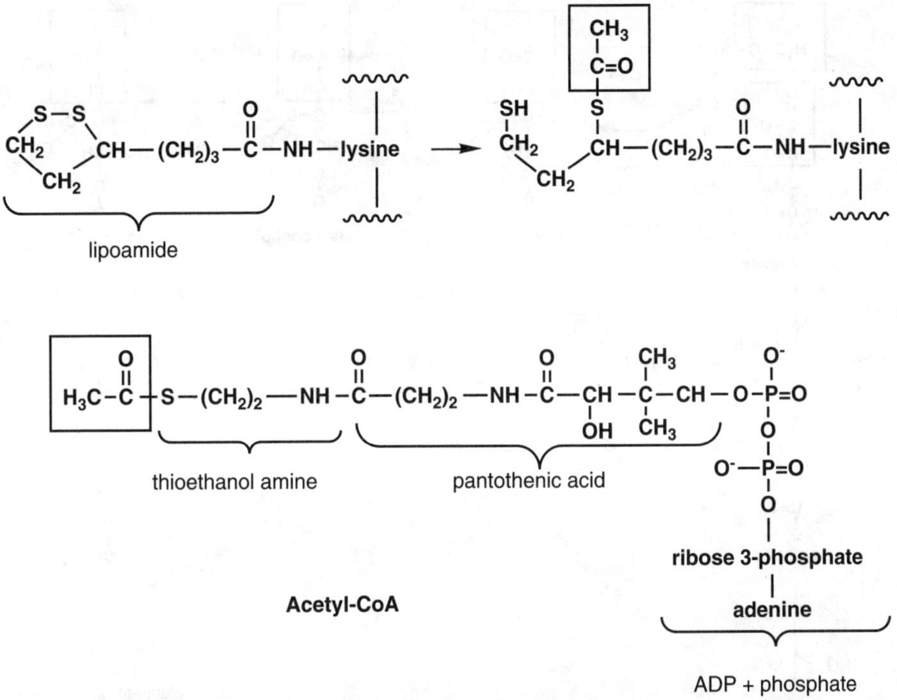

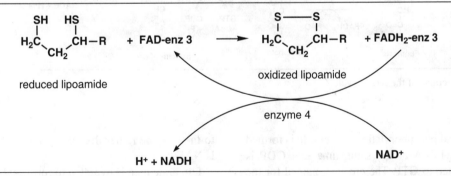

Fig. 14-2. The second and third reactions in the production of acetyl-CoA from pyruvate. The two-carbon acetyl group, shown in the box, is transferred from thiamine pyrophosphate to lipoamide. Then the acetyl group is transferred to the thio end of coenzyme A.

Fig. 14-3. Regeneration of oxidized lipoamide by NAD using an enzyme containing a bound FAD.

In the next reaction (4), isocitrate is dehydrogenated and decarboxylated to form α-**ketoglutarate** by the enzyme **isocitrate dehydrogenase.** NAD accepts the hydrogens in this reaction, producing NADH and H$^+$.

α-**Ketoglutarate dehydrogenase** catalyzes the next reaction (5) in which α-ketoglutarate is decarboxylated, dehydrogenated, and attached to coenzyme A to form **succinyl-CoA.** The mechanism of this irreversible reaction is similar to the one that formed acetyl-CoA from pyruvate. NADH is formed in regenerating the FAD-containing enzyme of the complex.

Reaction 6 is the only reaction in this pathway in which a high-energy phosphate is produced by

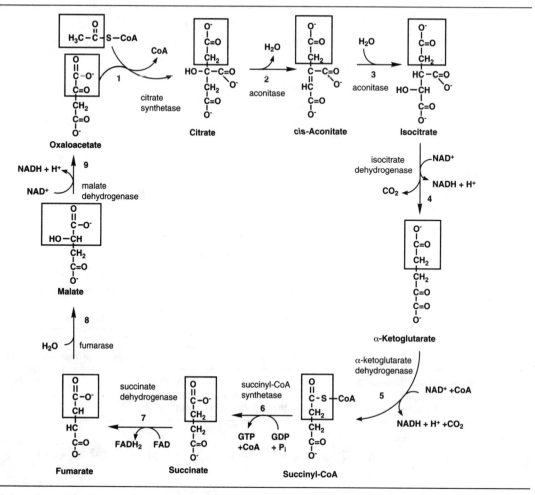

Fig. 14-4. Reactions of the Krebs cycle.

substrate-level phosphorylation. **Succinate** is formed from succinyl-CoA at the same time that GDP is phosphorylated to GTP. The enzyme, named for the reverse reaction, is **succinyl-CoA synthetase.** This enzyme is located on the inner membrane of the mitochondria instead of in the matrix as are all the other Krebs cycle enzymes. Because both anhydrides and thio esters are high-energy compounds, the succinyl-CoA synthetase reaction is reversible.

In the next reaction (7), succinate is dehydrogenated to **fumarate** by **succinate dehydrogenase.** The reducing agent is FAD, which is tightly bound

to the enzyme rather than being freely diffusible as is NAD.

Fumarate then is hydrated to form **malate** in reaction 8. The addition of water is stereospecific, with only L-malate being formed. The enzyme catalyzing this reversible reaction is **fumarase.**

The final reaction (9) regenerates **oxaloacetate.** Here, **malate dehydrogenase,** using NAD as the oxidizing agent, converts malate to oxaloacetate. This is also a reversible reaction.

The sum of the reactions of the Krebs cycle is as follows:

$$acetyl\text{-}CoA + 3\ NAD + FAD + GDP + P_i$$
$$+ 2\ H_2O \rightarrow 2\ CO_2 + 3\ NADH + FADH_2$$
$$+ GTP + 2\ H^+ + CoA$$

Note that the two carbons entering the cycle as acetyl-CoA are not the carbons lost as carbon dioxide on the first round of the cycle. They become the top two carbons of oxaloacetate at the end of one cycle. The first of these carbons is lost as carbon dioxide during the second round of the cycle. The second carbon is not lost until the fourth round. Verify this for yourself by tracing these carbons through successive cycles. Because of the need to regenerate NAD and FAD, the Krebs cycle can operate only when oxygen is available and when the mitochondrial electron transport chain is functioning.

Energetics

The conversion of pyruvate to acetyl-CoA is irreversible under cellular conditions. Once the carbons have been converted to acetyl-CoA, there is no mechanism for going back to pyruvate. Therefore, there can be no net synthesis of glucose or other carbohydrates starting from acetyl-CoA.

Under cellular conditions, all the reactions of the Krebs cycle have a negative ΔG with the exception of reaction 2. Because of this, the overall pathway has a large negative ΔG and goes in the direction as written in Figure 14-4. This may be surprising as the net energy yield is only one high-energy phosphate compound for the oxidation of two carbons to carbon dioxide. As we will see in the next chapter, the majority of energy is produced when hydrogens from NADH and FADH$_2$ are oxidized to water by the electron transport chain.

Regulation

The activity of the Krebs cycle is controlled at several points, as summarized in Figure 14-5. The first of these is the formation of acetyl-CoA. **Pyruvate dehydrogenase** is inhibited allosterically by GTP and is activated by AMP. However, the major control occurs by phosphorylation and dephosphorylation of a serine residue in the first enzyme of the complex.

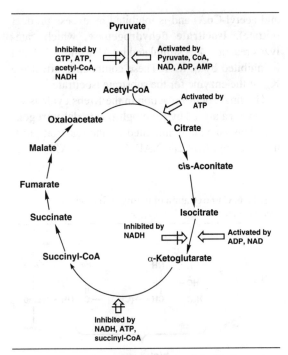

Fig. 14-5. Regulation of the enzymes of the Krebs cycle.

The phosphorylated form of the enzyme is inactive. Phosphorylation is enhanced by high levels of ATP, acetyl-CoA, and NADH. Phosphorylation of the enzyme is inhibited by pyruvate, CoA, NAD, AMP, and ADP. Pyruvate also activates a phosphatase that dephosphorylates and activates the enzyme.

Control of the production of acetyl-CoA from pyruvate coordinates the rate of activity of the Krebs cycle with the cell's need for energy. When the cell has ample energy, when the Krebs cycle is slowed by lack of NAD, or when acetyl-CoA has accumulated beyond the capacity of the Krebs cycle to use it, no more pyruvate is converted to acetyl-CoA. When there is excess pyruvate and the other conditions are not present, the conversion of pyruvate to acetyl-CoA is enhanced.

Several reactions of the Krebs cycle itself also are regulated. **Citrate synthetase,** enzyme 1, is inhibited by ATP. The nucleotide increases the K_m of the enzyme for citrate, which decreases the rate of the reaction. In addition, the enzyme activity is controlled by the availability of substrate (oxaloacetate and acetyl-CoA) and is inhibited by excess product (citrate). **Isocitrate dehydrogenase,** which catalyzes reaction 4, is stimulated by ADP and NAD and is inhibited by NADH. These compounds affect the K_m of the enzyme for the substrate isocitrate.

The final control reaction in the Krebs cycle is reaction 5, catalyzed by α-**ketoglutarate dehydrogenase.** This enzyme is inhibited by the reaction products succinyl-CoA and NADH as well as ATP.

Fig. 14-6. Carboxylation of pyruvate to produce oxaloacetate.

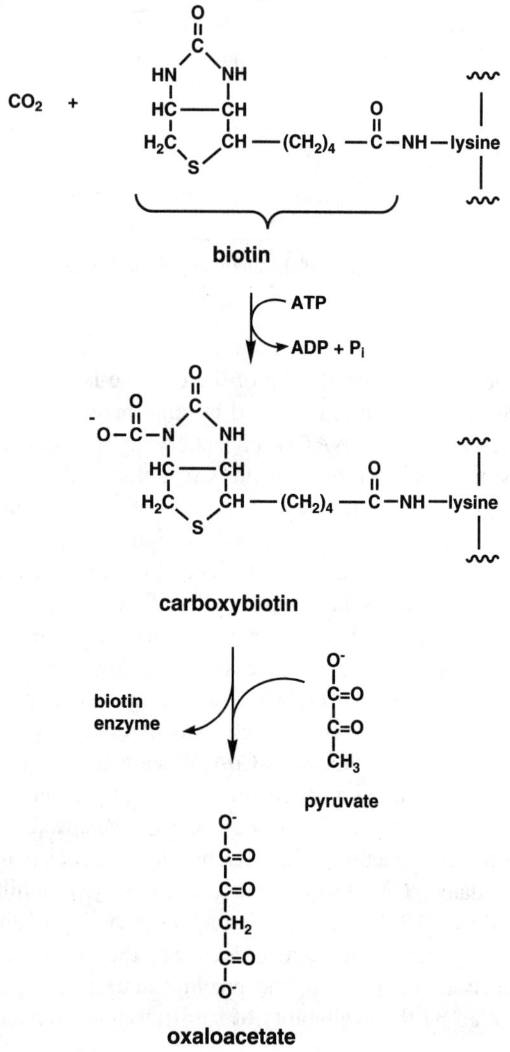

In addition to regulation of specific enzymes, the Krebs cycle is controlled by the availability of oxaloacetate. If the concentration of oxaloacetate is reduced by removal of Krebs cycle intermediates for other reactions, it must be regenerated before the cycle can continue. This is accomplished by carboxylation of pyruvate in the reaction outlined in Figure 14-6. The carboxylation uses biotin, another vitamin, as the carbon dioxide carrier and energy from the hydrolysis of ATP for the binding of the carbon dioxide to biotin.

A final control is exerted by the availability of the vitamin thiamine. A dietary deficiency of thiamine inhibits the decarboxylation of pyruvate as well as α-ketoglutarate. Under these conditions, the Krebs cycle cannot operate and NAD must be produced from NADH by conversion of pyruvate to lactate. Lactate accumulates, producing a serious metabolic acidosis.

Entry of Other Compounds

As shown in Figure 14-7, many compounds besides glucose are oxidized to carbon dioxide and water by the Krebs cycle. The degradation of the amino acids cysteine, alanine, glycine, serine, and threonine produce pyruvate. Aspartate and asparagine are converted to oxaloacetate. Leucine, isoleucine, lysine, and the aromatic amino acids are degraded to acetyl-CoA, as are the fatty acids. Tyrosine and phenylalanine yield fumarate. Succinyl-CoA can come from isoleucine, methionine, and valine. α-Ketoglutarate arises from glutamate, glutamine, histidine, proline, and arginine. The reactions that convert these amino acids to Krebs cycle intermediates will be discussed when we cover amino acid degradation. It is important for you to keep in mind that the Krebs cycle plays a major role in the oxidation of many classes of cellular compounds.

Inhibitors

There are four important inhibitors of the Krebs cycle—fluoroacetate, arsenite, mercury, and malonate. **Fluoroacetate,** found in some plants, is converted to fluoroacetyl-CoA and then condensed with oxaloacetate, producing **fluorocitrate.** Fluorocitrate then irreversibly poisons aconitase, the second en-

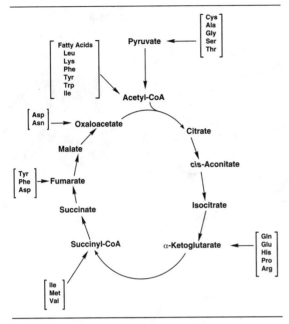

Fig. 14-7. Points at which amino acids and fatty acids enter into the Krebs cycle for oxidation.

zyme of the Krebs cycle. Fluoroacetate is called a **suicide inhibitor** because the cell must convert it from an inactive compound (fluoroacetate) to an active one (fluorocitrate). Only the activated form will lead to irreversible inhibition of an essential enzyme. You will encounter additional suicide inhibitors later in your study of metabolism.

$$
\begin{array}{ccc}
\text{F} & \text{O}^- & \text{O}^- \\
| & | & | \\
\text{CH}_2 & \text{C}=\text{O} & \text{C}=\text{O} \\
| & | & \text{F}-\text{CH} \quad \text{O}^- \\
\text{C}=\text{O} \quad + & \text{C}=\text{O} \longrightarrow & | \quad | \\
| & | & \text{HO}-\text{C}-\text{C}=\text{O} \\
\text{S} & \text{CH}_2 & | \\
| & | \quad \text{CoA} & \text{CH}_2 \\
\text{CoA} & \text{C}=\text{O} & | \\
& | & \text{C}=\text{O} \\
& \text{O}^- & | \\
& & \text{O}^-
\end{array}
$$

fluoroacetyl-CoA   **oxaloacetate**   **fluorocitrate**

Arsenite and **mercury** inhibit the Krebs cycle by binding to the SH group of lipoamine in the pyruvate and α-ketoglutarate dehydrogenase complexes. Binding of these inhibitors prevents the reoxidation of the SH group to a disulfide.

Malonate competitively inhibits succinate dehydrogenase because of its structural resemblance to succinate.

$$
\begin{array}{cc}
\text{O}^- & \text{O}^- \\
| & | \\
\text{C}=\text{O} & \text{C}=\text{O} \\
| & | \\
\text{CH}_2 & \text{CH}_2 \\
| & | \\
\text{C}=\text{O} & \text{CH}_2 \\
| & | \\
\text{O}^- & \text{C}=\text{O} \\
& | \\
& \text{O}^-
\end{array}
$$

malonate   **succinate**

Summary

The final oxidation of pyruvate begins by its entry into the mitochondria via a specific transport system. Then pyruvate is converted to acetyl-CoA by a combination of dehydrogenation, decarboxylation, and attachment to coenzyme A using a four-enzyme complex. These reactions require thiamine pyrophosphate, lipoamine, FAD, NAD, and coenzyme A. They are inhibited by arsenite and mercury.

The Krebs cycle itself consists of nine reactions in which two carbons are lost as carbon dioxide. First, acetyl-CoA condenses with oxaloacetate, forming citrate. This is isomerized to isocitrate with *cis*-aconitate as the intermediate. This reaction can be inhibited by fluoroacetate after the inhibitor is converted first to fluoroacetyl-CoA and then to fluorocitrate. In the next reaction, isocitrate is dehydrogenated and decarboxylated, producing NADH and α-ketoglutarate. α-Ketoglutarate then is dehydrogenated, decarboxylated, and attached to coenzyme A to form succinyl-CoA and another NADH. Succinate is formed, yielding GTP, the only high-energy phosphate produced by the Krebs cycle. Succinate is converted to fumarate, producing FADH$_2$. This reaction is inhibited by malonate. Finally, fumarate is converted to malate, which in turn is converted to oxaloacetate, completing the cycle. The total cycle has a large negative ΔG.

The control enzymes of the Krebs cycle are pyruvate decarboxylase, citrate synthetase, isocitrate dehydrogenase, and α-ketoglutarate dehydrogenase. Excess pyruvate can be converted to oxaloacetate by pyruvate carboxylase, a biotin-requiring enzyme. Degradation of fatty acids and amino acids produces acetyl-CoA as well as a number of intermediates of the cycle.

15 Oxidative Phosphorylation

Most ATP formed from the complete oxidation of glucose is produced by oxidative phosphorylation in the mitochondria. Here, electrons are passed between protein carriers that are attached to the inner mitochondrial membrane. At the same time, protons are pumped across the membrane, which is impermeable to most ions. The energy generated when the protons reenter the mitochondrial matrix is utilized to synthesize ATP. Electrons from NADH and $FADH_2$ produced in the cytoplasm, as well as those from the Krebs cycle, are funneled into the oxidative phosphorylation pathway.

Objectives

After completing this chapter, you should be able to

Relate $\Delta G^{\circ\prime}$ to $\Delta E^{\circ\prime}$.

Determine the direction in which electrons would pass in an oxidation-reduction couple, given either ΔG or ΔE.

Describe the components of the electron transport chain, including the points at which the various inhibitors act.

Discuss the mechanism of ATP production.

Diagram the shuttles for the transport of cytoplasmic reducing equivalents into the mitochondria.

List the sources and amounts of ATP produced by the complete oxidation of glucose.

Oxidation and Reduction

Oxidation is defined as the removal of electrons and **reduction** as the gain of electrons. When one compound is oxidized, another must be reduced. Oxidation-reduction reactions can be described by the difference in the electron donation potential (redox, or ΔE) between reactants and products. This is measured in volts relative to the hydrogen electrode. **Electronegative** compounds donate electrons to the hydrogen electrode, whereas **electropositive** compounds receive electrons from the hydrogen electrode. ΔE is related to ΔG by the following equation:

$$\Delta G = -nF\Delta E$$

where n equals the number of electrons transferred in the reaction and F is the Faraday constant 23.062 kcal/volt. Note that a negative ΔG produces a positive ΔE. For biological systems, the energy difference is given as $\Delta E^{\circ\prime}$, which is defined under the same standard conditions as $\Delta G^{\circ\prime}$, pH 7 with all other reactants at 1-M concentration. Obviously, this is not the concentration at which reactants are found in cells.

Table 15-1 gives the $\Delta E^{\circ\prime}$ for a number of cellular oxidation-reduction pairs. The $\Delta E^{\circ\prime}$ in the table is the potential in going from the oxidized member of the pair to the reduced member of the pair under standard conditions. At pH 1, the ΔE° for the hydrogen couple is set at zero. A positive $\Delta E^{\circ\prime}$ occurs

Table 15-1. Redox potentials of some pairs of the electron transport chain

Reduced form	Oxidized form	E
NADH	NAD$^+$	−0.32
FADH$_2$	FAD	−0.21
Coenzyme QH$_2$	Coenzyme Q	+0.10
2 Fe^{2+} cytochrome b	2 Fe^{3+} cytochrome b	+0.14
2 Fe^{2+} cytochrome c	2 Fe^{3+} cytochrome c	+0.44
H$_2$O	½ O$_2$	+0.82

when the reduced member of one pair gives electrons to (reduces) the oxidized member of another pair that has a more positive $\Delta E°'$. Using this table, you can see that NADH reduces FAD ($\Delta E°' = 0.11$), FADH$_2$ reduces coenzyme Q ($\Delta E°' = 0.32$), and molecular oxygen is reduced by all the other compounds listed.

The electron transport or respiratory chain functions by passing electrons from more electronegative compounds such as NADH and FADH$_2$ to more electropositive ones such as coenzyme Q and cytochrome c. The final electron recipient of the electrons of the respiratory chain is molecular oxygen. The $\Delta E°'$ for the entire chain is 1.14 volts or −52.6 kcal/mol.

Components of the Electron Transport Chain

The electron transport chain contains four protein complexes and two mobile electron carriers. The net result is the transfer of electrons from NADH and FADH$_2$ to oxygen-forming water. The protein complexes are an integral part of the inner mitochondrial membrane. As membrane-bound complexes, these proteins accept electrons from the preceding complex and pass them to the subsequent one. The order of the complexes of the electron transport chain is given in Figure 15-1.

The first protein complex (I) passes two electrons from NADH to coenzyme Q. This complex is called **NADH-CoQ reductase** and contains 25 different proteins including several nonheme iron proteins and a covalently bound flavin mononucleotide (FMN). Like FAD, FMN is derived from the vitamin riboflavin. It has the following structure:

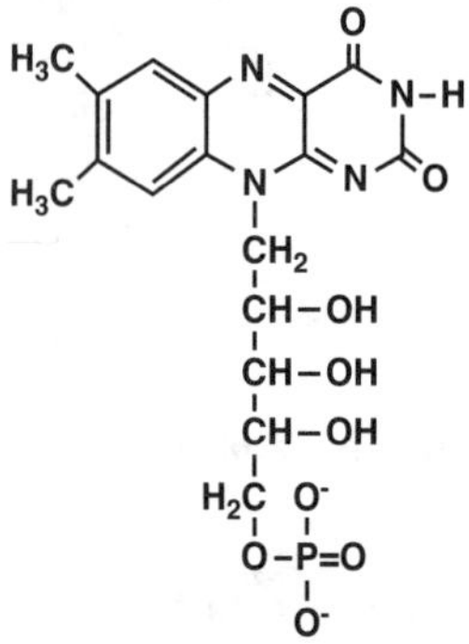

In complex I, FMN receives the two electrons from NADH and passes them to the nonheme iron proteins. In the reduction process, the iron atom changes from a +3 to a +2 charge.

The recipient of the electrons from complex I is **coenzyme Q.** This compound, also called **ubiquinone,** is capable of receiving either one or two

Fig. 15-1. Order of the protein complexes and mobile electron carriers of the mitochondrial electron transport chain and the points at which inhibitors act. Two protons are ejected from the inner mitochondrial membrane for each electron pair that passes through complexes I, III, and IV. ATP is synthesized when protons re-enter the mitochondrial matrix. (*CoQ* = coenzyme Q.)

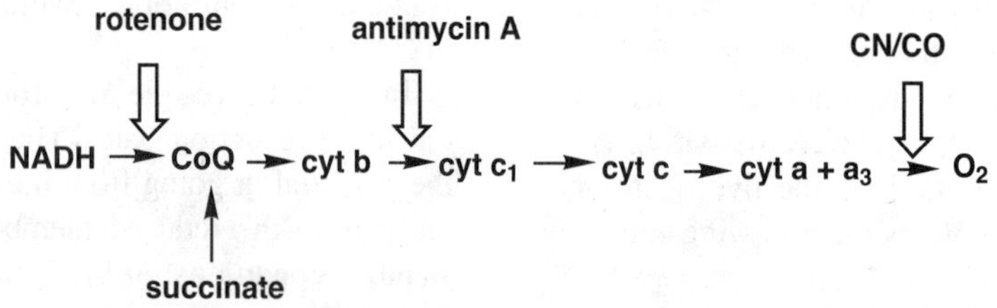

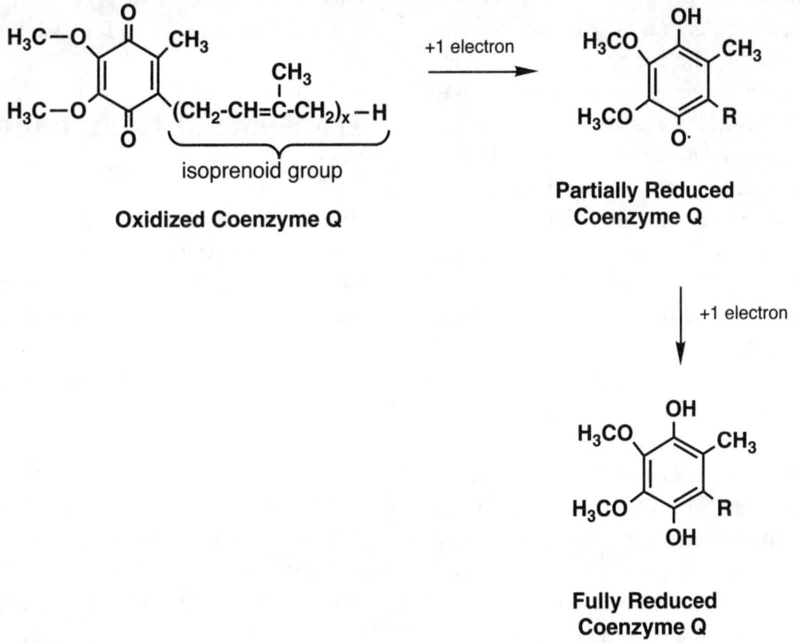

Fig. 15-2. Structure of oxidized, partially reduced, and fully reduced coenzyme Q.

electrons (Fig. 15-2). CoQ is not firmly attached to any protein and serves as a lipid-soluble mobile electron carrier. Membrane attachment for CoQ occurs through the isoprenoid tail, which varies in length depending on the species. The most common number of isoprenoid units in human CoQ is 10.

The second complex, **succinyl-CoQ reductase,** serves to link succinate, $FADH_2$, and enzymes containing bound $FADH_2$ to the rest of the electron transport chain. In complex II, two electrons are transferred from succinate to FAD, producing $FADH_2$. The electrons then are passed from $FADH_2$ through several iron-sulfur proteins and finally to CoQ. The $FADH_2$ used in this reaction can either be bound to the mitochondrial matrix or be covalently attached to an enzyme such as succinate dehydrogenase of the Krebs cycle.

Cytochrome c reductase is the third complex of the respiratory chain. It contains an iron-sulfur protein and cytochromes b and c_1, which are heme iron proteins. Like the iron in complex I, the heme iron changes from +3 to +2 on reduction. Complex III transfers two electrons from CoQ to **cytochrome c,**

the second mobile electron carrier. Unlike the large enzyme complexes, cytochrome c is a small protein containing only 104 amino acids. It is loosely attached to the inner mitochondrial membrane rather than being embedded in it. As with the other cytochromes, the iron atom of the heme in cytochrome c changes from +3 to +2 on reduction.

The final complex (IV) is **cytochrome c oxidase.** It transfers a single electron from each of two cytochrome c molecules to molecular oxygen. Like complex III, cytochrome oxidase contains heme iron as part of cytochromes a and a_3. In addition, it has a copper-containing enzyme that changes charge from +2 to +1 on reduction.

Production of ATP

The **Mitchell chemiosmotic hypothesis** is widely accepted as the mechanism for coupling electron transport and ATP production in mitochondria. When the electron transport chain is operating, protons are pumped to the outside of the inner mitochondrial membrane, which is impenetrable to positive ions. This makes the pH outside the mitochondrial membrane lower (higher hydrogen ion concentration) than that inside. The energy provided by the flow of protons back into the mitochondria is used to phosphorylate ADP to ATP.

As electrons are passed between NADH and CoQ, hydrogen ions are ejected from the inner mitochondrial membrane. These protons are not the same as those associated with NADH. Additional protons are ejected as electrons pass between CoQ and cytochrome c and between cytochrome c and oxygen. It is estimated that a total of six protons are ejected from the mitochondria as two electrons pass from NADH to oxygen. No protons are ejected as electrons pass from $FADH_2$ to CoQ.

Protons re-enter the mitochondrial matrix by flowing through a stalk-shaped protrusion on the inner membrane. This contains a hydrophobic proton channel, the F_0 subunit, and a spherical subunit, F_1, which is responsible for the synthesis of ATP. F_1 is an ATPase. In the presence of a proton gradient, the reaction is reversed and ATP is synthesized. The exact mechanism by which F_0 and F_1 couple proton transport with ATP synthesis remains unclear. No high-energy chemical intermediate has ever been isolated. However, it is estimated that one ATP is synthesized for each two protons passing through the complex.

This would give three ATPs synthesized for each NADH reduced and two for each $FADH_2$.

Transport of Compounds into the Mitochondria

Glycolysis and other cytoplasmic reactions produce NADH and $FADH_2$, which are then reoxidized by the electron transport chain. Because mitochondria are impermeable to NADH and $FADH_2$, the electrons from the cytoplasmic reducing equivalents enter the mitochondria by way of one of two shuttle systems. The NAD and FAD molecules remain in the cytoplasm; only the electrons are carried into mitochondria.

One of the shuttles is the **glycerol-phosphate shuttle,** outlined in Figure 15-3. Dihydroxyacetone phosphate from glycolysis reacts with NADH to produce glycerol 3-phosphate and NAD. Glycerol 3-phosphate is oxidized by a flavoprotein in the outer mitochondrial membrane, producing dihydroxyacetone phosphate. This dissociates from the mitochondria, completing the cycle. The net result is the conversion of cytoplasmic NADH into mitochondrial $FADH_2$. Only two ATPs are produced from cytoplasmic NADH when this shuttle is used.

Fig. 15-3. The glycerol-phosphate shuttle for the conversion of cytoplasmic NADH into mitochondrial $FADH_2$.

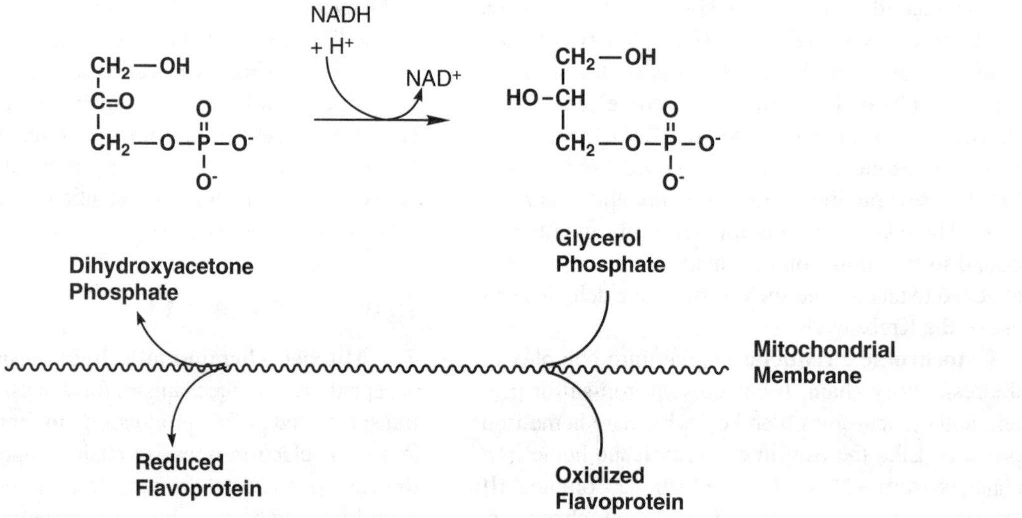

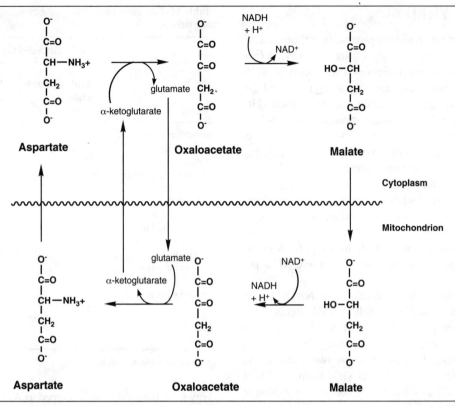

Fig. 15-4. The malate-aspartate shuttle, which converts cytoplasmic NADH into mitochondrial NADH.

A second shuttle that can transfer cytoplasmic reducing potential into the mitochondria is the **malate-aspartate shuttle,** shown in Figure 15-4. In this shuttle, oxaloacetate reacts with NADH, producing malate. This is a reversal of the last reaction of the Krebs cycle. However, the enzyme catalyzing the shuttle reaction is found in the cytoplasm, not in the mitochondria. Malate enters the mitochondria, where it is reoxidized to oxaloacetate by mitochondrial NAD. Oxaloacetate then is transaminated to aspartate, which leaves the mitochondria. This transamination is necessary because mitochondria are not permeable to oxaloacetate. In the cytoplasm, another transamination occurs and oxaloacetate is regenerated, completing the cycle. The net result is conversion of cytoplasmic NADH to mitochondrial NADH, with no loss of ATP production potential except for that produced by a possible proton exchange.

Another shuttle that must be completed for oxidative phosphorylation to occur is ADP for ATP. As these are highly charged compounds, a specific carrier is required for their transport across the mitochondrial membrane. The carrier binds ADP in the cytoplasm and transfers it into the mitochondria. Then it carries mitochondrial ATP back into the cytoplasm. This transport is an exchange reaction and will function only if ADP is present on one side of the membrane and ATP on the other. The shuttle can transport in both directions. In cells, the shuttle tends to move ADP in and ATP out of the mitochondria because of the greater negative charge on ATP. The energy needed for the ADP-ATP exchange is driven by the mitochondrial proton gradient. The phosphate needed for ATP phosphorylation enters the mitochondria in exchange for OH^- or malate.

Energy Yield

The complete oxidation of glucose to carbon dioxide and water using the glycolysis pathway, Krebs cycle, and oxidative phosphorylation yields 36 ATPs if the glycerol-phosphate shuttle is used and 38 ATPs if the malate-aspartate shuttle is used. The sources of the ATP are summarized in Table 15-2.

Glycolysis produces two ATPs by substrate-level phosphorylation and two NADHs. If these reducing equivalents enter the mitochondria by the glycerol-phosphate shuttle, they yield two ATPs per NADH for a total of four ATPs. If the reducing equivalents enter by the malate-aspartate shuttle, they yield three ATPs per NADH, for a total of six.

The conversion of pyruvate to acetyl-CoA yields two NADHs in the mitochondria or six ATPs per glucose.

The Krebs cycle produces three NADHs, one $FADH_2$, and one GTP. Each NADH yields three ATPs and each $FADH_2$ two ATPs when electrons from these compounds pass through the electron transport chain. Because there are two acetyl-CoAs produced per glucose, the net yield from the Krebs cycle is 24 ATPs, 22 of which come from oxidative phosphorylation.

With 6 or 8 ATPs from glycolysis (depending on which shuttle is used), 6 from acetyl-CoA production and 24 from the Krebs cycle, the total is 36 or 38 ATPs for each glucose oxidized completely to carbon dioxide and water.

Inhibitors

A number of compounds have been shown to inhibit oxidative phosphorylation, the electron transport chain, or both. These inhibitors can be classified as uncouplers, exchange inhibitors, or electron transport blockers. Use of various classes of inhibitors has been essential for elucidating the order and mechanism of oxidative phosphorylation.

Any agent that allows H^+ to pass through the mitochondrial membrane disrupts the proton gradient and prevents ATP production. However, such agents do not inhibit the transport of electrons in the respiratory chain. In their presence, NADH and $FADH_2$ are oxidized, heat is produced, but none of the en-

Table 15-2. Sources of ATP from the complete oxidation of glucose

Source	ATP produced per mole	
	With glycerol-phosphate shuttle	With malate-aspartate shuttle*
Substrate level		
Glycolysis	2	2
Krebs cycle	2	2
Oxidative phosphorylation		
Glycolysis— 2 NADHs shuttled into mitochondria	4	6
Acetyl-CoA synthesis— 2 NADHs	6	6
Krebs cycle		
6 NADHs	18	18
2 FADs	4	4
Total	36	38

*Assuming no proton exchange.

ergy from the oxidation is trapped as ATP. These two reactions, which are normally tightly coupled, are uncoupled in the presence of such inhibitors. Among the uncouplers of oxidative phosphorylation are **dicoumarol** and **dinitrophenol.** These compounds dissolve in the mitochondrial membrane and allow protons to pass into the mitochondria, thereby eliminating the proton gradient necessary for ATP production. Ionophores such as **valinomycin, gramicidin,** and **nigericin** can carry positive ions other than protons across the mitochondrial membrane. When positive ions are inside the mitochondria, they can be exchanged for hydrogen ions outside the mitochondria. This also dissipates the proton gradient and uncouples phosphorylation from oxidation. Uncoupling of electron transport and ATP production occurs naturally in the brown fat of newborn infants, where it is used as a mechanism for the production of heat.

Oligomycin interferes with ATP production by directly interacting with the proteins involved in ATP production. This prevents proton entry into the mitochondria and the synthesis of ATP. In the presence of oligomycin, neither ATP production nor the electron transport chain are functional. However, the addition of an uncoupler such as dinitrophenol to oligomycin-treated mitochondria relieves the inhibition of the electron transport chain but not the effect on ATP production.

Another group of inhibitors affects the electron transport chain directly. **Rotenone** and **barbiturate** block the transfer of electrons from complex I to CoQ. In their presence, NADH accumulates but $FADH_2$ is oxidized. Unless $FADH_2$ is regenerated, the electron transport chain will stop after one cycle. **Antimycin A** blocks the function of complex III so that neither NADH nor $FADH_2$ is oxidized. Complex IV is inhibited by **cyanide, carbon monoxide, sulfide,** and **azide.** In the presence of these inhibitors, all the respiratory chain intermediates remain in the reduced state. No oxygen is utilized and no ATP synthesized, because there is no electron flow or proton gradient to drive its production.

The rate of oxidative phosphorylation is controlled by the availability of NADH, $FADH_2$, oxygen, and ADP. To function, the electron transport chain needs electrons to transfer and oxygen to receive them. If either of these are in low concentration, the electron transport chain will operate very slowly. To a major extent, the rate of respiration and oxygen delivery to the mitochondria control the rate of electron transport.

In addition to oxygen and reducing equivalents, the electron transport chain requires ADP. If most of the cellular ADP is converted to ATP—that is, if the cellular energy charge is high—the electron transport chain will operate slowly even in the presence of an ample supply of oxygen and reducing equiva-lents. The requirement for ADP allows oxidative phosphorylation to be blocked by compounds such as **atractyloside** and **bonbrekic acid**, which inhibit the ADP-ATP exchange reaction.

Summary

In oxidation-reduction reactions, such as the electron transport chain, electrons flow from more electronegative compounds to more electropositive ones. $\Delta E^{\circ\prime}$ and $\Delta G^{\circ\prime}$ are related by the equation $\Delta G^{\circ\prime} = -nF\Delta E^{\circ\prime}$. In the mitochondrial electron transport chain, NADH reduces complex I, which reduces coenzyme Q. CoQ is reduced also by complex II, which is reduced by $FADH_2$. Two electrons are transferred in these reactions. CoQ passes one electron to complex III, which contains two heme iron proteins, cytochromes b and c_1. Complex III passes one electron to cytochrome c, which in turn passes it to complex IV and then to oxygen. Three ATPs are made for each NADH oxidized, by trapping the energy of a proton gradient produced when protons are ejected across the inner mitochondrial membrane. Two ATPs are synthesized for each $FADH_2$ that is oxidized. The energy yield from the complete oxidation of glucose is 36 ATPs if the glycerol-phosphate shuttle is used and 38 if the malate-aspartate shuttle is used.

Cytoplasmic reducing equivalents are transported into mitochondria by the glycerol-phosphate shuttle and the malate-aspartate shuttle. ADP-ATP and phosphate-OH enter by exchange reactions. Oxidative phosphorylation can be inhibited by agents that disrupt the pH gradient, ADP-ATP production, or the electron transport chain.

16 Gluconeogenesis and the Interconversion of Carbohydrates

Glucose is used as a major fuel for most cells and the exclusive fuel for cells of the central nervous system, exercising muscle, and erythrocytes. Circulating glucose can be produced from glycogen stores in the liver or by synthesis from noncarbohydrate sources. In addition to serving as a major metabolic fuel, glucose is used as the starting point for the synthesis of other important cellular compounds including lactose in the mammary gland, glycerol in adipose tissue, ribose for nucleic acid synthesis, and uronic and amino sugars of the glycoproteins and proteoglycans.

Objectives

After completing this chapter, you should be able to

Outline the reactions of gluconeogenesis and name the four enzymes that are unique to this pathway.

Discuss the control points of gluconeogenesis.

Describe the Cori cycle.

List the major reactions and products of the pentose phosphate and uronic acid pathways.

Recognize the reactions catalyzed by transaldolase and transketolase.

List the inherited defects affecting these pathways.

Gluconeogenesis

From Pyruvate

Glucose is synthesized from noncarbohydrate sources in the liver and, to a lesser degree, in the kidney. For the most part, gluconeogenesis utilizes a reversal of the cytoplasmic reactions of glycolysis. The reactions of gluconeogenesis are outlined in Figure 16-1.

The gluconeogenesis pathway begins in the mitochondria where pyruvate is carboxylated to form oxaloacetate by **pyruvate carboxylase.** This reaction, which uses biotin and ATP, was described in Chapter 14 as the mechanism for the production of oxaloacetate when intermediates from the Krebs cycle had been removed for other reactions. Pyruvate carboxylase is the first of the four enzymes of gluconeogene-

sis that are different from those in glycolysis. Two of these enzymes, pyruvate carboxylase and phosphoenolpyruvate carboxykinase, are needed to circumvent the irreversible conversion of phosphoenolpyruvate to pyruvate.

Mitochondria are impermeable to oxaloacetate, which can exit the mitochondria only after transamination to aspartate as part of the malate-aspartate shuttle (discussed in Chapter 15). Once in the cytoplasm, oxaloacetate is regenerated by a reversal of the transamination reaction. Next, **phosphoenolpyruvate carboxylase** catalyzes the decarboxylation and phosphorylation of oxaloacetate to yield phosphoenolpyruvate. GTP serves as the energy source and the phosphate donor for this reaction. The overall reaction converts mitochondrial pyruvate to cytoplasmic phosphoenolpyruvate. It requires one ATP

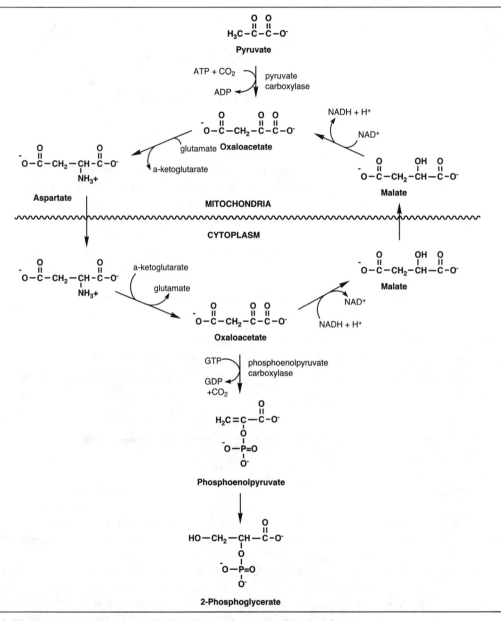

Fig. 16-1. The gluconeogenesis pathway. In this pathway, glucose is synthesized from pyruvate or glycerol. Only the enzymes that are different from those of glycolysis are given by name.

for the carboxylation reaction and one GTP for the phosphorylation. A relative deficiency in phosphoenolpyruvate in newborns affects their ability to perform gluconeogenesis and may be the cause of the hypoglycemia frequently observed in newborn infants.

The next series of reactions of gluconeogenesis constitute a reversal of the reactions of glycolysis. Phosphoenolpyruvate is converted to 2-phosphoglycerate, which then is converted to 3-phosphoglycerate. ATP transfers a phosphate to produce 1,3-diphosphoglycerate, which is converted to glyceraldehyde 3-phos-

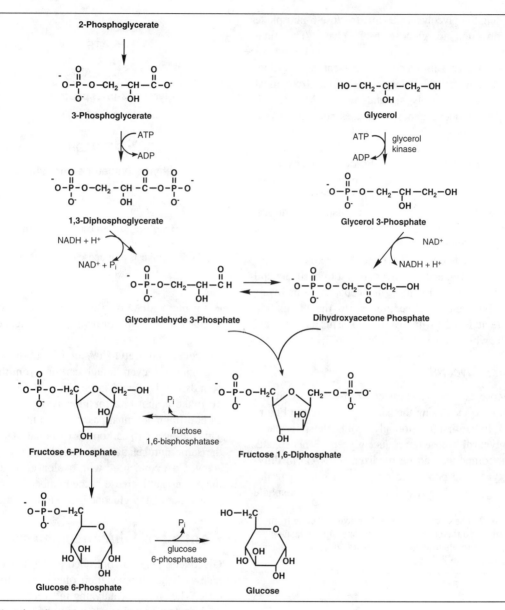

Fig. 16-1 (continued).

phate. Glyceraldehyde 3-phosphate, which is in equilibrium with dihydroxyacetone phosphate, condenses with that compound to give fructose 1,6-diphosphate.

The next step requires the function of the third enzyme unique to gluconeogenesis, **fructose 1,6-bisphosphatase.** Instead of transferring phosphate to ADP,

which would be a reversal of glycolysis, the phosphate merely is hydrolyzed, giving fructose 6-phosphate. Fructose 1,6-bisphosphatase is present in tissues that contain the enzymes of the gluconeogenesis pathway, such as liver, kidney, and striated muscle. The enzyme is absent from smooth and cardiac muscle.

Using the glycolysis enzyme, fructose 6-phosphate is converted to glucose 6-phosphate. The final new enzyme of gluconeogenesis is **glucose 6-phosphatase.** This enzyme is present in intestine, platelets, liver, and kidney. It is absent from most other tissues, notably skeletal muscle. Muscle can carry out gluconeogenesis but does not release free glucose to the circulation due to a lack of this enzyme.

Using the gluconeogenesis pathway, pyruvate, any intermediate of the Krebs cycle, or any compound that can be converted to pyruvate or a Krebs cycle intermediate can yield glucose. This is important for the metabolism of muscle, especially during strenuous exercise. Under these conditions, muscle converts glucose to lactate by the glycolysis pathway. The lactate is released into the circulation, travels to the liver, and is converted first to pyruvate and then to glucose by the gluconeogenesis pathway. The liver releases free glucose to the circulation where it is taken up by muscle. This is the **Cori cycle,** which is outlined in Figure 16-2.

From Glycerol

Glucose can be synthesized from glycerol in the liver or kidney using the pathway depicted in Figure 16-3. In the first reaction, glycerol is phosphorylated by **glycerol kinase** to produce glycerol 3-phosphate. This compound can be oxidized by NAD to dihydroxyacetone phosphate by glycerol phosphate dehydrogenase. Because dihydroxyacetone phosphate

Fig. 16-2. The Cori cycle. Muscle releases lactate from glycolysis to the circulation. Liver takes up the lactate, converts it to glucose, and releases the glucose to the circulation, where it is taken up by muscle.

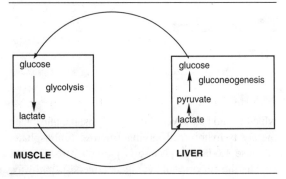

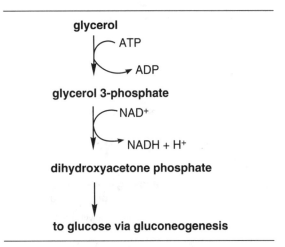

Fig. 16-3. Synthesis of glucose from glycerol.

is an intermediate of gluconeogenesis, glucose can be made from it by following the sequence of reactions given in Figure 16-1.

Acetate and acetyl-CoA are not substrates for gluconeogenesis, even though carbons from these compounds can be found in newly synthesized glucose. To make a substrate for gluconeogenesis, the two-carbon fragment must pass through the Krebs cycle. This removes two carbons as carbon dioxide, exactly the same number as were added to the cycle when acetyl-CoA condensed with oxaloacetate. Therefore, there is no net increase in the number of carbons and no net increase in gluconeogenesis substrates.

Control of Gluconeogenesis

Gluconeogenesis is controlled at the first step of the pathway, the carboxylation of pyruvate to oxaloacetate. The enzyme catalyzing this reaction, **pyruvate carboxylase,** is activated by acetyl-CoA and inhibited by ADP. An excess of acetyl-CoA occurs only if the Krebs cycle is proceeding at a maximal rate. Then pyruvate is carboxylated to oxaloacetate rather than decarboxylated to produce more acetyl-CoA. The oxaloacetate is converted into glucose, which either is stored as glycogen or is released to the circulation for use by other tissues. Under the condition of high levels of ADP, which indicates that the cell needs energy

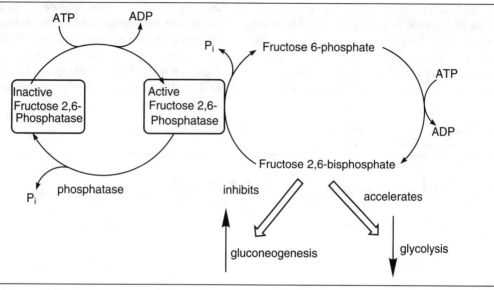

Fig. 16-4. Control of fructose 2,6-bisphosphatase by phosphorylation and dephosphorylation.

in the form of ATP, pyruvate is converted into acetyl-CoA for oxidation by the Krebs cycle.

The second control point of gluconeogenesis is the reaction catalyzed by **fructose 1,6-bisphosphatase.** This enzyme is activated by ATP and inhibited by AMP and fructose 2,6-bisphosphate. Fructose 2,6-bisphosphate increases the K_m of the enzyme for fructose 1,6-diphosphate, thus decreasing the reaction rate. There is also control at the level of phosphorylation-dephosphorylation of the enzyme fructose 2,6-bisphosphatase, as shown in Figure 16-4. Note that the control of fructose 1,6-bisphosphatase is exactly the opposite of that for the glycolysis enzyme phosphofructokinase.

Gluconeogenesis is controlled also at the transcriptional level by a number of hormones. In liver, insulin decreases the rate of transcription of the genes coding for the four enzymes unique to this pathway—pyruvate carboxylase, phosphoenolpyruvate carboxylase, fructose 1,6-bisphosphatase, and glucose 6-phosphatase. Cortisol, epinephrine, glucagon, starvation, and diabetes have the opposite effect on the transcription of these genes. Although gluconeogenesis takes place in kidney, neither insulin nor glucagon affects the renal enzyme levels.

Of the four enzymes, the concentration of pyruvate carboxylase is the least affected by changing hormonal and nutritional states. This is because the conversion of pyruvate to oxaloacetate is required for fatty acid biosynthesis and replenishment of the Krebs cycle as well as gluconeogenesis.

Pentose Phosphate Pathway

The **pentose phosphate pathway** or **hexose monophosphate shunt** produces NADPH outside of the mitochondria. NADPH is needed for the synthesis of fatty acids, cholesterol, and other steroids. The pathway is also the source of ribose for nucleotide biosynthesis. The reactions of this pathway occur in liver, adipose tissue, adrenal cortex, thyroid, erythrocytes, and lactating (but not nonlactating) mammary gland. In addition to producing NADPH and ribose, the pathway oxidizes glucose to carbon dioxide. However, it does not produce ATP by substrate-level phosphorylation. ATP can be produced if reducing equivalents generated by the pentose phosphate pathway are oxidized by the mitochondrial electron transport chain. Like glycolysis, all the enzymes of the pentose phosphate pathway are located in the cytoplasm. The reactions are given in Figure 16-5. The net result of the pentose phosphate pathway is the production of six NADPHs, three carbon dioxides, and one glyceraldehyde 3-phosphate for each glucose 6-phosphate passing through the pathway.

In the first reaction, glucose 6-phosphate is oxi-dized by NADP to 6-phosphogluconate by **glucose 6-phosphate dehydrogenase.** In the next reaction, 6-phosphogluconate is oxidized again by NADP and de-carboxylated to produce the five-carbon sugar **ribulose 5-phosphate.** Ribulose 5-phosphate can be isomer-ized to **ribose 5-phosphate,** the sugar required for nu-

Fig. 16-5. The pentose phosphate pathway. The structures of the sugars are drawn in their linear forms to better show the reactions.

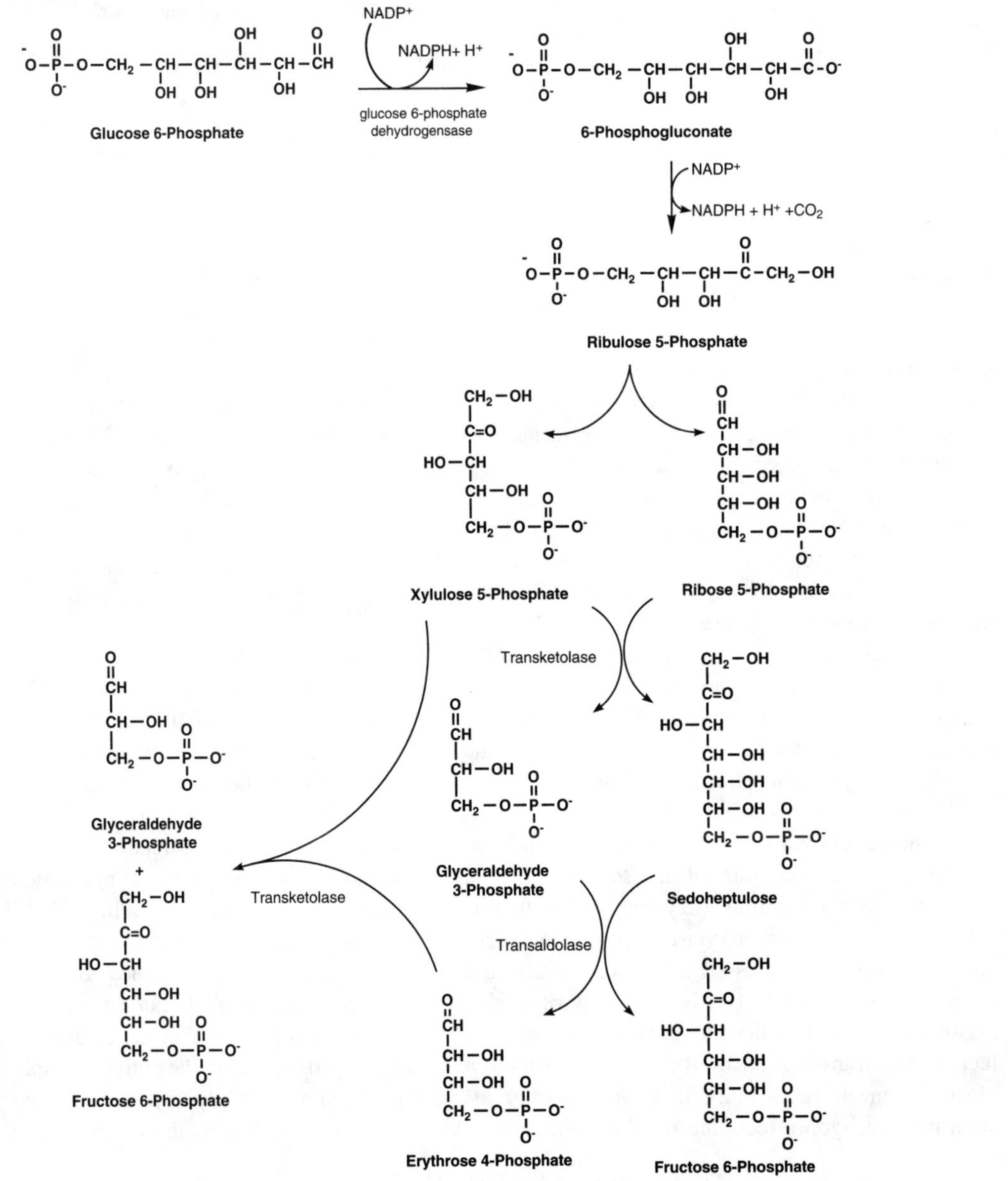

cleotide biosynthesis. In a different reaction, ribulose 5-phosphate can be epimerized at carbon 3 to yield xylulose 5-phosphate. These 2 five-carbon sugar phosphates form the start point for the next series of reactions catalyzed by transketolase and transaldolase. **Transketolase** transfers a two-carbon unit from a ketose to an aldose. **Transaldolase** transfers a three-carbon unit from an aldose to a ketose (Fig. 16-6).

The net result of the action of these two enzymes is the conversion of three pentose phosphates into one triose phosphate and two hexose phosphates. Because all these compounds are intermediates of glycolysis, these reactions link the pentose phosphate pathway with glycolysis and provide a pathway for the oxidation of excess ribose phosphate not needed for nucleotide biosynthesis.

In the pentose phosphate pathway, a two-carbon unit is transferred by transketolase from xylulose 5-phosphate to ribose 5-phosphate. The products are a three-carbon sugar, glyceraldehyde 3-phosphate, and a seven-carbon sugar, ketosedoheptulose. Using these two sugars, transaldolase can exchange a three-carbon unit, producing fructose 6-phosphate and erythrose 4-phosphate. Fructose 6-phosphate can enter glycolysis for further oxidation. In another reaction catalyzed by transketolase, xylulose 5-phosphate and erythrose 4-phosphate exchange a two-carbon unit, yielding glyceraldehyde 3-phosphate and fructose 6-phosphate.

Because there are no mitochondria and, thus, no Krebs cycle in red blood cells, reducing equivalents must come from the pentose phosphate pathway. Re-

Fig. 16-6. Reactions catalyzed by transaldolase and transketolase. A. Transaldolase transfers a three-carbon unit by binding the group to be transferred to an enzyme-bound amino group. B. Transketolase transfers a two-carbon unit by binding the group to be transferred to thiamine pyrophosphate (PP).

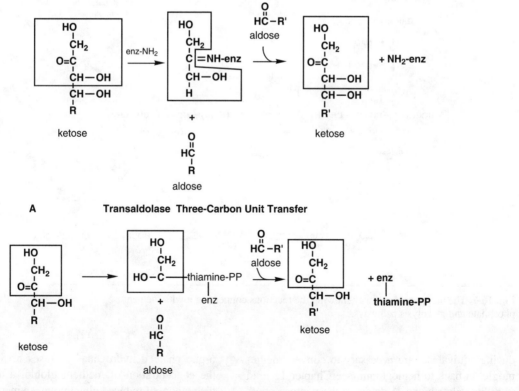

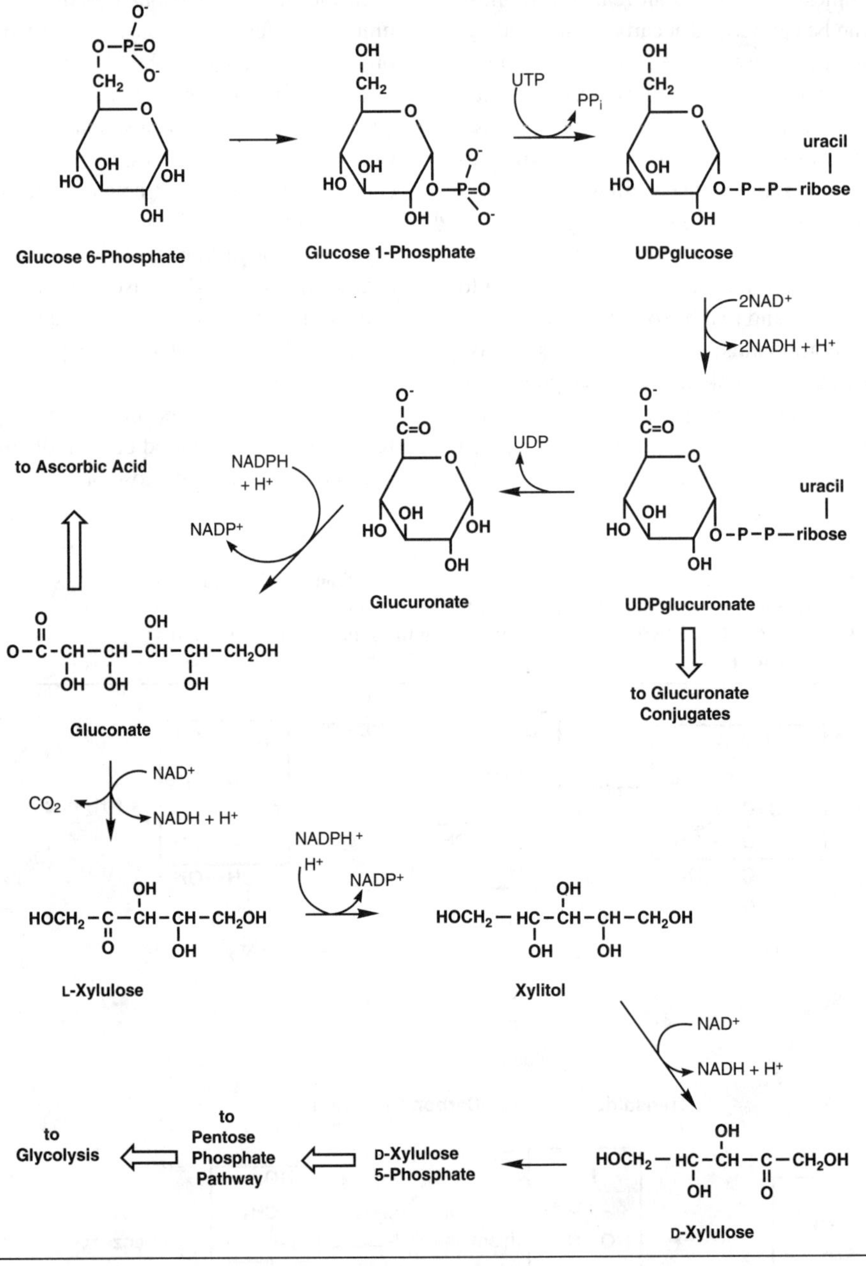

Fig. 16-7. The uronic acid pathway showing the reactions connecting it with the pentose phosphate and glycolysis pathways.

ducing equivalents are necessary to convert methemoglobin back to hemoglobin (see Chapter 12) and to reduce glutathione. A defect in the pentose phosphate pathway, especially in the first enzyme glucose

6-phosphate dehydrogenase, causes an increase in the concentration of methemoglobin, a decrease in the amount of reduced glutathione, an increase in hydrogen peroxide, and increased fragility of red blood cell membranes. This produces hemolysis, which is increased in the presence of oxidizing agents, including common over-the-counter drugs such as aspirin and antibiotics such as the sulfonamides.

Uronic Acid Pathway

The **uronic acid pathway,** as shown in Figure 16-7, converts glucose into glucuronic acid, which is necessary for the biosynthesis of glycoproteins and detoxification of drugs and other foreign compounds. Like the pentose phosphate pathway, the uronic acid pathway does not produce ATP by substrate-level phosphorylation. However, it does generate reducing equivalents, which can yield ATP via the electron transport chain.

In the first reaction of the uronic acid pathway, glucose 6-phosphate is isomerized to glucose 1-phosphate. This compound then reacts with UTP, yielding **UDPglucose.** Next, UDPglucose is oxidized to **UDPglucuronate** using two molecules of NAD. The UDPglucuronate can be further metabolized or used directly in the synthesis of polysaccharides and glycoproteins. UDPglucuronate is also important for the metabolism of drugs and other foreign substances. As glucuronide conjugates, these compounds are more water-soluble and their rate of metabolism, transport, and excretion is increased.

Because further reactions of glucuronate involve the sugar instead of the UDPsugar, the first step is the removal of the UDP, leaving free glucuronate. Glucuronate can be reduced by NADPH to yield gulonate. In all organisms except primates and guinea pigs, gulonate can be converted to **ascorbic acid** (vitamin C). All organisms can oxidize gulonate with NAD to yield 3-ketogulonate. Loss of carbon dioxide gives xylulose. Conversion of this to xylitol and then to xylulose 5-phosphate enables the uronic acid pathway to link with the pentose phosphate pathway and glycolysis. Lack of the enzyme needed to convert xylulose to xylitol produces **essential** or **idiopathic pentosuria** but no major clinical symptoms.

Interconversion of Carbohydrates

Glucose is the source of carbons for the biosynthesis of glycerol, fructose, lactose, and the amino sugars as well as ribose and the uronic acids.

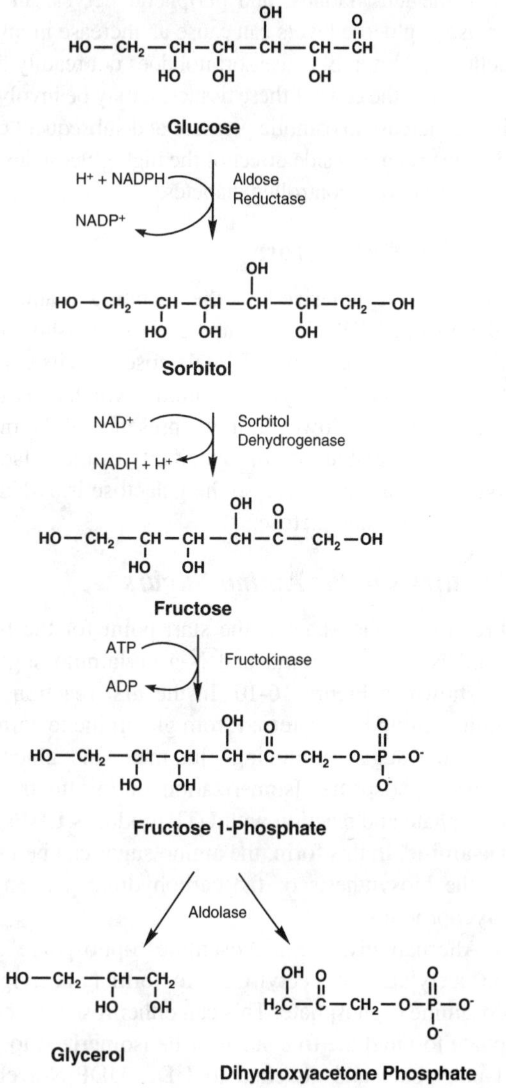

Fig. 16-8. Synthesis of glycerol. The structures of the sugars are drawn in their linear forms to better show the reactions.

Synthesis of Glycerol

As shown in Figure 16-8, glucose can be converted by aldose reductase and NADPH to yield sorbitol. Oxidation of sorbitol by NAD produces **fructose.** Reaction with fructokinase gives fructose 1-phosphate, which is cleaved by aldolase to form glyceraldehyde and dihydroxyacetone phosphate. The glyceraldehyde can then be reduced by NADH to yield **glycerol.**

In the lens, kidney, and peripheral nerves, an increase in glucose levels can cause an increase in intracellular sorbitol. Because sorbitol does not readily diffuse out of the cells of these tissues, it may be involved in the increase in osmotic pressure and subsequent cell damage seen as a side effect of the high glucose levels found in poorly controlled diabetes.

Synthesis of Lactose

Lactose is synthesized in the lactating mammary gland from UDPgalactose and glucose, as shown in Figure 16-9. Normally, UDPgalactose adds its galactose unit to an N-acetylglucosamine, which is part of a glycoprotein. However, in the presence of the milk protein α-lactalbumin, the specificity of the galactosyl transferase is altered so the galactose is added to glucose, forming lactose.

Synthesis of the Amino Sugars

Fructose 6-phosphate is the start point for the biosynthesis of the amino and N-acetylamino sugars, as shown in Figure 16-10. In the first reaction, an amino group is transferred from glutamine to carbon 2 of the sugar, producing glutamate and glucosamine 6-phosphate. Isomerization of this to the 1-phosphate and reaction with UTP produces **UDPglucosamine.** In this form, the amino sugar can be used in the biosynthesis of the carbohydrate portion of glycoproteins.

Alternatively, the glucosamine 6-phosphate can be acetylated by acetyl-CoA to form N-acetylglucosamine 6-phosphate. This can either lose the phosphate to yield the free sugar or be isomerized to the 1-phosphate and attached to UDP. UDP-N-acetylglucosamine can then be used directly in the biosynthesis of glycoproteins. Epimerization at carbon 4 produces UDP-N-acetyl-galactosamine, another sugar found in glycoproteins. Note that the epimerization can take place only when the sugar is attached to UDP.

Synthesis of Neuraminic Acid

N-acetylneuraminic acid (NANA), a **sialic acid,** is found in the apoproteins associated with lipid transport in the blood and the iron-binding protein trans-

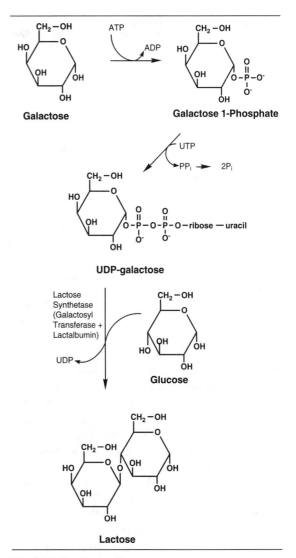

Fig. 16-9. Synthesis of lactose.

ferrin. It is synthesized from N-acetylglucosamine 6-phosphate in the liver by the pathway outlined in Figure 16-11.

First, an epimerase produces N-acetylmannosamine 6-phosphate. Then reaction with phosphoenolpyruvate gives the nine-carbon sugar N-acetylneuraminic acid 9-phosphate. In contrast to the other sugars, CDP, not UDP, is the carrier for NANA in attachment to proteins to form glycoproteins.

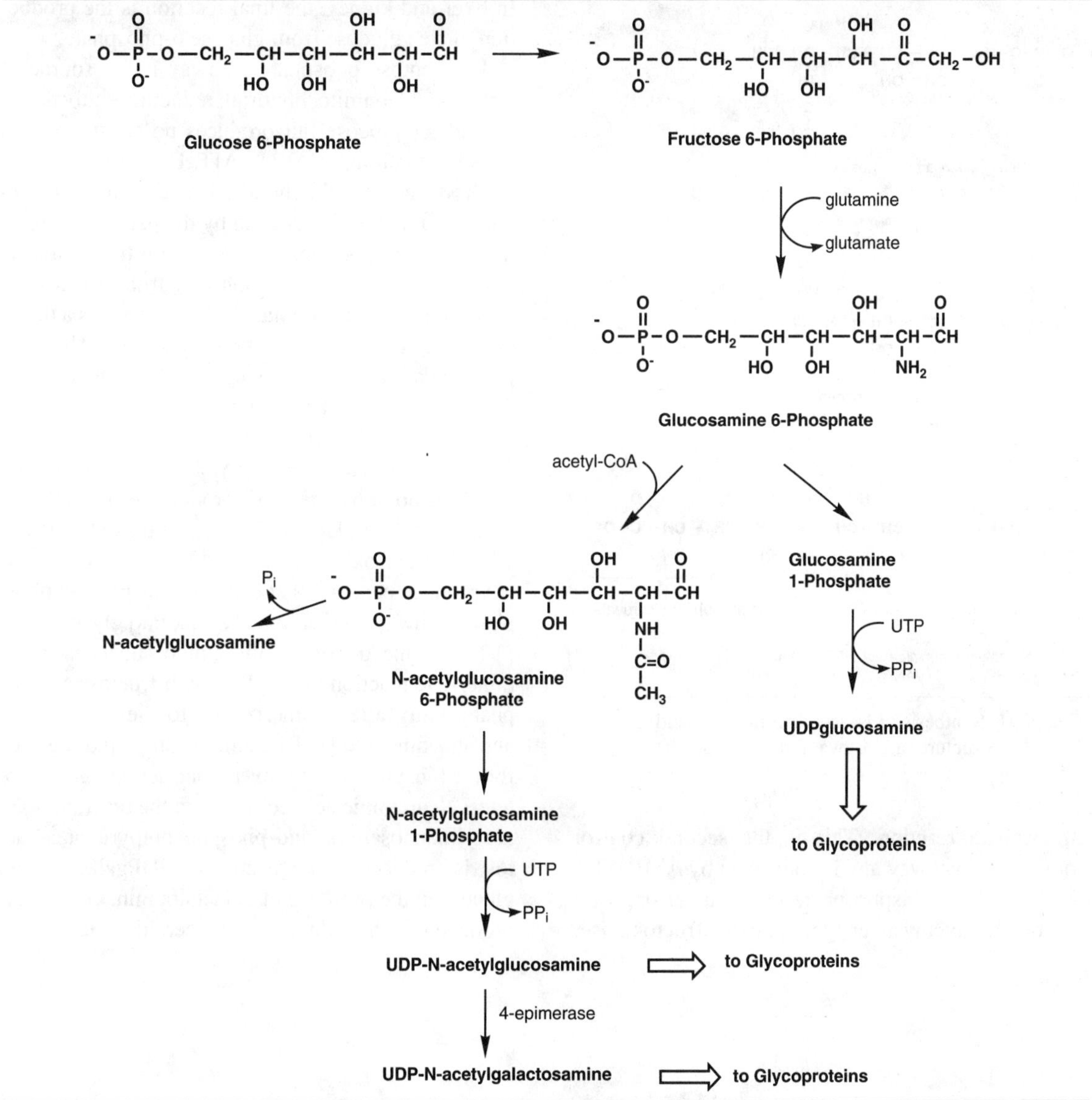

Fig. 16-10. Synthesis and interconversion of the amino sugars. The structures are drawn in their linear forms to better show the reactions.

Summary

Gluconeogenesis occurs in the cytoplasm by a reversal of glycolysis. Four enzymes unique to gluconeogenesis are necessary to reverse three steps of the glycolysis pathway. A combination of pyruvate car- boxylase and phosphoenolpyruvate carboxykinase are required to convert pyruvate in the mitochondria into phosphoenolpyruvate in the cytoplasm. Pyruvate carboxylase is the rate-controlling enzyme for gluconeogenesis. Fructose 1,6-bisphosphatase reverses the fructose 6-phosphate–to–fructose 1,6-

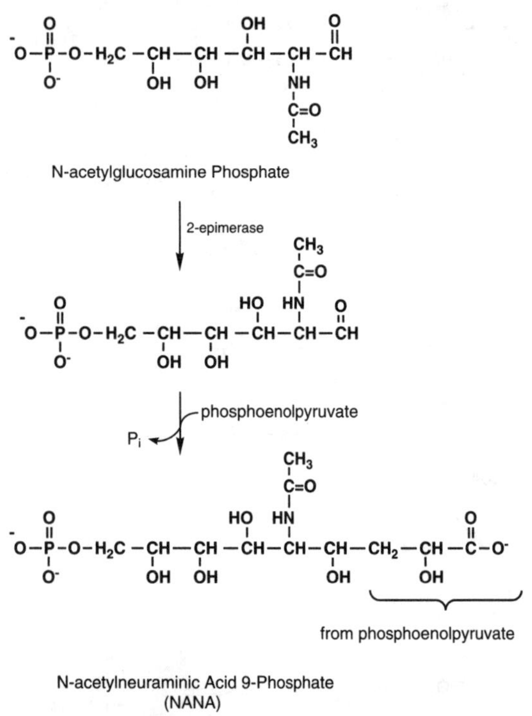

N-acetylglucosamine Phosphate

2-epimerase

N-acetylneuraminic Acid 9-Phosphate
(NANA)

Fig. 16-11. Synthesis of N-acetylneuraminic acid (NANA). Structures are drawn in their linear forms.

diphosphate reaction. This is the second control point of the pathway and is controlled by AMP, ATP, and fructose 2,6-bisphosphate in a manner opposite that of the glycolysis enzyme phosphofructokinase.

In liver and kidney, the final reaction is the production of free glucose from glucose 6-phosphate.

The pentose phosphate pathway is the source of ribose and nonmitochondrial reducing equivalents. It oxidizes glucose but produces no substrate-level phosphorylation of ADP to ATP. However, ATP can be produced by mitochondrial oxidation of the reducing equivalents produced by the pathway. Sugars in the pathway are interconverted by transketolase, which transfers a two-carbon unit from a ketose to an aldose, and transaldolase, which transfers a three-carbon unit from an aldose to a ketose. The final products are all glycolysis intermediates that can be further metabolized by that pathway.

Glucose is converted to gulonate by the uronic acid pathway. This pathway yields reducing equivalents but no substrate-level production of ATP. The oxidations take place while the sugar is attached to UDP. Further oxidation and the loss of carbon dioxide produce pentoses that can enter the pentose phosphate pathway and, ultimately, link into glycolysis.

The amino donor for the amino sugars is glutamine. The reaction takes place with fructose 6-phosphate. Only after isomerization to the 1-phosphate and attachment to UDP can these compounds be used for the biosynthesis of polysaccharides and glycoproteins. Neuraminic acid comes from the reaction of N-acetylmannosamine and phosphoenolpyruvate. Lactose is produced by the reaction of UDPgalactose and glucose in the presence of α-lactalbumin. Glycerol is produced by the reduction of glyceraldehyde.

17 Synthesis and Degradation of Glycogen and Other Polysaccharides

Glucose and other sugars metabolized by glycolysis come from dietary sources, from glucose produced by gluconeogenesis in the liver, and from the breakdown of glycogen. Glycogen is the major form of glucose storage in animal cells. It is an α-1,4 polymer of glucose, with an average of one α-1,6 branch for each 16 glucose residues. On degradation of glycogen, the liver releases glucose into the circulation for use by other tissues. Muscle also stores glucose as glycogen. However, when muscle degrades glycogen, it burns it as metabolic fuel rather than releasing it to the circulation.

Plants store glucose as amylose, an unbranched α-1,4 glucose polymer, or amylopectin, an α-1,4 glucose polymer with fewer branches than glycogen. Both amylose and amylopectin can be digested by humans. Cellulose, the other glucose polymer of plants, has β-1,4 linkages and cannot be digested by humans and most other animals. Undigested cellulose forms the dietary fiber needed for proper human nutrition. Other polysaccharides produced by eukaryotic cells include the glycoproteins, proteoglycans, and glycosaminoglycans. These polysaccharide-containing molecules have a structural role as well as being involved in cell-to-cell recognition, blood clotting, immune response, nutrient transport, and joint lubrication.

Objectives

After completing this chapter, you should be able to

Discuss the digestion of carbohydrates.

Describe the structure of glycogen.

List the reactions leading from glucose to glycogen and those necessary for glycogen breakdown.

Outline the control points for glycogen synthesis and breakdown, including the differences between liver and muscle.

Discuss the effects of cAMP, calcium, insulin, epinephrine, and glucagon on glycogen metabolism.

Match the glycogen storage diseases with the enzyme defects responsible for them.

List the components of hyaluronic acid, chondroitin sulfate, keratan sulfate, heparin, heparan sulfate, and dermatan sulfate.

Match the defects in the breakdown of these compounds with the affected enzyme and the name of the syndrome or disease.

Digestion of Carbohydrates

Most of the sugars metabolized by cells come from dietary sources. The breakdown of ingested carbohydrates begins in the mouth with the action of salivary α-**amylase** (or **ptyalin**). This enzyme hydrolyzes α-1,4 bonds in simple and complex carbohydrates, as shown in Figure 17-1. The products are oligosaccharides two or three residues long and compounds containing 1,6 branches (limit dextrans) that the α-amylase cannot break. Because the action of α-amylase ceases when food enters the acidic environment of the stomach, very little breakdown of carbohydrates actually occurs during the short time food is in the mouth.

There is no further digestion of carbohydrates until food passes into the intestine. Here, pancreatic α-amylase continues the digestion of the larger carbohydrates. Its specificity is similar to that of salivary α-amylase. Enzymes specific for the further digestion of the products of α-amylase action, such as maltose, isomaltose, sucrose, and lactose, are found on the mucosal surface of the intestine. After conversion to simple sugars, the monosaccharides are absorbed into the blood. Disaccharides cannot be absorbed.

The enzyme **lactase** is required to hydrolyze lactose to glucose and galactose. An absence of this enzyme is common in adults of all races but is especially frequent in blacks, Orientals, and Ashkenazi Jews. Because lactose, usually from milk products, cannot be hydrolyzed by persons who lack lactase, the ingested lactose passes unchanged into the large intestine. There, bacteria convert the lactose into organic acids that attract water by osmosis and cause diarrhea and bloating. **Lactose intolerance** can occur in normal adults due to intestinal diseases such as colitis, peptic ulcers, and gastroenteritis. Lactose intolerance is different from milk intolerance. The latter condition usually is due to sensitivity to β-lactoglobin, a major protein in cow's milk. Although more rare, **sucrose intolerance,** caused by an absence of sucrase, shows symptoms similar to those of lactose intolerance. Sucrose intolerance tends to occur in early childhood rather than developing later in life.

Cellulose, which contains only β-1,4 bonds, is not digested by any of the enzymes that hydrolyze other carbohydrates in the human digestive system. As cellulose passes unchanged through the digestive system, it forms the major source of dietary fiber.

Structure of Glycogen

Several hours after a carbohydrate meal, glycogen makes up 6% of the weight of the human liver. This amount decreases to less than 1% after an overnight fast. Most of the glucose from glycogen breakdown in the liver is released to the circulation. Muscle contains a lower and relatively constant amount of

Fig. 17-1. Digestion of carbohydrate by α-amylase.

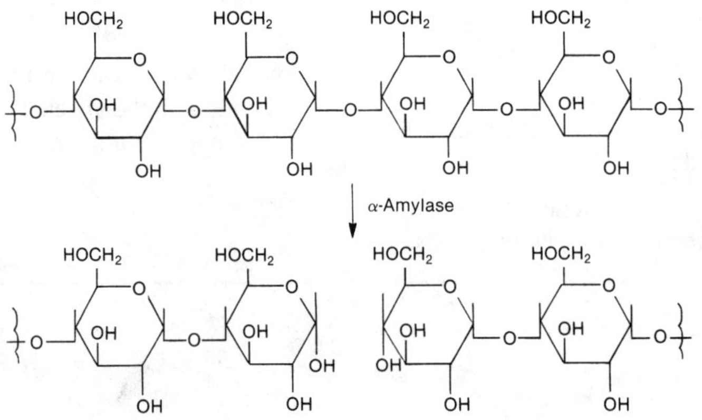

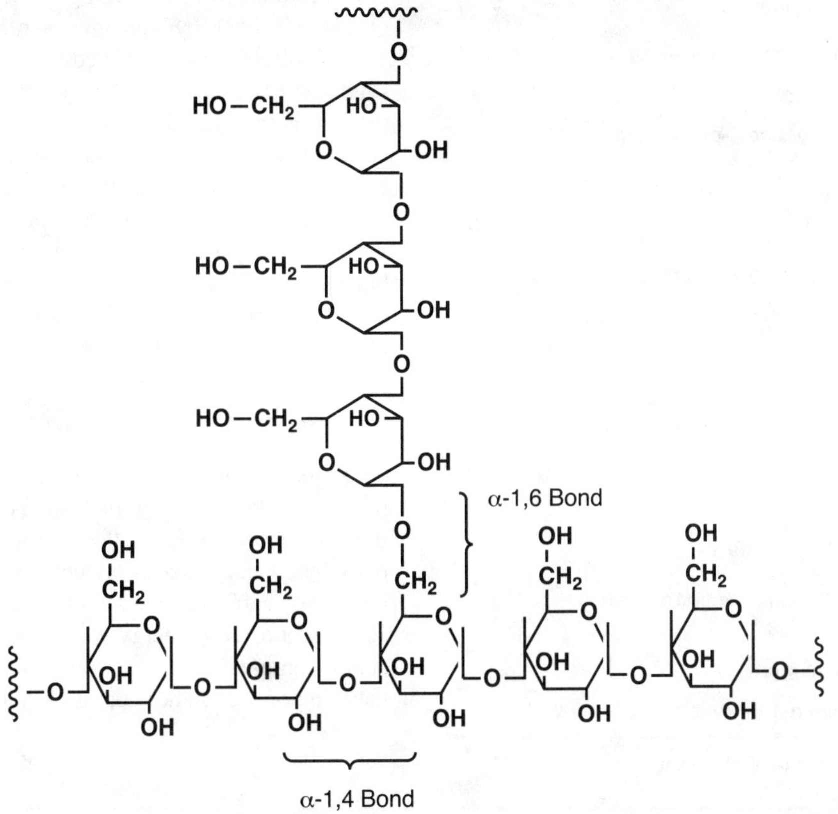

Fig. 17-2. Structure of a segment of glycogen showing α-1,4 bonds and an α-1,6 branch point.

glycogen, approximately 1% by weight. It is significantly decreased only after prolonged exercise. Then, if a high-carbohydrate diet is consumed, the amount of glycogen in muscle can rebound above its normal level. This is called *carbohydrate loading* and is practiced by many endurance athletes.

Glycogen is stored in both liver and muscle in granules of varying size. The enzymes required for synthesis and degradation and the factors necessary for the control of these enzymes are also found in the glycogen granules. The structure of glycogen is shown in Figure 17-2. The main bonding between glucose residues is α-1,4. Branches with α-1,6 bonds occur approximately once for every 16 glucose residues. Because of this compact branched structure, the size of glycogen granules can approach 2×10^7 daltons.

Glycogen Synthesis

Glycogen is synthesized by adding glucose residues, one at a time, to a preexisting glycogen primer. The reaction sequence is outlined in Figure 17-3. In liver, the first reaction is the phosphorylation of glucose by **glucokinase,** producing glucose 6-phosphate. The phosphate then is transferred to the 1 position by **phosphoglucomutase** with glucose 1,6-diphosphate as the intermediate. Because the donor of the second phosphate is ATP, this is an irreversible reaction.

Next, glucose 1-phosphate reacts with UTP, producing UDPglucose and pyrophosphate. As the pyrophosphate is hydrolyzed immediately to two molecules of inorganic phosphate, this also is an irreversible reaction.

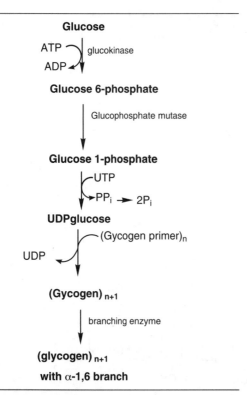

Fig. 17-3. Biosynthesis of glycogen.

In the final reaction, UDPglucose is added to a glycogen primer by **glycogen synthase.** This enzyme is also called UDPglucosyl transferase. In the synthesis reaction, C-1 of the UDPglucose attaches to C-4 of the primer in an α-1,4 linkage. After a chain of 6 to 11 residues has been synthesized, a branching enzyme transfers approximately 6 glucose residues to the 6 position of another glucose residue in the primer, as shown in Figure 17-4. Both branches continue to grow by the sequential addition of more glucose residues in α-1,4 linkage until another branch is created.

Glycogen Degradation

Glycogen is degraded by a set of three enzyme activities, as shown in Figure 17-5. The first enzyme, **phosphorylase,** breaks the α-1,4 linkage, adds phosphate, and releases glucose 1-phosphate. No nucleotide carrier is involved in the degradative reaction and no ATP energy is consumed. This degradation continues until only four glucose residues remain before a branch point. Then a transferase

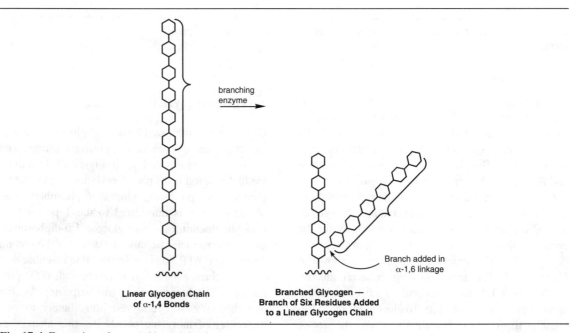

Fig. 17-4. Formation of an α-1,6 branch in glycogen.

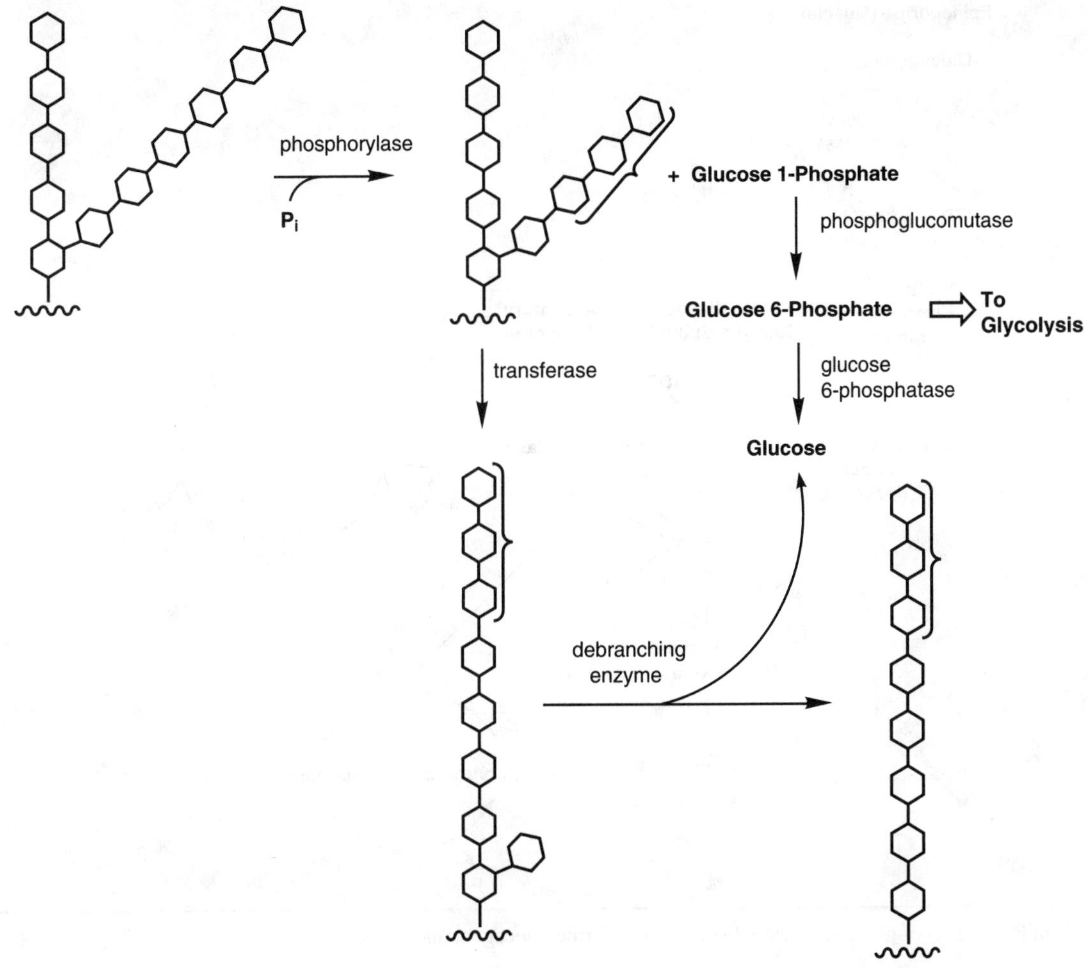

Fig. 17-5. Degradation of glycogen.

moves three of the remaining glucose residues to the end of another branch, leaving a single glucose residue in an α-1,6 linkage. A **debranching enzyme** releases this glucose as free glucose, not glucose 1-phosphate. Both the transferase and the debranching activities are associated with a single large protein.

Glucose 1-phosphate released by phosphorylase is isomerized to glucose 6-phosphate by **phosphoglucomutase.** As glucose 6-phosphate, it can enter glycolysis. Alternatively in liver and kidney, the phosphate can be removed by **glucose 6-phosphatase,** producing free glucose that then can leave the cell and enter the circulation.

Control of Glycogen Metabolism

The synthesis and degradation of glycogen is hormonally controlled by epinephrine, glucagon, and insulin. **Epinephrine** and **glucagon** both stimulate glycogen breakdown by controlling the phosphorylation state and, thus, the enzymatic activity of phosphorylase and glycogen synthase. The pathway is diagrammed in Figure 17-6. **Insulin** increases glycogen synthesis and inhibits its breakdown by a mechanism that is still not clearly understood. However, it probably also involves phosphorylation and dephosphorylation of enzymes.

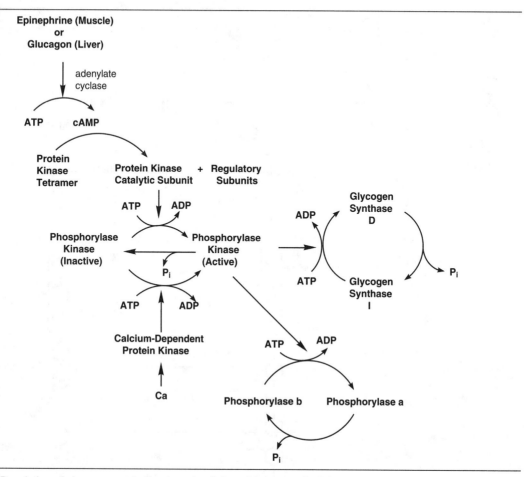

Fig. 17-6. Regulation of glycogen metabolism by epinephrine, glucagon, and calcium.

Epinephrine has its major effect in muscle, whereas glucagon primarily affects liver. Neither of these hormones enters the cell. Instead, they both act through specific membrane receptors that produce the intracellular second messenger **3′,5′-cAMP.** The catalytic subunit of cAMP-dependent protein kinase reacts with **phosphorylase kinase,** activating it by the addition of a phosphate group. This enzyme, in turn, phosphorylates **phosphorylase b,** producing **phosphorylase a,** the active enzyme for glycogen breakdown. Because the entire sequence of reactions is a cascade, a small amount of hormone can produce a large change in the amount of phosphorylated or dephosphorylated enzyme. ATP can bind to an allosteric site

on phosphorylase b and keep it in an inactive state. AMP can compete with ATP for this binding site and prevent the inhibition of phosphorylase b.

In contrast to phosphorylase, **glycogen synthase** is most active in the dephosphorylated state. Phosphorylation by phosphorylase kinase converts glycogen synthase from the active I form to the less active D form. The activity of the D (or dependent) form of glycogen synthase totally depends on the presence of high levels of glucose 6-phosphate. The I (or independent) form of glycogen synthase is active in the absence of glucose 6-phosphate. Because the synthetic and degradative enzymes are controlled in the opposite way by phosphorylation, a single hormonal stimulus can promote glycogen breakdown and prevent glycogen synthesis, or the reverse.

Phosphorylase phosphatase converts active phosphorylase a into inactive phosphorylase b. The rate of this reaction is enhanced in the presence of high levels of glucose. At the same time, phosphorylase phosphatase activates glycogen synthase by converting the inactive phosphorylated D form to the active nonphosphorylated I form.

In muscle, activation of phosphorylase kinase can occur also by a calcium-dependent protein kinase. The kinase phosphorylates inactive phosphorylase b, producing active phosphorylase a. This, in turn, catalyzes glycogen breakdown. Thus, the **calcium** released in muscle contraction can stimulate glycogen breakdown and provide the metabolic fuel necessary to continue the contractions.

Glycogen Storage Diseases

There are eight major types of defects in glycogen metabolism that produce profound clinical problems. These rare disorders are listed in Table 17-1.

In **von Gierke's disease,** glycogen cannot be metabolized to free glucose due to the absence of **glucose 6-phosphatase.** Under these conditions, phosphorylated glucose must be either catabolized or stored as glycogen. The liver and kidney become filled with glycogen, and glucose levels in the circu-

lation fall. Hyperlipidemia and hyperlactic acidemia also occur in von Gierke's disease.

The defective enzymes in **Pompe's disease** are the lysosomal α-1,4 and α-1,6 glucosidases. In the absence of these enzymes, the lysosomes and the cells become filled with undegraded glycogen. Death of affected individuals occurs in infancy or early childhood.

Type III glycogen storage disease (**Cori's disease**) is due to the absence of the debranching enzyme required for complete degradation of glycogen. Without this enzyme, glycogen can be degraded only up to the branch points. A lack of the branching enzyme produces type IV glycogen storage disease or **Andersen's disease,** in which only a linear, less dense glycogen product lacking the α-1,6 branches is produced. This altered glycogen takes up much more space in the liver than does normal glycogen containing the same number of glucose residues.

The fifth glycogen storage disease, also known as **McArdle's disease,** is due to the absence of muscle but not liver phosphorylase. Muscle contains a high concentration of glycogen but, because it cannot be broken down to glucose, there is a low tolerance for sustained exercise. The liver phosphorylase is normal, and so there is no hypoglycemia. A defect in liver phosphorylase produces type VI glycogen storage disease or **Hers' disease.** This syndrome has symptoms similar to those of type I glycogen storage disease, but they are much less severe.

Table 17-1. Glycogen storage diseases

Disease	Enzyme defect	Symptoms
Type I (von Gierke's)	Glucose 6-phosphatase	Increased liver glycogen, liver enlargement, hypoglycemia
Type II (Pompe's)	Lysosomal α-1,4 and α-1,6-glucosidase	Increased liver glycogen, cardiac and respiratory failure, early death
Type III (Cori's)	Debranching enzyme	Glycogen with short branches, liver enlargement
Type IV (Andersen's)	Branching enzyme	Glycogen with long outer branches, liver cirrhosis, early death
Type V (McArdle's)	Muscle phosphorylase	Increased muscle glycogen, exercise intolerance
Type VI (Hers')	Liver phosphorylase	Increased liver glycogen
Type VII	Muscle phosphofructokinase	Exercise intolerance
Type VIII	Liver phosphofructokinase	Liver enlargement, hypoglycemia

Types VII and VIII glycogen storage disease are due to defects in muscle (type VII) and liver (type VIII) phosphofructokinase. The muscle defect produces exercise intolerance. The liver defect causes liver enlargement and hypoglycemia.

Synthesis and Structure of Glycoproteins and Proteoglycans

Polysaccharides other than glycogen can be found as free molecules or covalently attached to proteins. **Glycoproteins** consist of proteins to which less than 100 saccharide residues have been added. Included in the glycoproteins are the blood-clotting factors, immunoglobulins, and hormone receptors of the plasma membrane. The carbohydrate of glycoproteins is first attached to a nucleotide carrier and then added one residue at a time while the protein is in the Golgi complex. The enzymes responsible for the synthesis of the carbohydrate portion of the molecule, **glycosyl transferases,** are specific for particular carbohydrates. The carbohydrate is attached to the protein through an O-glycosidic linkage to serine, threonine, hydroxylysine, or hydroxyproline or an N-glycosidic linkage to asparagine.

Proteoglycans or **mucopolysaccharides** contain a protein chain with a repeating polysaccharide unit. The repeating unit consists of either N-acetylglucosamine or N-acetylgalactosamine and glucuronic or iduronic acid. Modification by the attachment of sulfate or removal of the acetyl group of some of the carbohydrate occurs after the structure has been completely synthesized. The sulfate donor for proteoglycan synthesis is **3-phosphoadenosine 5-phosphosulfate** (PAPS).

Proteoglycans are found extracellularly between the collagen and elastin fibers in connective tissue and on the surface of cells. If the protein portion of the proteoglycan is removed, the remaining carbohydrate is called a **glycosaminoglycan.** Glycosaminoglycans are synthesized while attached to protein and then are cleaved from it. The different classes of proteoglycans and glycosaminoglycans are differentiated by the repeating disaccharide units. The major classes are summarized in Table 17-2.

Hyaluronic acid is found in synovial fluid and in the vitreous humor of the eye. Its repeating unit is glucuronic acid and N-acetylglucosamine in a β-1,4 linkage. The carbohydrate residues are added sequentially from UDP carriers. Unlike the other polysaccharides, hyaluronic acid has no sulfate. In addition, there is some question as to whether it is first synthesized on a protein primer.

Chrondroitin sulfate, dermatan sulfate, heparan sulfate, and heparin are all synthesized while joined to a protein. A xylose residue is attached first to a serine of the protein. Then two galactose residues are added before the alternating carbohydrate structure is synthesized. In **chondroitin sulfate,** the repeating residues are glucuronic acid and N-acetylgalactosamine. Sulfate is found predominantly on the 4 position of the N-acetylgalactosamine in chondroitin 4-sulfate and on the 6 position in chondroitin 6-sulfate. Both of these are found in cartilage, tendon, and ligaments.

Dermatan sulfate is synthesized similarly except that, after synthesis, many of the glucuronic acid residues are epimerized at the 5 position to form iduronic acid. In addition, sulfate groups are added to the two hydroxyl groups of the iduronic acid residues in dermatan sulfate.

Heparan sulfate is modified after synthesis by deacylation of the amine and the addition of sulfate at N-2 and O-6. In **heparin,** as in dermatan sulfate, the glucuronic acid is epimerized to iduronic acid and sulfate added at position 2. Although heparan sulfate and heparin have similar structures, only heparin demonstrates anticoagulant activity. Heparin is found inside mast cells. Heparan sulfate is found on the surface of most cells, in blood vessel walls, and in brain tissue.

Keratan sulfate has a repeating unit of galactose and N-acetylglucosamine. Sulfate occurs on C-6 of the amine residue. Keratan sulfate, found in the cornea, is linked to protein by an N-glycosidic bond.

Table 17-2. Major classes of proteoglycans and glycosaminoglycans

Hyaluronic Acid

β-glucuronic acid   N-acetyl–β-glucosamine

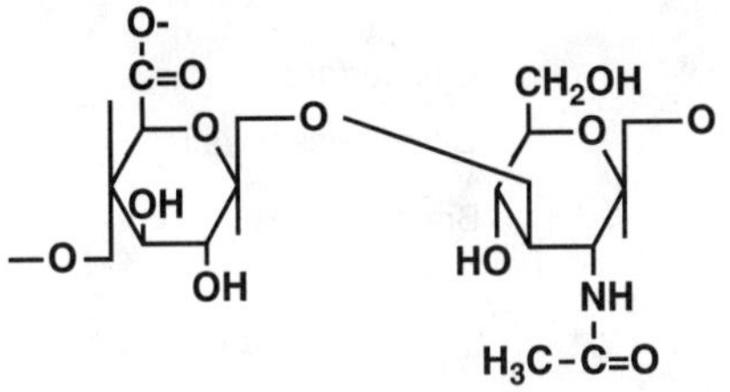

Cell surface

Synovial fluid

Vitrous humor

Chrondroitin Sulfate

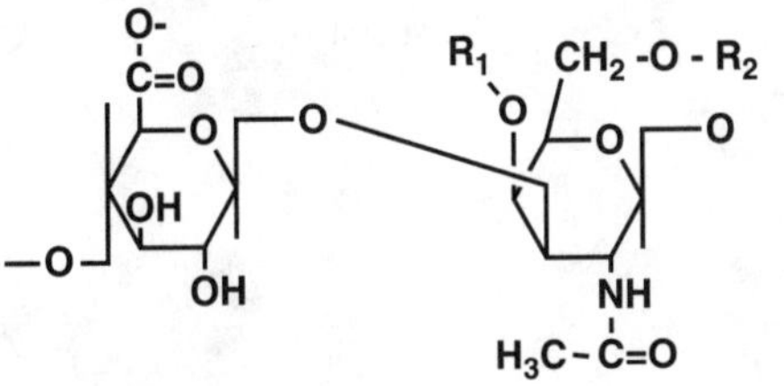

β-glucuronic acid   N-acetyl–β-galactosamine

Cartilage

Tendon

Bone

$R_1 = SO_3$, $R_2 = H$ for chondroitin 4 sulfate

$R_1 = H$, $R_2 = SO_3$ for chondroitin 6 sulfate

Keratin Sulfate

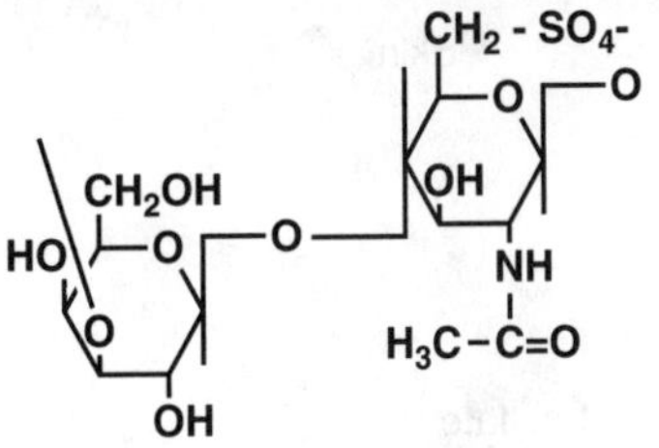

β-galactose   N-acetyl–α-glucosamine
6-sulfate

Cornea

Cartilage

In cartilage, it is linked by an O-glycosidic bond to a serine of a different protein.

Defects in Proteoglycan Metabolism

The carbohydrate portion of proteoglycans is degraded one saccharide unit at a time by enzymes specific for both the saccharide and the particular bond. Defects in these enzymes can produce an accumulation of the affected compound and profound, sometimes lethal, effects. With the exception of iduronate sulfatase, whose gene is on the X chromosome, the genes coding for these enzymes are autosomal and the syndromes are inherited as autosomal recessives. The most common of the defects are summarized in Table 17-3.

Table 17-2 (continued).

Heparan sulfate

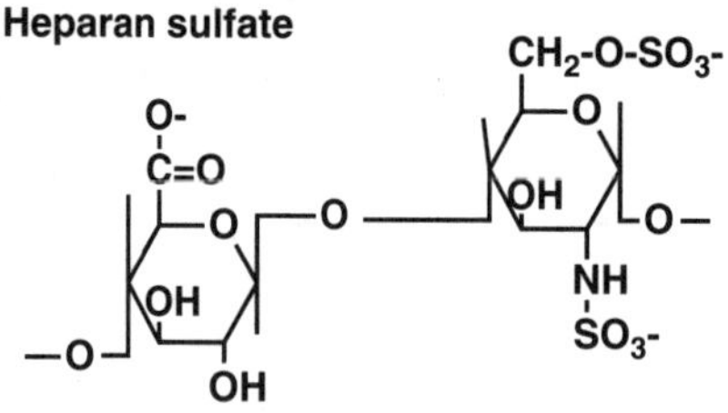

β-glucuronic acid   β-glucosamine N,6-sulfate

Cell surface

Blood vessel wall

Brain

Heparin

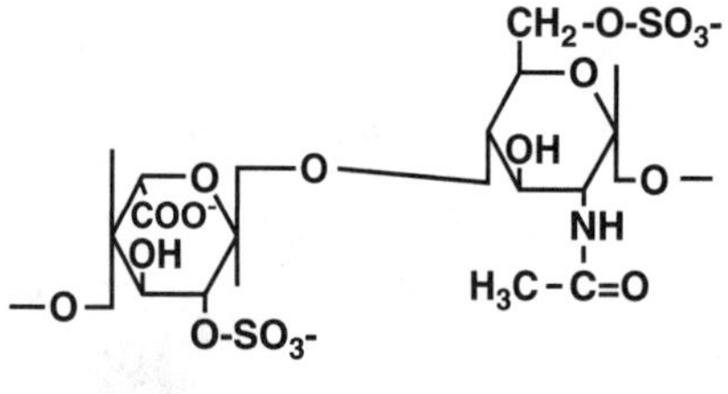

β-iduronic acid 2-sulfate   N-acetyl–α-glucosamine 6-sulfate

Mast cell

Dermatan sulfate

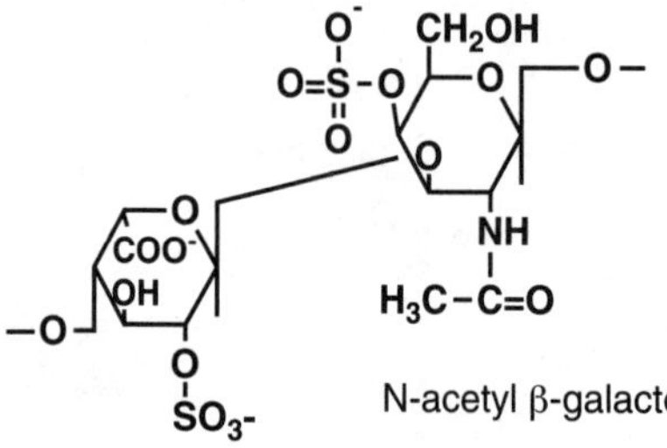

N-acetyl β-galactosamine 4-sulfate

β-iduronic acid 2-sulfate

In **Hurler's** and **Scheie's syndromes,** iduronidase is low or absent. Hurler's syndrome produces an accumulation of dermatan and heparan sulfates in the urine, tissues, and amniotic fluid. This results in mental retardation, skeletal deformities, and corneal opacity. The milder Scheie's syndrome shows only dermatan sulfate accumulation in the urine and corneal opacity as the major clinical symptoms.

Hunter's syndrome is characterized by deafness, mental retardation, and skeletal abnormalities. It is due to a lack of iduronic acid sulfatase and to the resulting accumulation of dermatan and heparan sulfates.

Defects in the ability to remove the sulfate attached to position 4 or 6 of N-acetylgalactosamine produces **Maroteaux-Lamy** and **Morquio syndromes.** In both one sees skeletal deformities. In ad-

Table 17-3. Inherited defects in proteoglycan metabolism

Syndrome	Enzyme defect	Symptoms
Hurler's	Iduronidase	Mental retardation, skeletal deformity, corneal opacity
Hunter's	Iduronate sulfatase	Mental retardation, skeletal deformity, deafness
Scheie's	Iduronidase	Corneal opacity
Maroteaux-Lamy	N-acetylgalactosamine 4-sulfatase	Skeletal deformity, corneal opacity
Morquio	N-acetylgalactosamine 6-sulfatase	Skeletal deformity
Sanfilippo's		
A	N-acetylglucosaminidase	Mental retardation, skeletal deformity
B	Sulfamidase	
C	Acetyl transferase	

dition, Maroteaux-Lamy syndrome involves corneal opacity.

There are three types of **Sanfilippo's syndrome,** and in all there are mental retardation and skeletal abnormalities. Type A is due to a lack of N-acetyl-glucosaminidase, type B to a lack of sulfamidase, and type C to a lack of acetyl transferase.

Summary

Carbohydrates are partially broken down by salivary α-amylase, and the breakdown is continued by pancreatic α-amylase. The products of these enzymes are limit dextrans and two- and three-unit oligosaccharides. These compounds are hydrolyzed to monosaccharides by enzymes on the intestinal mucosal membrane and then are absorbed into the circulation. Lack of the enzymes required to hydrolyze lactose or sucrose produces intolerance for these sugars, with associated bloating and diarrhea.

Glycogen is an α-1,4 glucose polymer with α-1,6 branches. It is found in abundance in liver, kidney, and muscle. The synthesis of glycogen begins with the phosphorylation of glucose to glucose 6-phosphate. This then is isomerized to glucose 1-phosphate, activated by reaction with UTP, and added to a glycogen primer. Branches are formed by a branching enzyme that transfers a segment of six glucose residues to the 6 position of another glucose of the polymer.

In degradation, phosphorylase splits the α-1,4 linkage, releasing glucose 1-phosphate. When four residues remain on a branch, a transferase moves three glucose residues to another branch. Then, a debranching enzyme splits the α-1,6 bond, releasing the final glucose as free glucose. The glucose 1-phosphate can be isomerized to glucose 6-phosphate and either enter glycolysis or, in liver, be hydrolyzed to free glucose by the enzyme glucose 6-phosphatase.

Insulin increases glycogen synthesis. Epinephrine and glucagon have the opposite effect. These latter two hormones act through a cAMP cascade that phosphorylates phosphorylase kinase, which in turn phosphorylates the glycogen-degrading enzyme phosphorylase. Phosphorylase is active in the phosphorylated state (a form) and inactive in the dephosphorylated state (b form). In contrast, glycogen synthase is active in the dephosphorylated I form and inactive in the phosphorylated D form. Defects in the enzymes responsible for glycogen synthesis and degradation result in various syndromes, the symptoms of which vary from exercise intolerance to early death.

In glycoproteins, the carbohydrate is attached to a protein at serine, threonine, or asparagine. Proteoglycans contain a protein core and a repeating unit of carbohydrate. The various classes are differentiated by the repeating unit (usually an N-acetyl-glucosamine or N-acetylgalactosamine and a uronic acid) and by the number and location of sulfate groups. Proteoglycans, the protein-free glycosaminoglycans, are found in cartilage, skin, and cell membranes. Defects in the degradation of these compounds often produce mental retardation and skeletal deformities.

Part IV Questions: Metabolism of Carbohydrates

1. The rate-controlling reaction for glycolysis is
   A. glucokinase.
   B. inhibited by AMP.
   C. activated by citrate.
   D. phosphofructokinase.
   E. pyruvate kinase.

2. The net energy produced when fructose 6-phosphate is converted to lactate by the glycolysis pathway is
   A. 0.
   B. 1.
   C. 2.
   D. 3.
   E. 6.

3. The following structure would be classified as

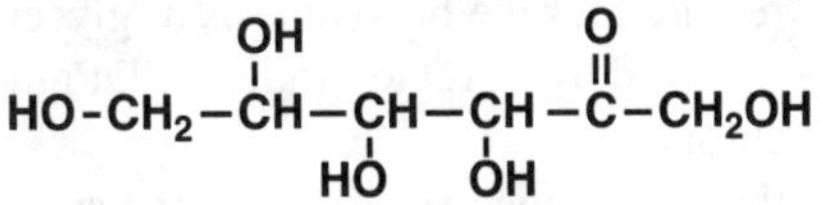

$$HO-CH_2-\overset{\overset{\displaystyle OH}{|}}{CH}-\overset{\overset{\displaystyle}{|}}{\underset{\underset{\displaystyle HO}{|}}{CH}}-\overset{\overset{\displaystyle}{|}}{\underset{\underset{\displaystyle OH}{|}}{CH}}-\overset{\overset{\displaystyle O}{\|}}{C}-CH_2OH$$

   A. a pentose.
   B. a structure with four asymmetrical carbons.
   C. a ketose.
   D. a pyranose.
   E. an intermediate of glycolysis.

4. In the formation of 2,3-diphosphoglycerate during glycolysis,
   A. two ATPs are formed.
   B. a substrate-level phosphorylation is lost.
   C. NADH is produced.
   D. ATP is hydrolyzed.
   E. a high-energy phosphate group is produced.

5. Intolerance to galactose can be caused by a lack of
   A. galactokinase.
   B. glucokinase.
   C. phosphofructokinase.
   D. pyruvate kinase.
   E. enolase.

6. Which Krebs cycle intermediate accumulates in the presence of malonate?
   A. Oxaloacetate
   B. Malate
   C. Succinate
   D. Fumarate
   E. Citrate

7. Excess ATP inhibits which of the Krebs cycle enzymes?
   A. Oxaloacetate dehydrogenase
   B. Succinyl-CoA synthetase
   C. Isocitrate dehydrogenase
   D. Pyruvate dehydrogenase
   E. Fumarase

8. The Krebs cycle has a large negative ΔG because
   A. it produces GTP.
   B. hydrogens are transferred to NAD.
   C. most of the individual reactions have a negative ΔG.
   D. products are being continually removed.
   E. several reactions of the cycle are reversible.

9. Conversion of α-ketoglutarate to succinate involves all of the following *except*
   A. thiamine pyrophosphate.
   B. lipoamide.
   C. coenzyme A.
   D. NAD.
   E. biotin.

10. The rate-controlling reactions for the Krebs cycle include
    A. pyruvate dehydrogenase.
    B. aconitase.
    C. succinate dehydrogenase.
    D. malate dehydrogenase.
    E. fumarase.

11. How many moles of ATP are produced when an electron pair is passed from complex III to oxygen in coupled oxidative phosphorylation?
    A. 0
    B. 1
    C. 2
    D. 4
    E. 6

12. In oxidative phosphorylation, electrons are transferred to coenzyme Q directly from
    A. NADP.
    B. complex II.
    C. FADH$_2$.
    D. cytochrome oxidase.
    E. cytochrome c reductase.

13. The evidence that the chemiosmotic hypothesis of Mitchell may account for oxidative phosphorylation is that
    A. all the intermediates postulated by the hypothesis have been isolated.
    B. structural changes have been observed in mitochondria during respiration.
    C. there is a pH gradient across the mitochondria during oxidative phosphorylation.
    D. mitochondria contain an ATPase.
    E. mitochondria are freely permeable to positive ions.

14. 2,4-Dinitrophenol increases body temperature because there is
    A. an increase in both oxygen uptake and ATP formation.
    B. a decrease in both oxygen uptake and ATP formation.
    C. an increase in ATP formation but no change in oxygen uptake.
    D. a decrease in ATP formation but no change in oxygen uptake.
    E. a decrease in ATP formation and an increase in oxygen uptake.

15. Reduced NADH generated in the cytoplasm is able to pass its electrons to oxygen through the mitochondrial electron transport system by which of the following mechanisms?
    A. A specific transport system enables NADH to pass through the mitochondrial membrane.
    B. NADH reduces FAD to FADH$_2$, which then passes through the mitochondrial membrane.
    C. NADH reduces dihydroxyacetone phosphate to glycerol phosphate, which can pass through the mitochondrial membrane.
    D. NADH reduces membrane-bound NADP to NADPH, which passes electrons to coenzyme Q.
    E. NADH oxidizes aspartate to oxaloacetate, which can enter the mitochondria.

16. Which of the following enzymes is unique to the gluconeogenesis pathway?
    A. Pyruvate kinase
    B. Enolase
    C. Malate dehydrogenase
    D. Pyruvate carboxylase
    E. Glucokinase

17. UDPglucose is
    A. a major intermediate in the pentose phosphate pathway.
    B. directly required for the production of lactose.
    C. required for the biosynthesis of glycerol.
    D. oxidized by NAD to produce UDPglucuronate.
    E. the direct precursor to the amino sugars.

18. N-acetylneuraminic acid (NANA)
    A. requires phosphoenolpyruvate for its biosynthesis.
    B. is synthesized from galactosamine 6-phosphate.
    C. is a seven-carbon amino sugar.
    D. is an intermediate in the uronic acid pathway.
    E. is produced by the action of transketolase.

19. Gluconeogenesis is
    A. activated by high concentrations of acetyl-CoA.
    B. inhibited by ATP.
    C. activated by glucagon in the kidney.
    D. predominantly a mitochondrial pathway.
    E. stimulated by insulin in the liver.

20. Essential pentosuria
    A. is due to a defect in the synthesis of amino sugars.

B. produces hypoglycemia.

C. is caused by an inability to convert xylulose to xylitol.

D. causes fragility of red blood cells.

E. prevents the synthesis of ribose 5-phosphate for nucleotide biosynthesis.

21. The branch points in glycogen have which type of linkage?

A. α-1,4

B. β-1,4

C. α-2,6

D. α-1,6

E. β-1,6

22. An abnormal glycogen structure can be produced by

A. a lack of debranching enzyme.

B. a lack of liver phosphorylase.

C. von Gierke's disease.

D. Hers' disease.

E. a lack of glucokinase.

23. In glycogen breakdown in the liver,

A. glucose is attached first to a nucleotide carrier.

B. glucose from branch points is released as free glucose.

C. a high-energy phosphate bond is produced.

D. the α-1,4 bonds are split by phosphorylase kinase.

E. most of the glucose is released as glucose 6-phosphate.

24. All of the following are true about proteoglycans *except*

A. the carbohydrate can be attached to a protein by O-glycosidic bonds to serine or threonine residues of a protein primer.

B. is sulfated by PAPS after the carbohydrate has been synthesized.

C. usually includes repeating units of a uronic acid and an N-acetylamino sugar.

D. the carbohydrate for heparin is iduronic acid and N-acetylglucosamine sulfate.

E. if the carbohydrate is cleaved, the remaining protein is called a glycosaminoglycan.

25. As opposed to children, adults would be more likely to have difficulty digesting

A. casein.

B. lactose.

C. sucrose.

D. galactose.

E. fructose.

Part
V Lipid Metabolism

18 Lipid Biosynthesis

Lipids make up approximately 15% of the body. They are predominantly hydrocarbon and have little affinity for water. Included in lipids are fatty acids, triglycerides, steroids, prostaglandins, and the fat-soluble vitamins. In addition to being an important fuel store, lipids are also important for the structure of membranes, to transport fat-soluble vitamins, as carriers of oligosaccharides, and as hormones. Most fatty acid synthesis occurs in the cytoplasm of the liver. Then the fatty acids are released to the circulation and taken up by other tissues. Adipose tissue synthesizes triglycerides from glycerol and fatty acids. Phospholipids, necessary for membrane biosynthesis, are produced by most tissues. The biosynthesis of two other kinds of lipids, steroids and the eicosanoids, will be discussed in Chapter 20.

Objectives

After completing this chapter, you should be able to

Discuss the major chemical characteristics of fatty acids, triglycerides, phospholipids, and sphingolipids.

Diagram the synthesis of fatty acids starting from mitochondrial acetyl-CoA.

Write the reaction for elongation and desaturation of palmitoyl-CoA.

List the essential fatty acids.

Outline the biosynthesis of triglycerides and phospholipids.

Identify the structures of sphingolipids and plasmalogens.

Structure of Lipids

The simplest of the lipids are the fatty acids. Most fatty acids are straight-chain compounds containing an even number of carbon atoms. They are composed entirely of hydrocarbons except for the polar acid group at one end. Because one end of the molecule is polar and the other is not, fatty acids are said to be **amphipathic.**

If there are no double bonds, the fatty acid is saturated. If there are double bonds, they usually are in the *cis* configuration (the hydrogen atoms are on the same side of the double bond). The position of double bonds can be designated in two ways. The symbol $\Delta 9$ indicates that a double bond is located between carbons 9 and 10 when the carbonyl carbon is counted as 1. With the ω designation, counting begins from the methyl end of the chain. A ω-9 fatty acid has a double bond between the ninth and tenth carbon atoms counting from the methyl end. Examples of both types of nomenclature are shown here.

$$CH_3-(CH_2)_4-CH=CH-(CH_2)_3-COO^-$$

Δ^5 or ω-6 fatty acid

Fatty acids usually are found esterified through the carboxyl carbon to various alcohols such as glycerol, ethanolamine, or choline. Glycerol plus three fatty acids produces a **triglyceride,** the main form of stored fuel for humans. The combination of glycerol, two fatty acids, and phosphoric acid produces **phospholipids.** If there is no other group attached to the phosphoric acid, the compound is a **phosphatidic acid.** The addition of another small hydroxyl-containing molecule, such as choline, serine, or ethanolamine, creates more complex phospholipids. Phospholipids that contain a long-chain fatty alcohol rather than a fatty acid at carbon 1 of glycerol are **plasmalogens.** The **sphingolipids** are composed of the amino alcohol sphingosine plus carbohydrate. They are important constituents of the membranes of the central and peripheral nerves. Examples of structures of these types of lipids are given in Figure 18-1.

Fig. 18-1. Structures of some of the important cellular lipids.

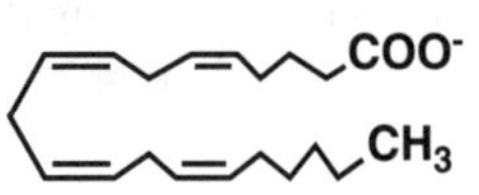

arachidonic acid

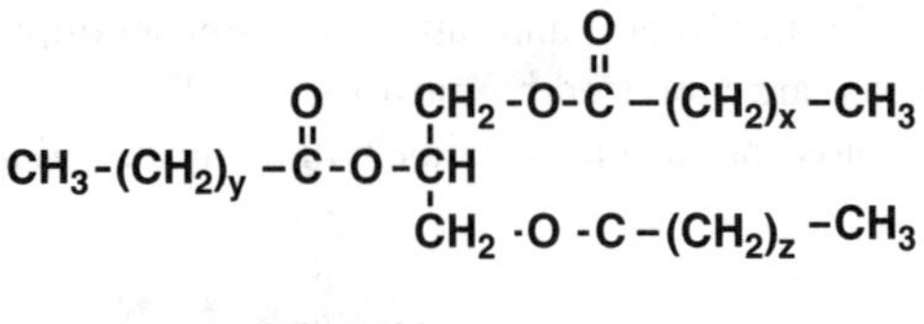

triglyceride

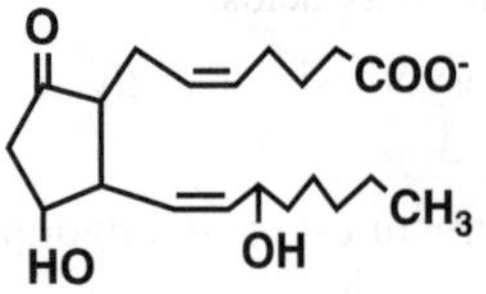

prostaglandin E_2

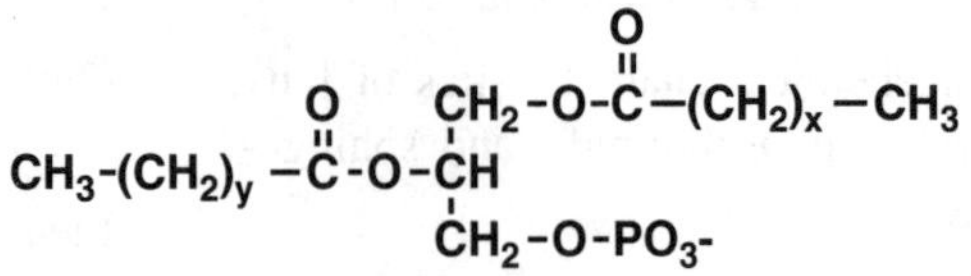

phospholipid

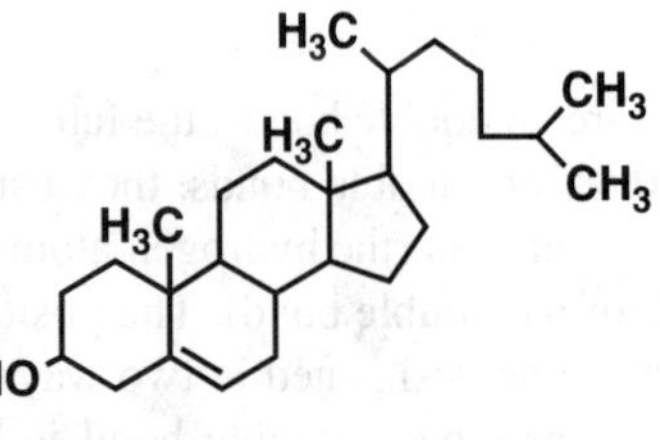

cholesterol

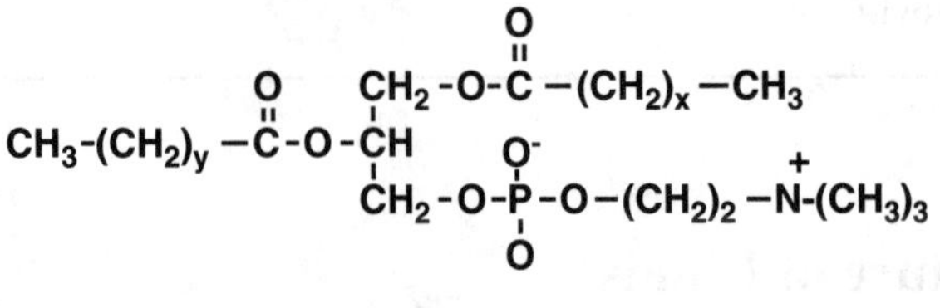

phosphatidylcholine

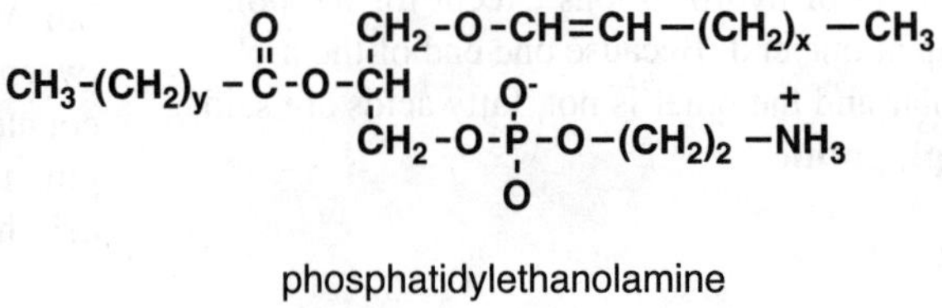

phosphatidylethanolamine

The steroids are hydrocarbons containing four fused rings. The important steroids for humans are all derived from cholesterol, which is found primarily as part of the plasma membrane. Other steroids of physiological significance include the mineralocorticoids, glucocorticoids, sex hormones, bile acids, and vitamin D.

Biosynthesis of Fatty Acids

Production of Cytoplasmic Acetyl-CoA

The enzymes for the initial biosynthesis of fatty acids occur in the cytoplasm of liver and adipose tissue. Because the reaction sequence begins with acetyl-CoA, it must first be transported out of the mitochondria using the sequence of reactions given in Figure 18-2. The **citrate shuttle** transport system uses the first reaction of the Krebs cycle, the condensation of oxaloacetate with acetyl-CoA, to produce citrate. The citrate then leaves the mitochondria. Once in the cytoplasm, citrate can be cleaved by **citrate lyase** to produce oxaloacetate and acetyl-CoA. This reaction uses energy from the hydrolysis of ATP, and so it is irreversible in the cell. The transport cycle is completed by conversion of oxaloacetate to malate and then malate to pyruvate. Pyruvate then re-enters the mitochondria and is converted

into oxaloacetate by carboxylation. This reaction also requires the energy from hydrolysis of ATP. The total energy requirement for converting mitochondrial acetyl-CoA into cytoplasmic acetyl-CoA using the citrate shuttle is two ATPs.

Synthesis of Malonyl-CoA

The first reaction committed to fatty acid biosynthesis is the conversion of acetyl-CoA into malonyl-CoA by **acetyl-CoA carboxylase.** This rate-limiting step of fatty acid biosynthesis is outlined in Figure 18-3. First, carbon dioxide is covalently bound to biotin using the energy from the hydrolysis of ATP. Then, using the same enzyme, the carbon dioxide is transferred to acetyl-CoA, producing **malonyl-CoA.**

The rate of fatty acid biosynthesis is controlled both by the activity and the amount of acetyl-CoA carboxylase. The enzyme is activated allosterically by high concentrations of citrate and is inactivated by long-chain fatty acids, especially palmitic acid. Citrate, a positive allosteric effector, controls enzyme activity by promoting polymerization of the enzyme. The negative effectors act in the opposite manner by promoting depolymerization of the enzyme. Activity of acetyl-CoA carboxylase also is controlled by phosphorylation, which decreases the enzymatic activity, and dephosphorylation, which

Fig. 18-2. Citrate shuttle for transferring mitochondrial acetyl-CoA into the cytoplasm.

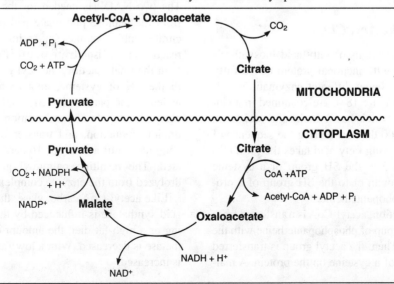

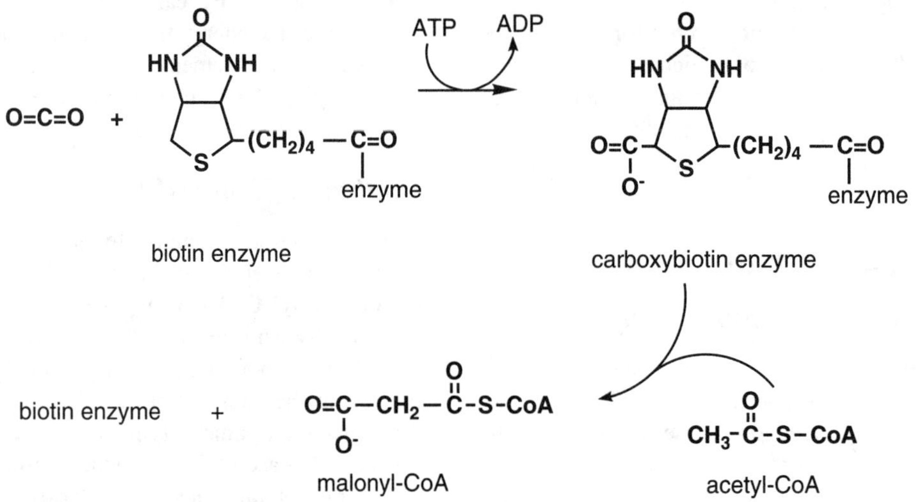

Fig. 18-3. Biosynthesis of malonyl-CoA from acetyl-CoA.

increases it. Phosphorylation can be catalyzed by either a cAMP-dependent or a calcium-dependent protein kinase.

Changes in diet affect the amount of fatty acid biosynthesis by affecting the amount of acetyl-CoA carboxylase. A diet rich in carbohydrate or low in fat increases the biosynthesis of the enzyme by affecting the rate of transcription. Starvation or a diet high in fat has the opposite effect and reduces the rate of synthesis of acetyl-CoA carboxylase.

Synthesis of the Acyl Chain

The rest of the reactions of fatty acid biosynthesis take place on a multifunctional protein called **fatty acid synthetase.** Seven different enzymatic activities, outlined in Figure 18-4, are contained in a single polypeptide, which has a molecular weight in humans of 250,000 daltons. During the sequence of reactions, the growing fatty acid takes the form of a thio ester attached to the SH group of a cysteine residue of the protein or to the SH group of a protein-bound phosphopantetheine.

In the first reaction, acetyl-CoA is transferred from CoA to the SH group of phosphopantetheine with the release of CoA. Then, the acetyl group is transferred to the SH group of a cysteine on the protein. A mal-

onyl group from malonyl-CoA is added to the pantetheine SH group just vacated by the acetyl group.

In the next step, the acetyl group and the malonyl groups are condensed, with the release of carbon dioxide. This forms acetoacetate attached to the pantetheine SH group. Using NADPH, the ketone is reduced to a hydroxyl group. Then the compound is dehydrated and the double bond is reduced by NADPH. This series of reactions forms a four-carbon acyl group still attached to the phosphopantetheine. The two NADPHs needed for the reductions come from the pentose phosphate pathway and from the citrate shuttle. Another source for NADPH for these reactions is the **isocitrate shuttle** (Fig. 18-5).

In the final reaction, the acyl group is transferred to the SH of cysteine, another malonyl group is added to the pantetheine SH, and the cycle begins again. This cycle of condensation, reduction, dehydration, reduction, and transfer of the acyl group continues until a chain of 16 carbons has been created. The resulting palmitoyl group then is hydrolyzed from the enzyme complex.

Like acetyl-CoA carboxylase, the amount of fatty acid synthetase is influenced by the diet. With fasting or a high-fat diet, the amount of fatty acid synthetase is decreased. With a low-fat diet, the amount is increased.

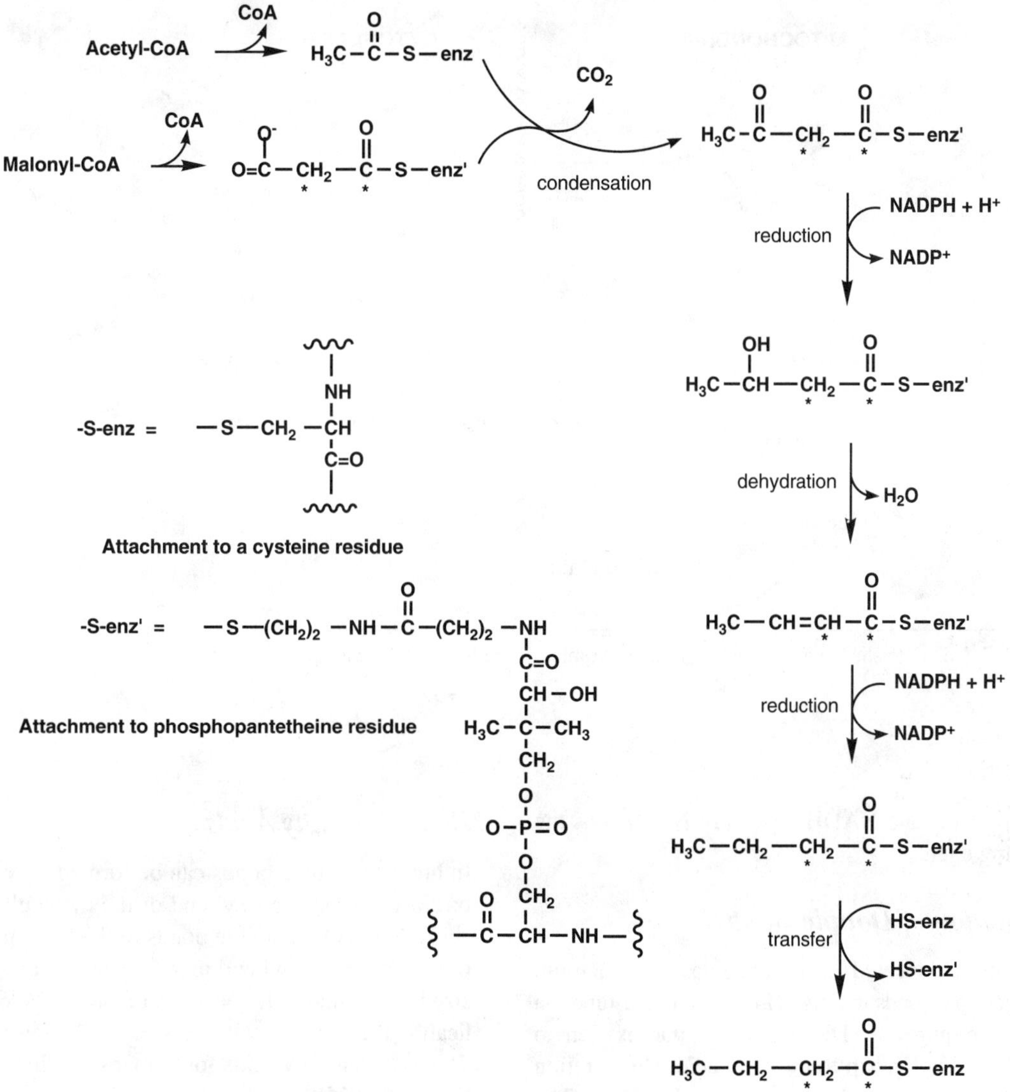

Fig. 18-4. Reaction catalyzed by fatty acid synthetase. The carbons contributed by malonyl-CoA are indicated by asterisks. The release of the completed 16-carbon fatty acid is catalyzed by a hydrolase function of the same fatty acid synthetase enzyme.

Modification of Fatty Acids

Elongation

Elongation of fatty acids to produce chains longer than 16 carbons takes place in the endoplasmic reticulum and in the mitochondria. In the endoplasmic reticulum, the reactions are similar to those of fatty

acid synthesis. Malonyl-CoA is condensed with palmitoyl-CoA. The product is reduced, dehydrated, and then reduced again. In most tissues, the major fatty acid from the elongation process is stearoyl-CoA (18 carbons). In the brain, elongation can yield fatty acids with up to 24 carbons. The elongation process in mitochondria uses acetyl-CoA instead of

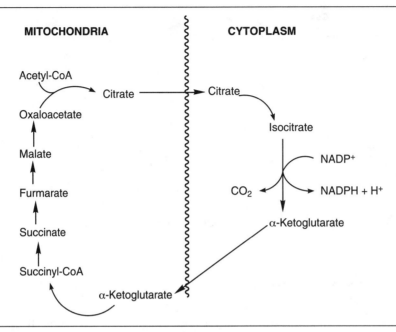

Fig. 18-5. Isocitrate shuttle for the production of cytoplasmic reducing equivalents.

malonyl-CoA and NADH instead of NADPH as the reducing agent.

Formation of Double Bonds

A desaturase located in the endoplasmic reticulum forms double bonds in fatty acids. This mixed-function oxidase requires NADH and molecular oxygen to add a hydroxyl group to the fatty acid. Dehydration follows to form a nonconjugated double bond. The major products are palmitoleic and oleic acids.

$$CH_3 - (CH_2)_5 - CH=CH - (CH_2)_7 - COO^-$$

palmitoleic acid

$$CH_3 - (CH_2)_7 - CH=CH - (CH_2)_7 - COO^-$$

oleic acid

Essential Fatty Acids

In humans, double bonds can be formed only at C-9 or closer to the carboxyl end of the molecule. Fatty acids that contain double bonds in the *trans* position or nearer the methyl end than C-9 cannot be synthesized by humans. If they are required for optimal health, they are classified as essential fatty acids. The essential fatty acids for humans are **linoleic** and **linolenic acids.**

$$H_3C - (CH_2)_4 - CH=CH - CH_2 - CH=CH - (CH_2)_7 - COO^-$$

linoleic acid

$$H_3C - CH_2 - CH=CH - CH_2 - CH=CH - CH_2 - CH=CH - (CH_2)_7 - COO^-$$

linolenic acid

These fatty acids are necessary for normal membrane structure. In addition, linoleic acid serves as the precursor for arachidonic acid, from which the eicosanoids are formed. Arachidonic acid is pro-

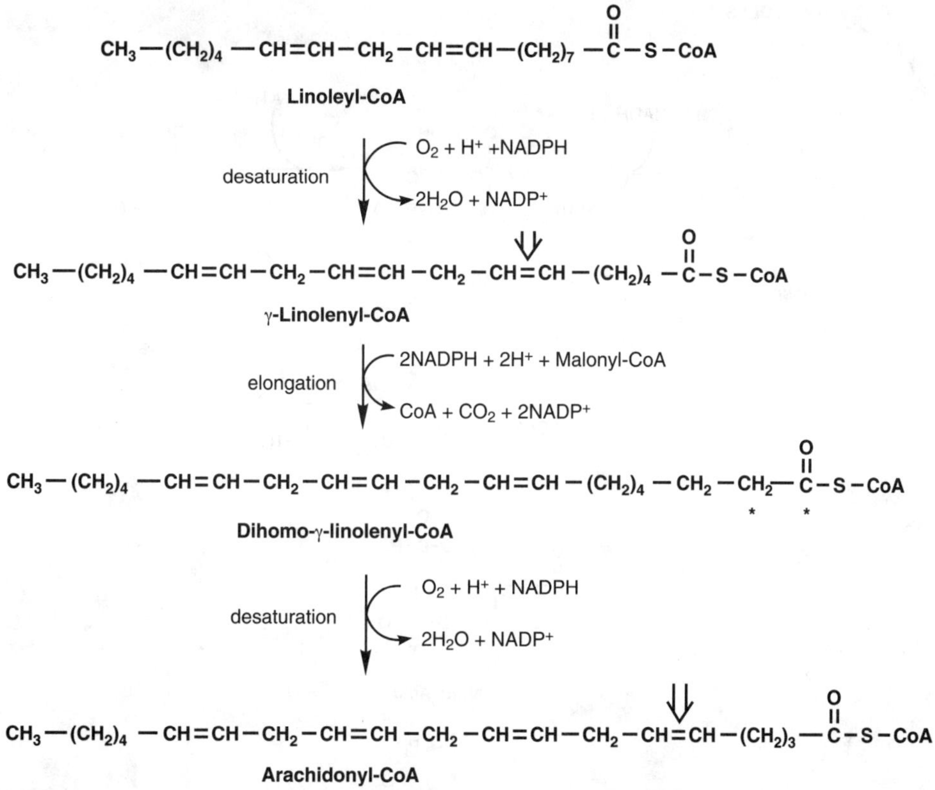

$$CH_3-(CH_2)_4-CH=CH-CH_2-CH=CH-(CH_2)_7-\overset{\overset{\displaystyle O}{\|}}{C}-S-CoA$$

Linoleyl-CoA

desaturation

$O_2 + H^+ + NADPH$

$2H_2O + NADP^+$

$$CH_3-(CH_2)_4-CH=CH-CH_2-CH=CH-CH_2-CH=CH-(CH_2)_4-\overset{\overset{\displaystyle O}{\|}}{C}-S-CoA$$

γ-Linolenyl-CoA

elongation

$2NADPH + 2H^+ + Malonyl\text{-}CoA$

$CoA + CO_2 + 2NADP^+$

$$CH_3-(CH_2)_4-CH=CH-CH_2-CH=CH-CH_2-CH=CH-(CH_2)_4-CH_2-CH_2-\overset{\overset{\displaystyle O}{\|}}{C}-S-CoA$$

Dihomo-γ-linolenyl-CoA

desaturation

$O_2 + H^+ + NADPH$

$2H_2O + NADP^+$

$$CH_3-(CH_2)_4-CH=CH-CH_2-CH=CH-CH_2-CH=CH-CH_2-CH=CH-(CH_2)_3-\overset{\overset{\displaystyle O}{\|}}{C}-S-CoA$$

Arachidonyl-CoA

Fig. 18-6. Conversion of linoleic acid to arachidonic acid by a combination of desaturation and elongation. Double bonds formed in the desaturation reactions are marked with an arrow. The carbons donated by malonyl-CoA in the elongation reaction are indicated by asterisks.

duced by elongation and the addition of two double bonds, as shown in Figure 18-6.

Synthesis of Triglycerides

Biosynthesis of triglycerides, the storage form of fats in the body, begins with glycerol 3-phosphate, as shown in Figure 18-7. Glycerol 3-phosphate can come from a reduction of dihydroxyacetone phosphate from glycolysis or from direct phosphorylation of glycerol by ATP. This latter reaction cannot take place in adipose tissue because it lacks the necessary enzyme, **glycerol kinase.**

In the endoplasmic reticulum, acyl transferases add fatty acids to the 1 and 2 positions of glycerol 3-

phosphate, thus producing a phosphatidic acid. Next, the phosphate at position 3 is removed, yielding 1,2-diacylglycerol. Finally, a third fatty acid is added, producing a triglyceride. All the transfers of fatty acids come from CoA derivatives. Free fatty acids can enter triglycerides only by being attached first to CoA and then being transferred to glycerol.

Synthesis of Phospholipids

Phosphatidic acid (glycerol plus two fatty acids plus phosphate) is the starting point for the biosynthesis of phospholipids, a major constituent of membranes. The addition of a hydroxyl-containing compound, such as choline or ethanolamine, to the phosphate

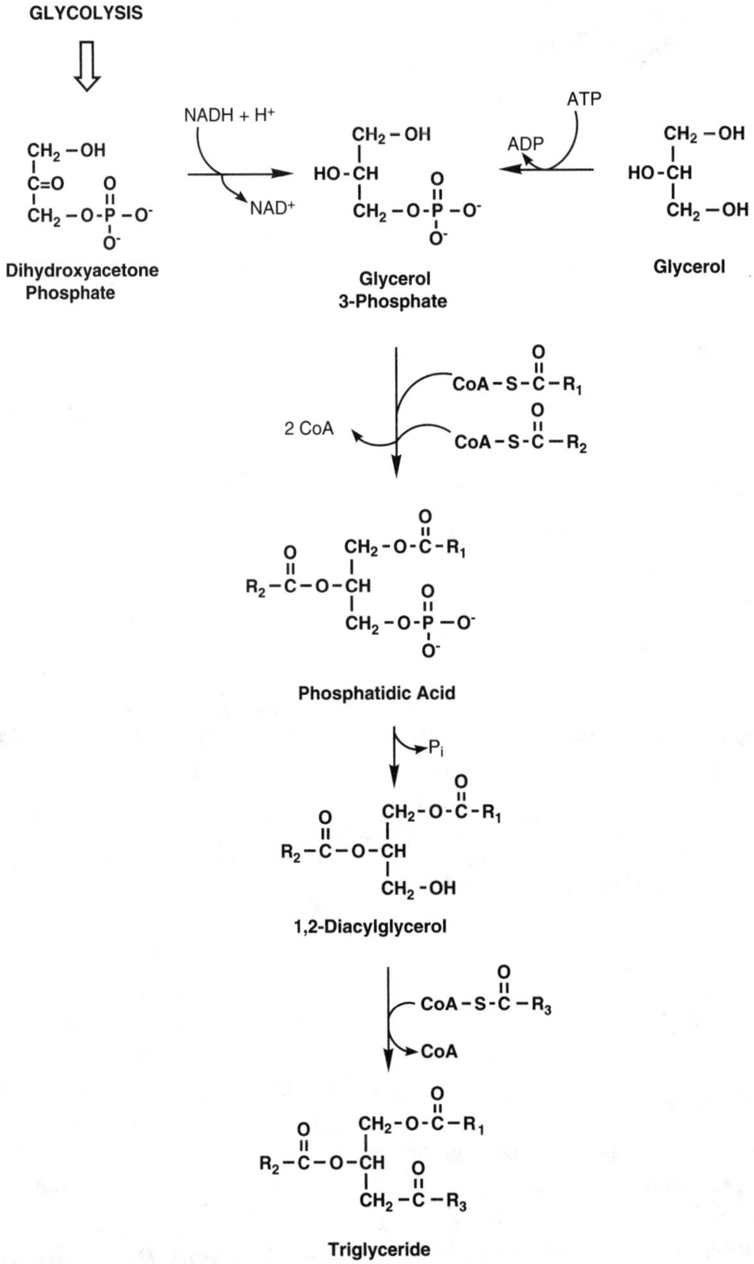

GLYCOLYSIS

Fig. 18-7. Biosynthesis of triglycerides.

group on a phosphatidic acid can occur by activation of either the compound to be added or the phosphatidic acid. In both cases, activation occurs by reaction with cytidine triphosphate (CTP).

For the synthesis of **phosphatidylcholine** or **lecithin,** phosphatidic acid reacts with CTP, producing cytidine diphosphate (CDP)-diacylglycerol. Condensation with choline with the loss of cytidine monophosphate (CMP) yields phosphatidylcholine, as shown in Figure 18-8. Alternatively, choline can

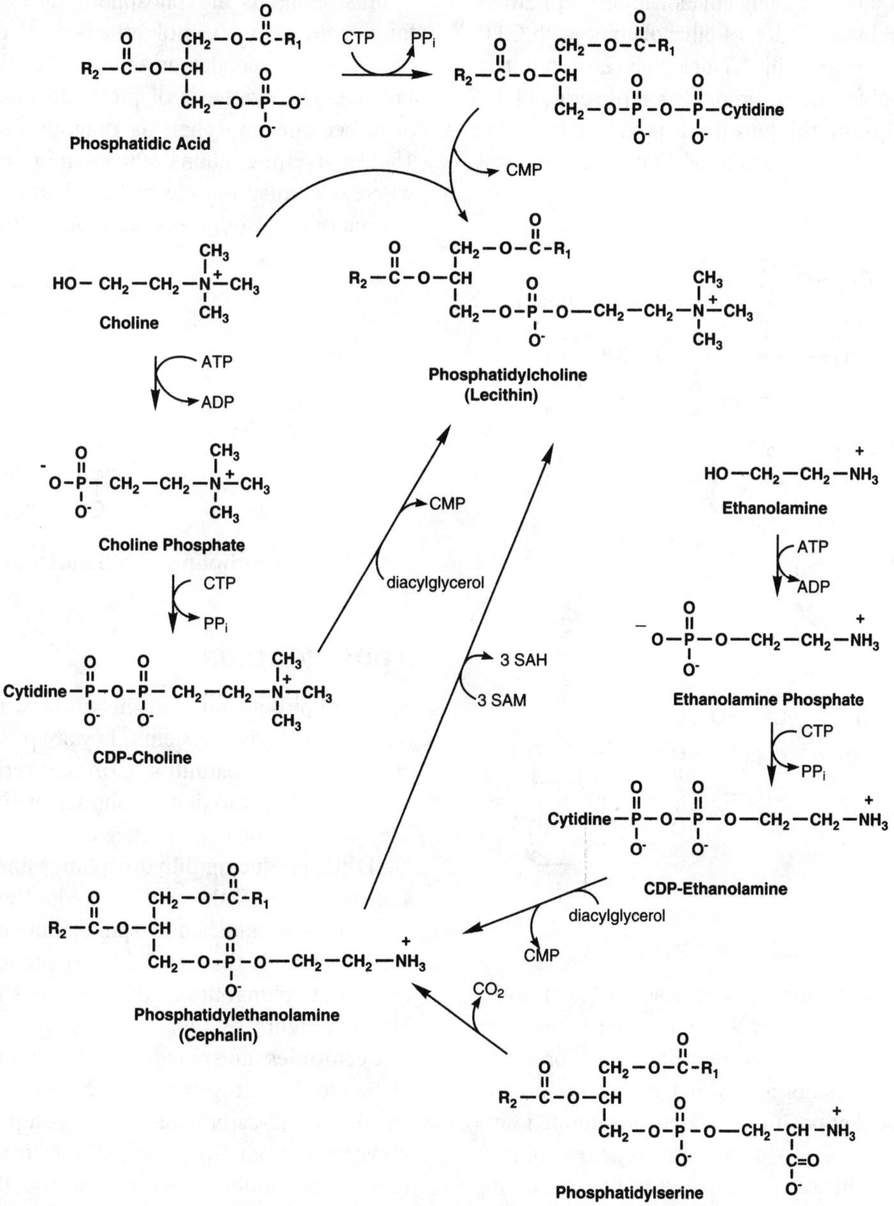

Fig. 18-8. Biosynthesis and interconversion of phospholipids (*SAH* = S-adenosyl homocysteine.)

be phosphorylated by ATP and then activated by reaction with CTP. The resulting CDP-choline can react with 1,2-diacylglycerol to form phosphatidylcholine. In a third mechanism, phosphatidylcholine is produced by repeated methylation of phosphatidylethanolamine by **S-adenosylmethionine.** Extensive synthesis of phosphatidylcholine occurs in the lung shortly before birth. Measurement of its level indicates lung maturity and the ability to utilize oxygen from the air.

Phospholipids containing ethanolamine (**cephalins**) are produced by activation of ethanolamine with CTP followed by reaction with 1,2-diacylglycerol. A minor pathway involves the decarboxylation of serine phospholipids. **Phosphatidylinositol** is produced by a single mechanism: the reaction of CDP-diacylglycerol with inositol.

CDP-diacylglycerol

phosphatidyl inositol

The fatty acids originally attached to the 1 and 2 positions of the glycerol are not necessarily the same as those found in the final phospholipid. For example, phosphatidylcholine normally contains a saturated fatty acid at position 1 and an unsaturated fatty acid at position 2. To permit an exchange of fatty acids, phospholipases must first remove the existing fatty acid. Then a new fatty acid can be added by direct acylation with a CoA derivative of the fatty acid or by exchange with another phospholipid, as shown in Figure 18-9.

The addition of phosphatidic acid groups to the 1 and 3 positions of glycerol produces **cardiolipin** (Fig. 18-10). This phospholipid is a major constituent of the inner mitochondrial membrane.

Plasmalogens are phospholipids that contain a long-chain fatty alcohol attached to carbon 1 of glycerol. This creates an ether rather than an ester linkage. The synthesis of plasmalogens from dihydroxyacetone phosphate is diagrammed in Figure 18-11. Myelin contains ethanolamine plasmalogen, whereas cardiac muscle has predominantly choline plasmalogens. Both types are found in the mitochondrial membranes.

R$_3$ = choline or ethanolamine

Sphingolipids

Sphingolipids are components of the white matter of the central nervous system. They are produced by the condensation of palmitoyl-CoA and serine, with the loss of carbon dioxide as shown in Figure 18-12. The ketone then is reduced to a hydroxyl by NADPH, producing dihydrosphingosine. This is the start point for all the sphingolipids. Remember that the fatty acid attached to sphingosine can vary. All the compounds discussed in this section constitute classes of sphingolipids, the structures of which are given in Figure 18-13.

Ceramides are sphingolipids with a fatty acid attached to the nitrogen at carbon 2. This fatty acid is usually a 22-carbon saturated compound and is transferred from CoA. In addition, ceramides contain a *trans* double bond in what was the fatty acid portion of the molecule. Transfer of galactose from UDPgalactose to ceramide produces galactosyl ceramide or **cerebroside**. The addition of a sulfate to position 3 of the galactose of a cerebroside produces a **sulfatide. Gangliosides** are ceramides containing a polysaccharide chain of glucose, galactose, and sialic acid (e.g., NANA) attached to carbon 1. **Sphingomyelin** is the only sphingolipid that contains a

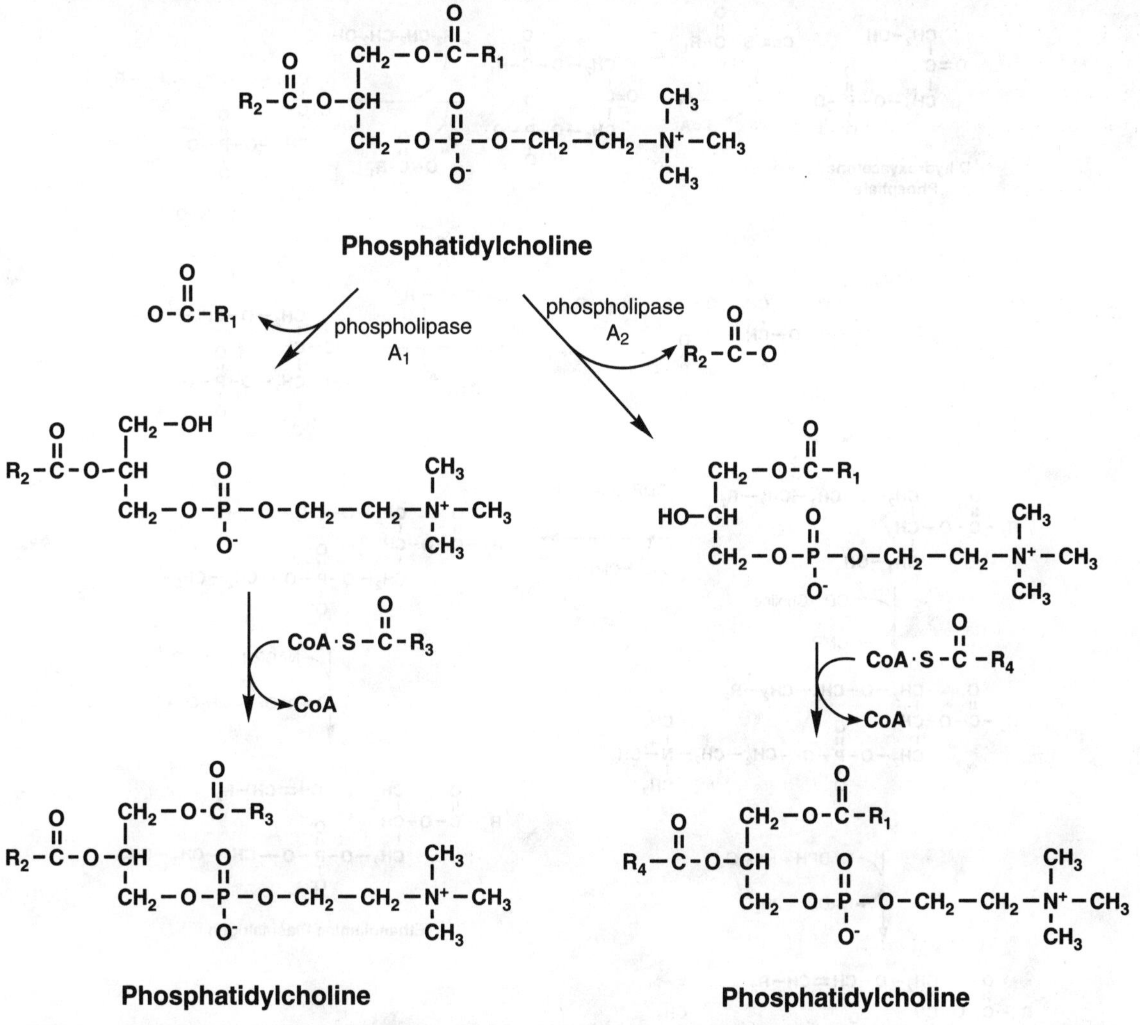

Phosphatidylcholine

phospholipase A₁

phospholipase A₂

CoA·S−C−R₃

CoA

CoA·S−C−R₄

CoA

Phosphatidylcholine

Phosphatidylcholine

Fig. 18-9. Exchange of acyl groups by action of phospholipases and either reacylation from fatty acid CoA or exchange with another phospholipid.

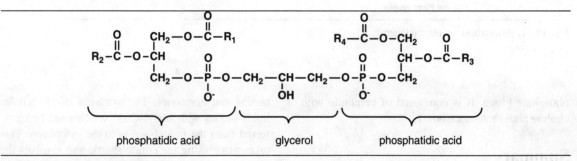

phosphatidic acid   glycerol   phosphatidic acid

Fig. 18-10. Structure of cardiolipin.

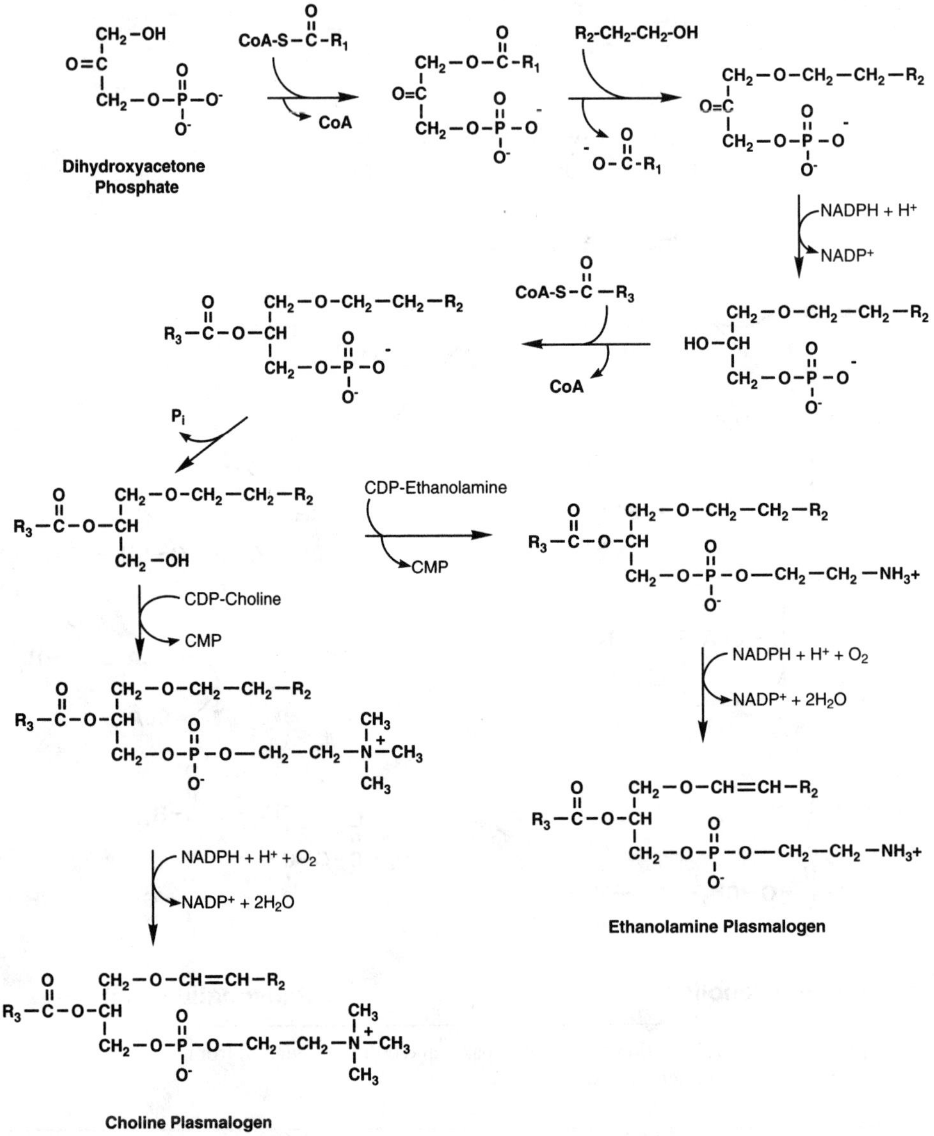

Fig. 18-11. Biosynthesis of plasmalogens.

phosphate group. It is composed of ceramide with choline phosphate at carbon 1.

Summary

Lipids contain primarily hydrocarbon and are important as cellular fuel, in membrane structure, and as vitamins and hormones. The synthesis of all cellular lipids begins with acetyl-CoA, which must be transferred from the mitochondria to the cytoplasm. This takes place using the citrate shuttle and involves the expenditure of two ATPs. For the synthesis of fatty acids, acetyl-CoA is carboxylated to malonyl-CoA in the rate-limiting reaction of fatty acid biosynthesis.

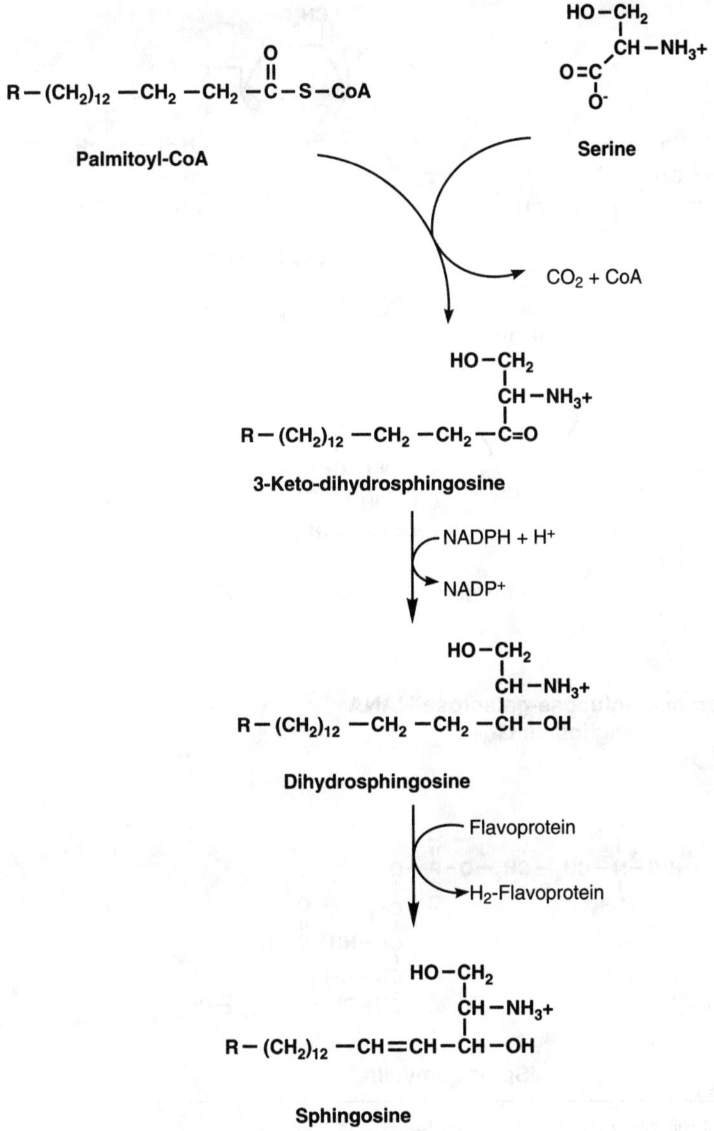

R—(CH$_2$)$_{12}$—CH$_2$—CH$_2$—C=O $\overset{O}{\underset{||}{}}$—S—CoA

Palmitoyl-CoA

HO—CH$_2$
CH—NH$_3$+
O=C—C
O$^-$

Serine

CO$_2$ + CoA

HO—CH$_2$
CH—NH$_3$+
R—(CH$_2$)$_{12}$—CH$_2$—CH$_2$—C=O

3-Keto-dihydrosphingosine

NADPH + H$^+$

NADP$^+$

HO—CH$_2$
CH—NH$_3$+
R—(CH$_2$)$_{12}$—CH$_2$—CH$_2$—CH—OH

Dihydrosphingosine

Flavoprotein

H$_2$-Flavoprotein

HO—CH$_2$
CH—NH$_3$+
R—(CH$_2$)$_{12}$—CH=CH—CH—OH

Sphingosine

Fig. 18-12. Synthesis of sphingosine from palmitoyl-CoA and serine.

Fatty acid synthetase catalyzes the remaining seven reactions required to produce palmitic acid. Malonyl-CoA and acetyl-CoA are transferred to SH groups and then condensed. The product is reduced, dehydrated, and reduced again. The cycle continues until a chain of 16 carbons is made. The reducing equivalents for fatty acid biosynthesis come from the pentose phosphate pathway and the citrate and isocitrate shuttles.

Fatty acids can be modified by elongation, producing fatty acids with up to 24 carbons. The modification involves a mechanism similar to that of fatty acid biosynthesis. Desaturation requires a mixed-function oxidase and yields nonconjugated *cis* double bonds nine carbons or closer to the carboxyl end of the molecule. Linoleic and linolenic acids, which have double bonds nearer the methyl end, cannot be

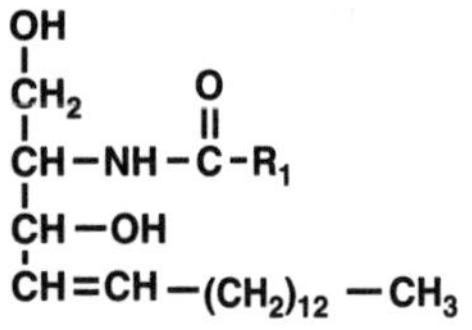

Ceramide

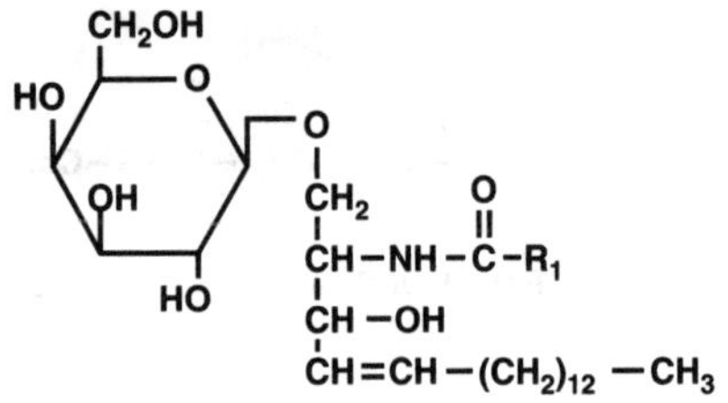

Cerebroside

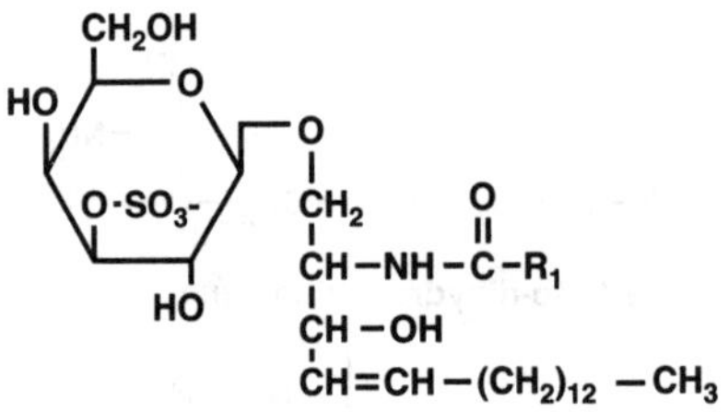

Sulfatide

Ceramide-glucose-galactose-NANA
Ganglioside G_{M3}

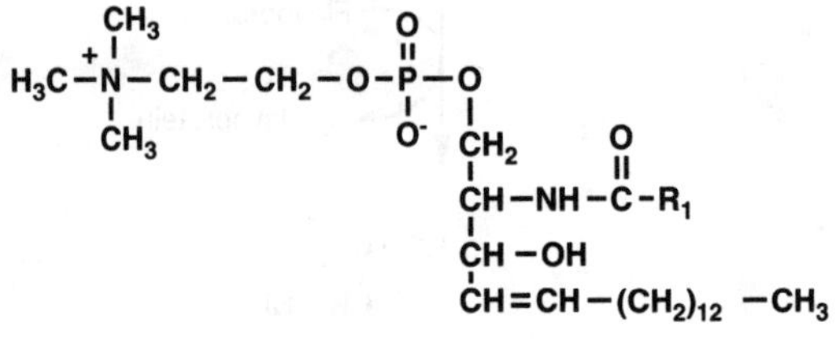

Sphingomyelin

Fig. 18-13. Structures of the major classes of sphingolipids.

formed by humans but are essential fatty acids necessary for normal membrane structure.

Triglycerides are formed by acylation of glycerol 3-phosphate, removal of the phosphate, and acylation of the final hydroxyl group. All the fatty acids are transferred from coenzyme A.

Phospholipids contain 1,2-diacylglycerol phosphate attached with an ester bond to choline, ethanolamine, serine, or inositol. They are formed from CDP carriers of the glycerol- or the hydroxyl-containing portions of the molecule. Phosphatidylcholine can also be produced by methylation of phosphatidylethanolamine. Cardiolipin is a phospholipid consisting of two phosphatidic acids attached to the 1 and 3 positions of glycerol. Plasmalogens contain choline or ethanolamine and an acyl glycerol phosphate with a fatty alcohol at position 1.

The condensation of serine and palmitoyl-CoA begins the pathway to the sphingolipids. This class of lipids contains the ceramides, cerebrosides, sulfatides, gangliosides, and sphingomyelin.

19 Lipid Degradation

Lipids are released from triglycerides or phospholipids whenever there is a need for energy. Because they can be converted into acetyl-CoA with the hydrolysis of only two ATP molecules, fatty acids, not carbohydrates, are the major energy source for the body when at rest. In response to hormones, fatty acids are released from adipose tissue into the circulation, where they travel to other tissues for oxidation as metabolic fuels. After conversion to acetyl-CoA by the β-oxidation pathway, saturated straight-chain fatty acids can be further oxidized by the Krebs cycle or converted into ketone bodies. Additional enzymes are required for the degradation of fatty acids containing branched chains, unsaturation (double bonds), or an odd number of carbon atoms. Failure to degrade lipids properly can produce serious clinical symptoms.

Objectives

After completing this chapter, you should be able to

List the contents of the circulating lipoprotein particles.

Diagram the steps of β-oxidation of fatty acids from release from triglycerides through the production of acetyl-CoA.

Calculate the energy yield from the complete oxidation of a fatty acid.

Describe the additional steps required for the oxidation of fatty acids containing an odd number of carbon atoms, unsaturation, or branches. Explain the effect this has on the energy yield.

Name the ketone bodies and outline their production.

Match the defects in sphingolipid degradation with the defective enzymatic step.

Digestion of Lipids

Digestion of ingested lipids begins in the intestine. Here, pancreatic lipase hydrolyzes fatty acids from the 1 and 3 positions of triglycerides. The intestine also contains an esterase that hydrolyzes cholesterol esters, producing more free fatty acids. Still more come from the action of pancreatic phospholipases, which release fatty acids from phosphatidylcholine (lecithin). Because fatty acids are not soluble in water, they are emulsified by combination with bile salts, lecithin, and monoglycerides. The results are micelle particles 200 to 500 μm in diameter. In these micelles, the hydrophilic hydroxyl and amino groups are on the outside of the particle and the hydrophobic fatty acids on the inside. As micelles, lipids are absorbed into the brush border of the intestinal mucosal cells.

From here, fatty acids of 12 carbons or fewer are bound to albumin and transported directly into the portal circulation. Fatty acids that contain more than 12 carbons are re-esterified to glycerol or cholesterol. They enter the lymphatic circulation combined with lipoproteins and phospholipids in a 100- to 1000-μm particle called a *chylomicron.*

Chylomicrons contain 90% triglyceride and only 1 to 2% protein. They serve as the means for transporting triglycerides to the tissues, especially liver and adipose tissue. While in the particle, lipoprotein lipase hydrolyzes the triglycerides to glycerol and fatty acids. After entering adipose cells, these compounds are re-esterified and stored.

A smaller, 30- to 50-μm particle, the **very low-density lipoprotein** (VLDL), serves to transport triglycerides from the liver to other tissues. VLDL can be broken down to **low-density lipoproteins** (LDLs), which are 20 to 25 μm in diameter. These particles contain 10% triglyceride, 30% phospholipids, 20% protein, and 40% cholesterol and cholesterol esters. The function of the LDLs is to transport cholesterol and phospholipids to tissues to be used in the synthesis of membranes.

High-density lipoproteins (HDLs), which measure only 8 to 10 μm, are produced by both the liver and the intestine. These particles contain 45% protein, 5% triglyceride, 30% phospholipid, and 20% cholesterol. A high concentration of HDLs is associated with resistance to atherosclerosis. The makeup of the various lipoprotein particles is summarized in Table 19-1.

Several clinical syndromes are associated with defects in the construction or metabolism of the lipoprotein particles. In **Tangier disease,** no HDLs are formed because of a lack of α-lipoprotein, one of the proteins in the particle. This produces a hypolipidemia with abnormal tissue deposition of cholesterol and its esters. Lack of β-lipoprotein, one of the proteins of chylomicrons, VLDLs, and LDLs, results in no lipid absorption from the intestine. Fats remain in the bowel, and fatty stools result. Because of a lack of lipid absorption, the lipid-soluble vitamins also remain unabsorbed. This syndrome is known as **hypobetalipoproteinemia** or **Bassen-Kornzweig syndrome.** An overproduction or improper metabolism of the lipoproteins can produce a variety of hyperlipoproteinemia syndromes. In type IIa, also called **familial hypercholesterolemia,** there is a defect in the surface receptors for LDLs in the liver and a resulting increase in the level of serum cholesterol. This defect produces a very high risk for atherosclerosis in childhood. In type V, the defect has been localized to a lack of lipoprotein lipase needed to hydrolyze the circulating triglycerides in the chylomicrons.

Degradation of Triglycerides

To be oxidized, fatty acids must first be released from storage as triglycerides in adipose tissue, as seen in Figure 19-1. This release is a hormonally sensitive event. In response to norepinephrine, epinephrine,

Table 19-1. Characteristics of the circulating lipoprotein particles

| Class | Composition (%) | | | | Function |
	Triglyceride	Cholesterol and esters	Phospholipid	Protein	
Chylomicron	90	2	6	2	Transport triglycerides from intestine to adipose tissue
VLDL	55	18	20	7	Transport triglycerides from liver to extrahepatic tissue
LDL	10	40	30	20	Transport cholesterol and phospholipids to tissues
HDL	5	20	30	45	Transport cholesterol from tissues to liver

VLDL = very low-density lipoprotein; LDL = low-density lipoprotein; HDL = high-density lipoprotein.

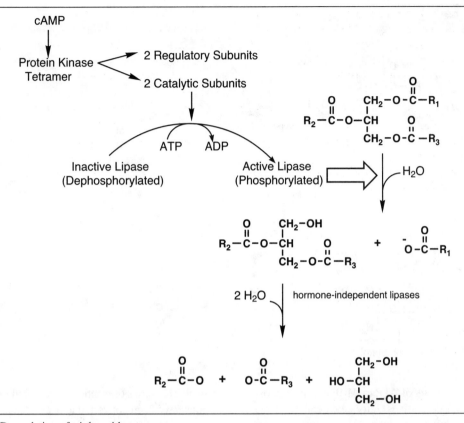

Fig. 19-1. Degradation of triglycerides.

adrenocorticotropic hormone (ACTH), or glucagon, the cAMP level of the adipose tissue rises. This produces the cAMP cascade discussed earlier. The cAMP-dependent protein kinase phosphorylates a hormone-sensitive lipase, which then removes the first fatty acid from a triglyceride. The other fatty acids are hydrolyzed by hormonally insensitive lipases. The free fatty acids leave the adipose tissue and enter the circulation bound to albumin. Glycerol also is released to the blood for oxidation by other tissues.

Degradation of Phospholipids

Phospholipids can be degraded into their constituent parts by a series of four enzymes, as shown in Figure 19-2. Phospholipase A_1 removes fatty acids attached to carbon 1 of glycerol. Similarly, phospholipase A_2 hydrolyzes the fatty acid at carbon 2. Phospholipase

C breaks the bond between glycerol and the phosphate. Phospholipase D removes the ethanolamine, choline, serine, or inositol from a phospholipid, leaving phosphatidic acid. After action by all four phospholipases, the fatty acids can be oxidized by pathways that will be discussed next. The phosphate, choline, serine, and inositol groups are excreted or reused. Glycerol is phosphorylated and either reused in triglyceride biosynthesis or oxidized via the glycolysis pathway.

Fatty Acid Oxidation

Fatty acids are oxidized by the β-oxidation pathway, which is located in the mitochondria. In the first step of the oxidation pathway, fatty acids are converted into CoA derivatives by an enzyme located on the outer mitochondrial membrane. This reaction re-

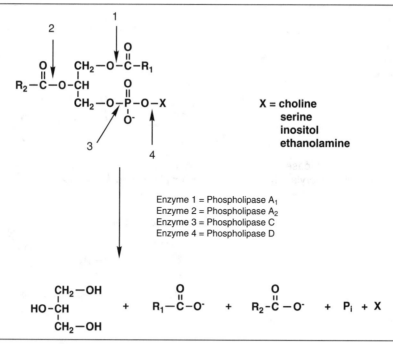

Enzyme 1 = Phospholipase A_1
Enzyme 2 = Phospholipase A_2
Enzyme 3 = Phospholipase C
Enzyme 4 = Phospholipase D

Fig. 19-2. Degradation of phospholipids.

quires the hydrolysis of ATP to AMP and pyrophosphate. As CoA derivatives, fatty acids can enter the mitochondria but cannot penetrate the inner membrane. To do so, they must be transferred to the hydroxyl group of another carrier molecule, **carnitine.**

$$CH_3 \overset{+}{\underset{CH_3}{\overset{CH_3}{\underset{|}{\overset{|}{N}}}}}-CH_2-\underset{OH}{\underset{|}{CH}}-CH_2-COO^-$$

Two enzymes catalyze the transfer of fatty acids to carnitine. One is specific for long-chain fatty acids and the other for short-chain acids. Once inside the inner mitochondrial membrane, the fatty acids are transferred back to CoA. Because β-oxidation occurs inside the mitochondria, the reducing equivalents that are produced can enter the electron transport chain directly without the need for a shuttle system.

The reactions of the β-oxidation pathway are given in Figure 19-3. First, the fatty acid is dehydrogenated using FAD as the hydrogen acceptor. Next, the double bond is hydrated, giving a β-hydroxy compound. Reaction with NAD produces a ketone. Finally, two carbons are split off in a thiolase reaction with CoA. This leaves a fatty acyl-CoA that is two carbons shorter than the

original. This series of reactions is repeated until all the carbons have been converted to acetyl-CoA.

The major differences between fatty acid biosynthesis and degradation are the cellular location and the identity of the hydrogen acceptor. Biosynthesis occurs in the cytoplasm and uses NADPH. Degradation occurs in the mitochondria and uses NAD and FAD. Degradation requires the function of several different enzyme proteins. In contrast, synthesis is catalyzed by a single multifunctional protein.

Energy Yield

Initial activation of the fatty acid (reaction with CoA) requires the expenditure of two ATP equivalents of energy. Each reduction by FAD will give two ATPs when the electrons pass through the electron transport chain. Similarly, each NADH produces three ATPs. When the acetyl-CoA units are oxidized by the Krebs cycle, they yield an additional 12 ATPs for each acetyl-CoA oxidized. The net energy produced is 17 ATPs for every two-carbon unit, except for the last one, which yields only 12 ATPs. This is because the thiolase cleavage of acetoacetyl-

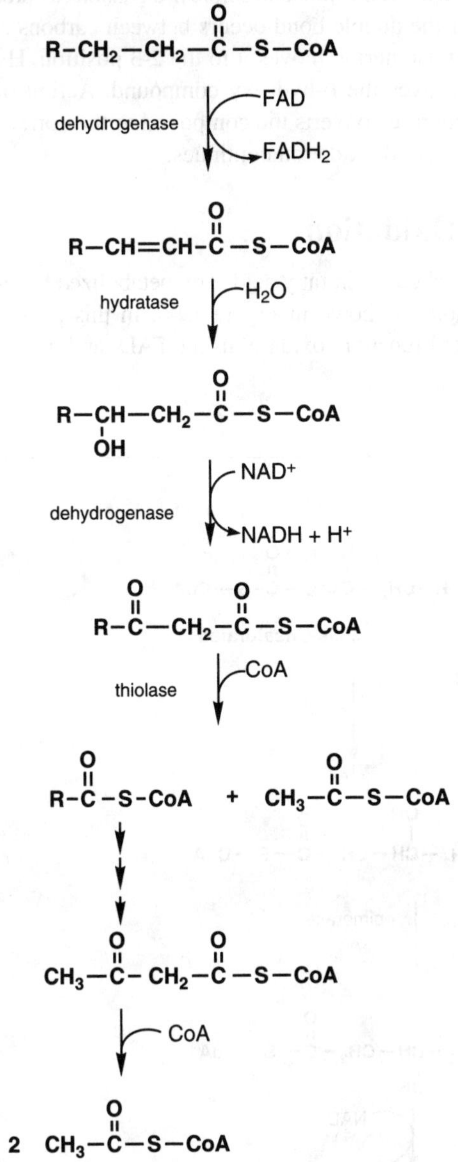

Fig. 19-3. Reactions of β-oxidation.

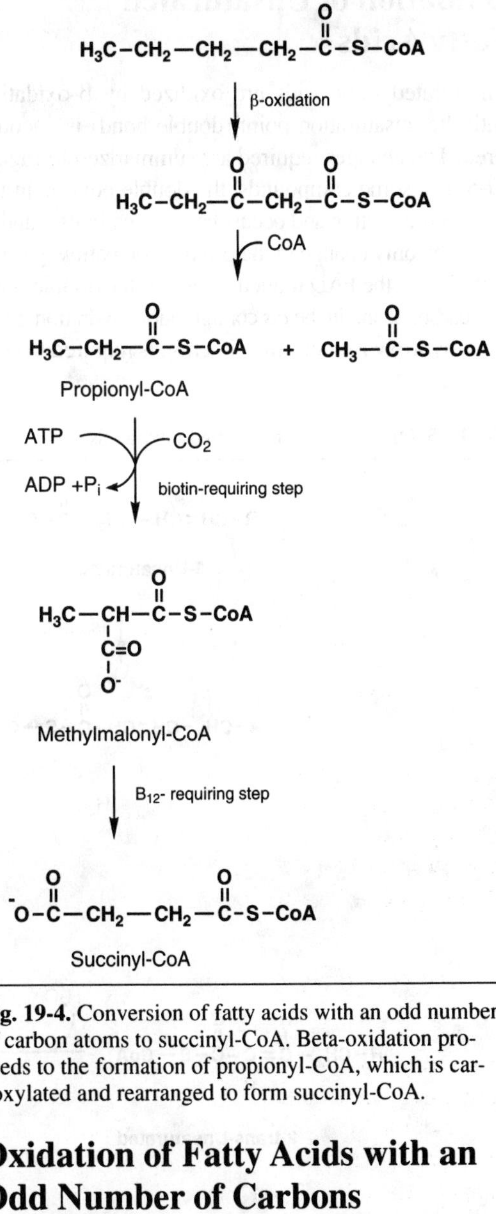

Fig. 19-4. Conversion of fatty acids with an odd number of carbon atoms to succinyl-CoA. Beta-oxidation proceeds to the formation of propionyl-CoA, which is carboxylated and rearranged to form succinyl-CoA.

Oxidation of Fatty Acids with an Odd Number of Carbons

Oxidation of fatty acids with an odd number of carbon atoms proceeds by β-oxidation up to the production of the three-carbon compound propionyl-CoA. Propionyl-CoA is carboxylated in a reaction that requires biotin and ATP. The product is methylmalonyl-CoA. Rearrangement by a reaction requiring vitamin B_{12} gives succinyl-CoA. This is an intermediate of the Krebs cycle and can be further oxidized by that route. These reactions are summarized in Figure 19-4.

CoA directly produces two acetyl-CoA units. The final two carbons are not subjected to the β-oxidation pathway, and no $FADH_2$ or NADH is produced for passage through the electron transport chain. For a fatty acid of 12 carbons, the energy yield would be (5 acetyl-CoA × 17 ATP) + (1 acetyl-CoA × 12 ATP) – 2 ATPs (for activation), or 95 ATPs.

Oxidation of Unsaturated Fatty Acids

Unsaturated fatty acids are oxidized by β-oxidation until the unsaturation point (double bond) is encountered. The changes required are summarized in Figure 19-5. For some compounds, the double bond is in the *trans* configuration and occurs between carbons 2 and 3. Thus, the only change in the β-oxidation pathway is the omission of the FAD reduction step. If the double bond is at carbon 2 and in the *cis* configuration, hydration gives the D-hydroxy isomer. An epimerase is required to con-

vert it to the L-compound before β-oxidation can proceed.

If the double bond occurs between carbons 3 and 4, an isomerase moves it to the 2-3 position. Hydration gives the D-hydroxy compound. Action of the epimerase converts the compound to the correct isomer, and β-oxidation continues.

α-Oxidation

Branched-chain fatty acids are metabolized by α-oxidation, as shown in Figure 19-6. In this pathway, a mixed-function oxidase using FAD and molecular

Fig. 19-5. Oxidation of unsaturated fatty acids.

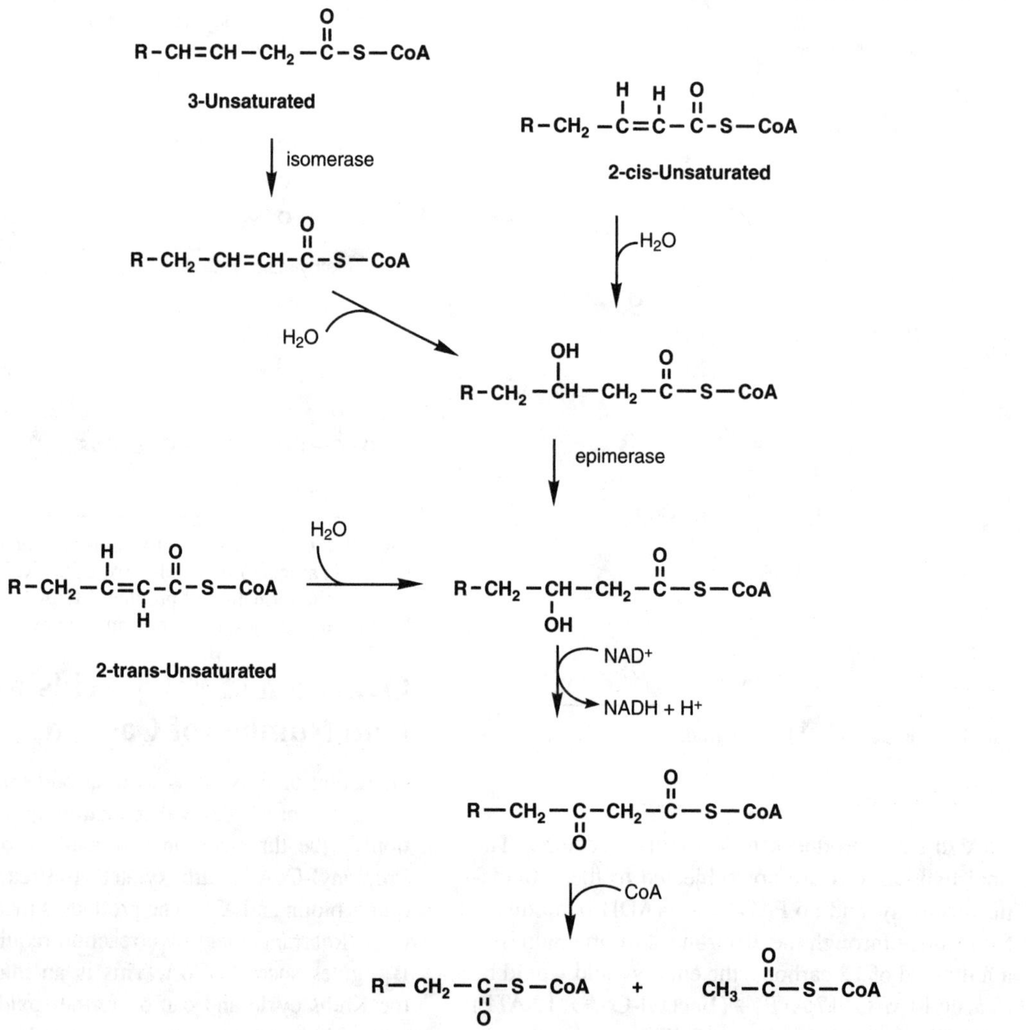

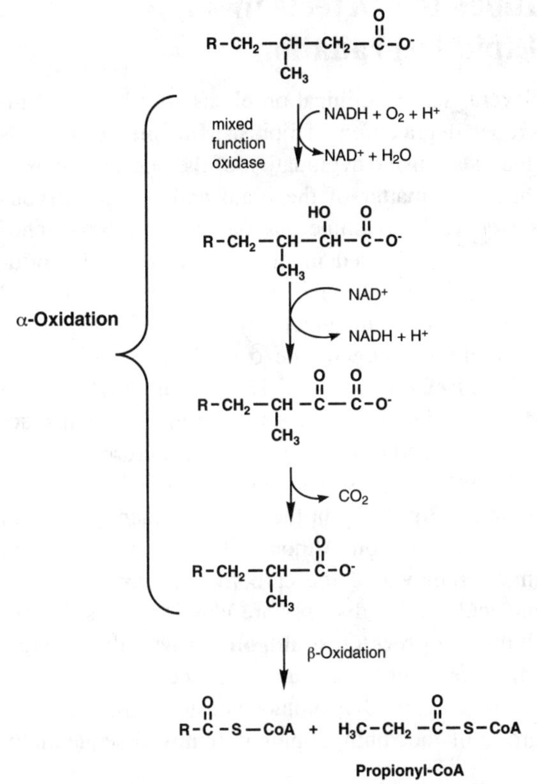

Fig. 19-6. Alpha-oxidation of a branched-chain fatty acid to produce a product for β-oxidation.

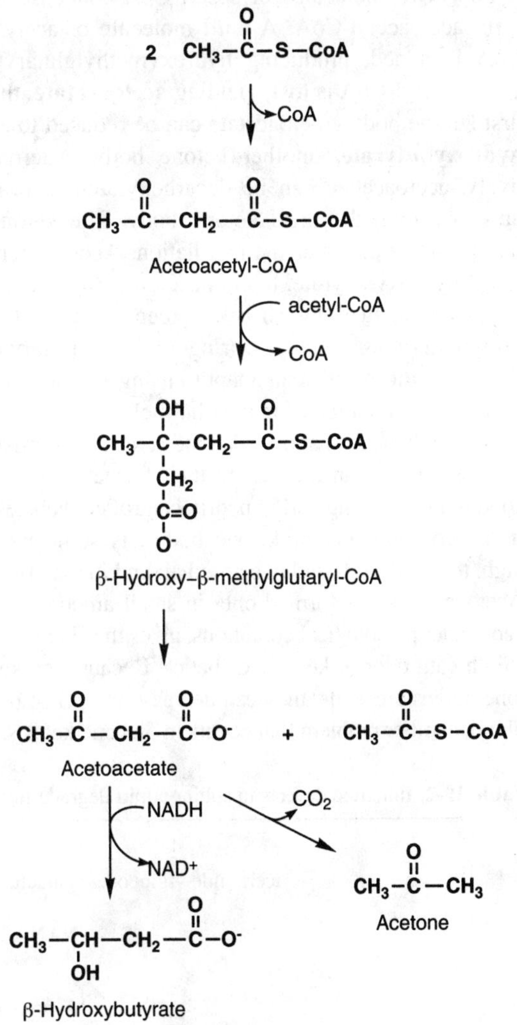

oxygen adds a hydroxyl group to the α-carbon of a fatty acid, with a branch at the β-position. The hydroxyl group is oxidized to a ketone and then decarboxylated. The resulting carboxylic acid is a substrate for β-oxidation, eventually yielding propionyl-CoA containing the branch point carbon. The main function of α-oxidation is to bypass branch points at the β-carbon so the remainder of the fatty acid can continue in the β-oxidation pathway.

ω-Oxidation

A third, but minor, pathway for fatty acid oxidation is ω-oxidation. Here, the methyl carbon of the chain is oxidized to a carboxyl group, which is decarboxylated as in β-oxidation. After attachment to CoA, β-oxidation can then continue from both ends of the molecule. ω-oxidation is one mechanism for removing branches containing multiple carbons.

Fig. 19-7. Synthesis of ketone bodies from acetyl-CoA.

Ketone Bodies

Ketone bodies are formed under conditions of carbohydrate starvation. Fatty acids released from adipose tissue are transported to the liver and converted into mitochondrial acetyl-CoA. If there is little carbohydrate present in the cell, most of the Krebs cycle oxaloacetate will have been converted into glucose, in which case little will remain for condensation with acetyl-CoA in the first step of the Krebs cycle. When this happens, acetyl-CoA is converted into ketone bodies by the pathway shown in Figure 19-7.

First, two molecules of acetyl-CoA condense to form acetoacetyl-CoA. A third molecule of acetyl-CoA is added, producing hydroxymethylglutaryl-CoA. Acetyl-CoA is lost, yielding **acetoacetate,** the first ketone body. Acetoacetate can be reduced to **3-hydroxybutyrate,** another ketone body. Alternatively, acetoacetate can be decarboxylated to produce **acetone,** the third ketone body. The ketone bodies are released to the circulation. Acetoacetate and 3-hydroxybutyrate, but not acetone, can be used as metabolic fuels by all cells except those of the central nervous system. During prolonged starvation, even those cells can adapt to using ketone bodies as a major source of metabolic fuel.

Although the formation of ketone bodies is a normal process, there is an increase in their formation under conditions of fasting and in poorly controlled diabetes. If the concentration of ketone bodies is sufficiently high, they will be found in expired air and in the urine. Acetone, which is formed only in small amounts except under pathological conditions, gives the characteristic breath odor of ketonic diabetics. Because the ketone bodies are acids, they can decrease the pH of the blood, another problem that occurs in diabetic ketosis.

Inherited Defects in Lipid Degradation

Several serious clinical problems are caused by improper degradation of lipids. **Multiple sclerosis** is due to extensive degradation of the phospholipids in the white matter of the brain and central nervous system. Ethanolamine plasmalogen is the phospholipid most affected in multiple sclerosis. Unfortunately, the events precipitating the excessive lipid degradation are unknown.

Nonfunctioning of the α-oxidation pathway occurs in **Refsum's disease.** Here dietary phytanic acid (from the degradation of chlorophyll from plants) accumulates and produces neurological problems.

Defective lysosomal degradation of the sphingolipids, especially in the brain, frequently leads to mental retardation. Although the enzyme defects are in the removal of the carbohydrate portion of the molecule, these diseases are classified as **sphingolipodystrophies** or **sphingolipidoses** (Table 19-2). Most are inherited as autosomal recessive traits and can be diagnosed by amniocentesis. Each saccharide group of the sphingolipids is removed sequentially

Table 19-2. Inherited defects in sphingolipid degradation

$$\text{ceramide} \xleftarrow{4} \text{glucose} \xleftarrow{3} \text{galactose} \xleftarrow{2} \text{N-acetylgalactosamine} \xleftarrow{1} \text{galactose}$$
$$\downarrow$$
$$\text{NANA}$$

$$\text{ceramide} \xleftarrow{6} \text{galactose} \xleftarrow{5} \text{sulfate}$$

$$\text{ceramide} \xleftarrow{7} \text{phosphate} - \text{choline}$$

$$\text{ceramide} - \text{glucose} - \text{galactose} \xleftarrow{8} \text{galactose}$$

Disease	Enzyme defect	Symptoms
GM$_1$ gangliosidosis	Enzyme 1 (β-galactosidase)	Mental retardation, spasticity
Sandhoff's or Tay-Sachs variant	Enzyme 2 (hexosamidase A and B)	Death by age 3
Tay-Sachs	Enzyme 2 (hexosamidase A)	Mental retardation, cherry-red spot in the eye, early death
Lactosyl ceramidosis	Enzyme 3 (β-galactosidase)	Brain degeneration
Gaucher's	Enzyme 4 (β-glucosidase)	Mental retardation
Metachromatic leukodystrophy	Enzyme 5 (sulfatidase)	Mental retardation
Krabbe's	Enzyme 6 (β-galactosidase)	Abnormal myelin
Niemann-Pick	Enzyme 7 (sphingomylinase)	Early death
Fabry's	Enzyme 8 (α-galactosidase)	Kidney failure

by a separate enzyme. If the enzyme is absent or defective, the precursor will accumulate.

Summary

Simple and complex lipids are broken down in the intestine to their constituent parts: fatty acids, cholesterol, glycerol, and phosphate. These are emulsified with bile salts and absorbed into the brush border of the intestinal mucosal cells. Small fatty acids enter the portal circulation bound to albumin. Larger fatty acids are re-esterified to glycerol or cholesterol and enter the lymphatic system in chylomicrons. Chylomicrons and the smaller VLDLs transport fatty acids to the tissues. These lipoprotein particles can be broken down to LDLs, which contain a high proportion of cholesterol, cholesterol esters, and phospholipids. Further breakdown produces HDLs. Defects in the proteins and in the metabolism of these particles can cause serious clinical symptoms.

When energy is required by the body, fatty acids are released from triglycerides and phospholipids in adipose tissue by the action of hormonally sensitive and insensitive lipases. The fatty acids enter the circulation and travel to other tissues for oxidation. Once in cells, the fatty acids are attached to CoA at the expense of ATP. The major pathway for oxidation of fatty acids is mitochondrial β-oxidation. In this pathway, acetyl-CoA is produced by a combination of dehydration by FAD, hydration, oxidation of the hydroxyl group by NAD, and cleavage of a two-carbon unit by CoA.

Unsaturated fatty acids require two additional enzymes to produce *trans* double bonds at position 2 and to convert D-hydroxy compounds into the required L-isomer. Fatty acids containing an odd number of carbon atoms are oxidized to the three-carbon level by β-oxidation. The resulting propionyl-CoA is converted in two steps to succinyl-CoA. Branches can be removed by the α- and ω-oxidation pathways.

Under conditions of starvation or diabetes, acetyl-CoA is converted into acetoacetate, 3-hydroxybutyrate, and acetone—the ketone bodies. Acetoacetate and 3-hydroxybutyrate can be used as primary metabolic fuels by all tissues except those of the central nervous system.

Improper degradation of lipids leads to serious clinical consequences such as multiple sclerosis or the mental retardation associated with the sphingolipidoses.

20 Specialized Lipids

There are two additional classes of lipids whose characteristics and biosynthesis were not covered previously. One class is derived from isoprene units and includes the steroids, ubiquinone, dolichol, and vitamins A, D, E, and K. The isoprene units are synthesized from acetyl-CoA. The second group of lipids is derived from the essential fatty acids. They are grouped together as the eicosanoids and include prostaglandins, thromboxanes, leukotrienes, and prostacyclins. These compounds occur in all tissues and function as hormones, producing profound physiological effects at minute concentrations.

Objectives

After completing this chapter, you should be able to

Recognize the structures of mevalonic acid, squalene, and cholesterol.

List the factors that control the rate-controlling reactions for the synthesis of these compounds and the factors that regulate these reactions.

Recognize the major intermediates in steroid biosynthesis.

Describe the relationship between cholesterol and atherosclerosis.

Describe the function of the bile salts and cite the rate-controlling reaction for the biosynthesis of these salts.

Characterize the major structural features and biological activities of the prostaglandins, leukotrienes, thromboxanes, and prostacyclins. List the precursors and the rate-contolling reactions for the biosynthesis of these compounds.

Describe the effects of aspirin and indomethacin on eicosanoid biosynthesis.

Steroids

Steroids are lipids that contain a characteristic structure composed of four fused nonaromatic rings (Fig. 20-1). The major steroid found in humans is cholesterol. It is found in the plasma membranes of animal, but not plant, cells. The glucocorticoids, mineralocorticoids, sex hormones, and bile acids are all steroids. The biosynthesis of cholesterol and the bile acids will be discussed here. The conversion of cholesterol into hormones is covered in Chapter 29.

Fig. 20-1. The four-fused-ring structure characteristic of steroids.

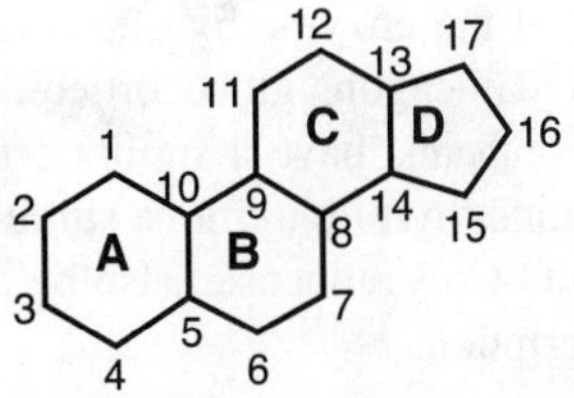

In this chapter, the synthesis of ubiquinone and dolichol, two other compounds synthesized from isoprene units, also is discussed. The function of vitamins A, D, E, and K is covered in Chapter 30.

Synthesis of Mevalonic Acid

The biosynthesis of cholesterol occurs in the cytoplasm of the liver and intestinal epithelium. The first important precursor is **mevalonate.** Its synthesis from acetyl-CoA is outlined in Figure 20-2. In the first reaction, two acetyl-CoA units are condensed to form acetoacetyl-CoA. The reaction is the same as that which occurs in fatty acid biosynthesis, but it uses a different enzyme. For steroid biosynthesis, the keto group is not reduced; instead, a third acetyl-CoA residue is added to it. This yields β-hydroxy-α-methylglutaryl-CoA, a compound that loses coenzyme A and is reduced to the alcohol using two molecules of NADPH, thus producing **mevalonic acid.**

The final reaction producing mevalonic acid is catalyzed by **hydroxymethylglutaryl-CoA reductase** (HMG-CoA reductase). This enzyme is located in the smooth endoplasmic reticulum and is the rate-controlling reaction for cholesterol biosynthesis. The activity of HMG-CoA reductase is regulated by a large number of factors, including cAMP, cholesterol, glucagon, glucocorticoids, thyroid hormone, and insulin. cAMP increases the activity of HMG-CoA reductase by phosphorylation of the enzyme, using the familiar cAMP cascade. Dephosphorylation decreases the enzymatic activity. High concentrations of cholesterol also decrease enzymatic activity. The mechanism of the inhibition by cholesterol is not a simple feedback inhibition as the addition of cholesterol to the isolated enzyme has no effect on its activity.

Hormones and diet affect the amount, rather than the activity, of HMG-CoA reductase. High concentrations of cholesterol in the diet decrease the biosynthesis of the enzyme by affecting the rate of transcription. Glucagon, glucocorticoids, and fasting, but not diabetes, have a similar effect. In contrast, insulin and thyroid hormone stimulate the synthesis of HMG-CoA reductase, also by affecting the rate of transcription.

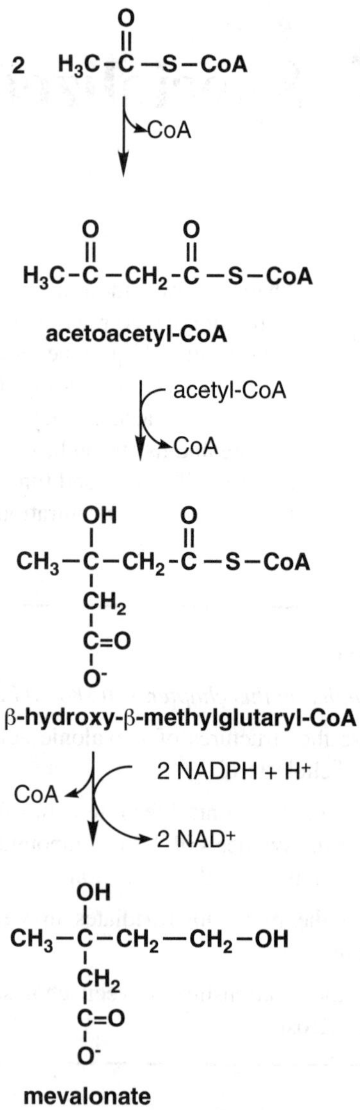

Fig. 20-2. Synthesis of mevalonate from three acetyl-CoA units.

Synthesis of Squalene

In the next steps of cholesterol biosynthesis, mevalonate is converted to active isoprene and then is polymerized to form squalene. These reactions are diagrammed in Figure 20-3. First, mevalonate is phosphorylated in two separate steps to form the pyrophosphate. It then is dehydrated and decarboxy-

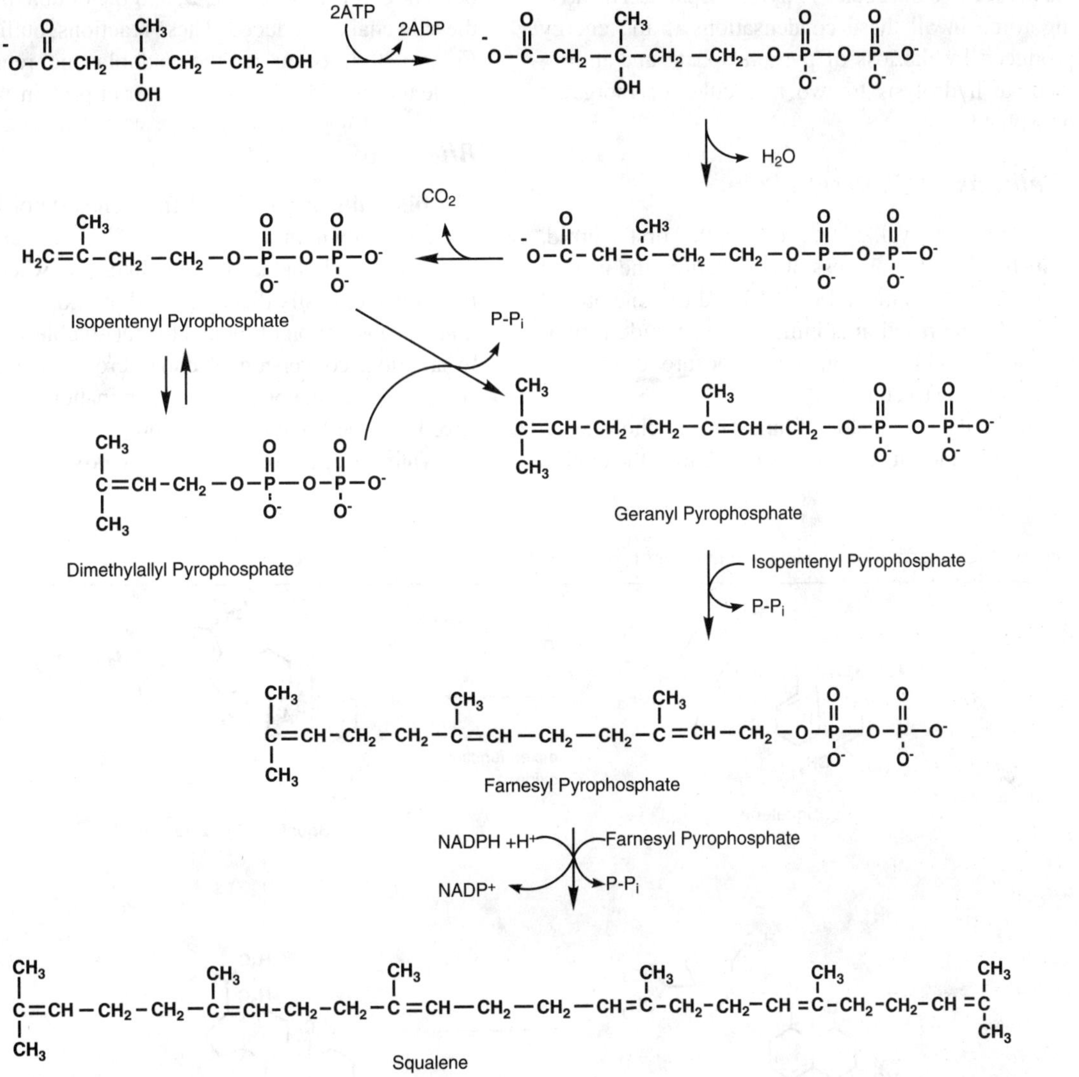

Fig. 20-3. Conversion of mevalonate to squalene.

lated to form isopentenyl pyrophosphate or so-called **active isoprene.** Isopentenyl pyrophosphate is in equilibrium with dimethylallyl pyrophosphate. These two compounds can condense, forming **geranyl pyrophosphate.** This is a head-to-tail condensation in which the pyrophosphate tail of dimethylallyl pyrophosphate adds to the methylene head of isopentenyl pyrophosphate. An additional head-to-

tail condensation of geranyl pyrophosphate with isopentenyl pyrophosphate gives **farnesyl pyrophosphate,** which is the precursor of cholesterol, ubiquinone, and dolichol.

The final reaction in the synthesis of squalene involves a head-to-head condensation of two molecules of farnesyl pyrophosphate to yield **squalene.** This reaction requires a reduction by NADPH and the loss of two molecules of pyrophosphate. The driving force in all these condensations is the energy produced by the loss of pyrophosphate and its immediate hydrolysis to two molecules of inorganic phosphate.

Synthesis of Cholesterol

Squalene can cyclize to produce the first steroid, **lanosterol.** In the process of cyclization, the methyl group at C-14 is moved to C-13 and the one at C-8 to C-14. The reaction is initiated by epoxide formation at C-3, with subsequent ring closure to form the basic steroid structure.

During the conversion of lanosterol to cholesterol, three molecules of carbon dioxide are lost, the double bond at C-8 is moved to C-5, and the double bond in the side chain is reduced. These reactions, outlined in Figure 20-4, occur in the endoplasmic reticulum while the steroid is bound to a carrier protein.

Bile Salts

The **bile salts** are produced from cholesterol by the liver, as shown in Figure 20-5. They are excreted through the gallbladder into the intestine, where they function to emulsify dietary fat and promote the digestion and absorption of fat and the fat-soluble vitamins. In addition, conversion of cholesterol into bile salts constitutes the major route for elimination of cholesterol from the body. The rate-controlling reaction for the synthesis of the bile salts is 7-hydroxylase. **Tauro-**

Fig. 20-4. Conversion of squalene to cholesterol.

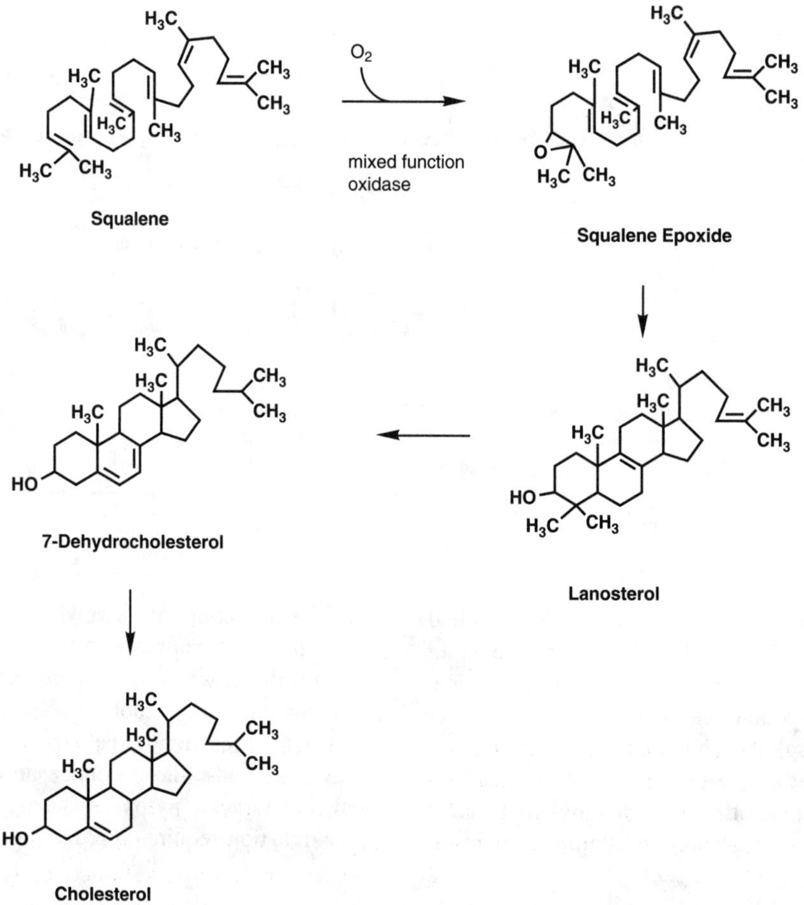

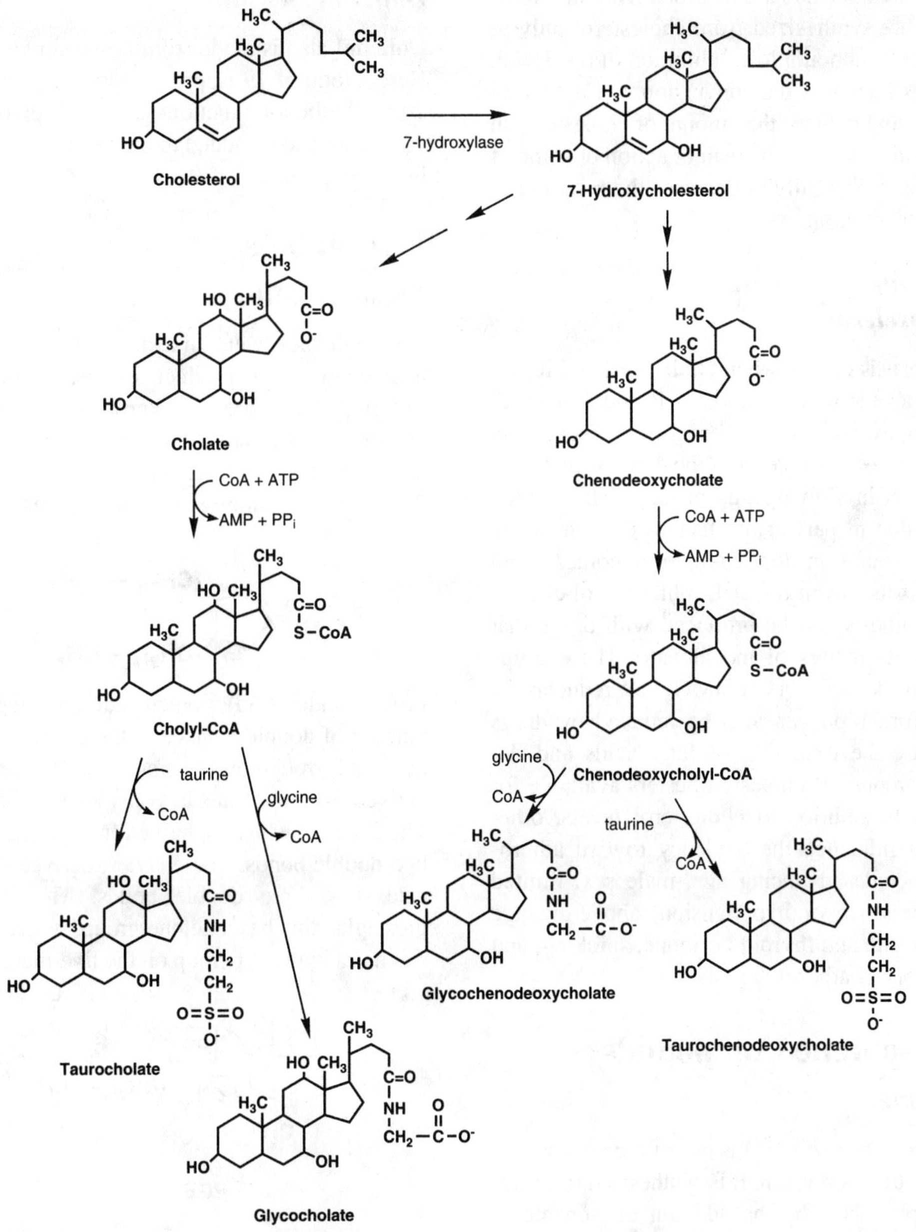

Fig. 20-5. Synthesis of the bile salts from cholesterol. Glycocholate is the major bile salt in humans.

cholate, taurochenodeoxycholate, glycochenode-oxycholate, and glycocholate are the primary bile salts. The secondary bile salts, deoxycholate and lithocholate, are formed by the action of intestinal bacteria on the primary bile salts.

Bile salts are sometimes called bile acids. Be-

cause bile is an alkaline fluid, these compounds occur as salts, not as free acids, and should be called bile salts. In addition to the bile salts, bile contains a large amount of cholesterol. Cholesterol is relatively insoluble in aqueous solutions, so it is cholesterol, not the bile salts, that precipitates to form gallstones.

Most of the bile salts are resorbed from the intestine. They are synthesized from cholesterol only to make up for the amount lost. Inhibition of resorption of bile salts increases the conversion of cholesterol to bile salts and reduces the amount of cholesterol in the body. This is the mechanism of action of some of the drugs used clinically to decrease the serum cholesterol concentration.

Relationship of Cholesterol to Atherosclerosis

Atherosclerosis occurs when cholesterol and its esters are deposited in arterial walls. Blood flow is restricted and, eventually, clots are formed. Reduction of the amount of cholesterol in the diet has met with some success in slowing this process. The limited success is due in part to the fact that much of the cholesterol found in the circulation comes from biosynthesis, not from the diet. Inhibition of cholesterol biosynthesis can be produced with drugs that are structural analogs of mevalonate. These drugs act as feedback inhibitors of HMG-CoA reductase.

Plaque formation can also be reduced by drugs that increase the oxidation of fatty acids and decrease the amount of cholesterol esters available for deposition. In addition to cholesterol levels, other factors that influence the tendency toward atherosclerosis include advancing age, male sex, limited exercise, presence of hypertension, above-average levels of insulin and thyroid hormone, smoking, and family history of atherosclerosis.

Other Isoprene Compounds

Ubiquinone

Ubiquinone or **coenzyme Q** is one of the members of the electron transport chain. It is synthesized from farnesyl pyrophosphate by the addition of isopentenyl pyrophosphate units. The number of such units is variable, but 10 is the most common number in mammals.

Dolichol

Dolichol also is made from farnesyl pyrophosphate. Here a total of 19 isoprene units are combined (Fig. 20-6). Dolichol functions as a carrier of carbohydrate that is to be added to the nitrogen of asparagine in glycoproteins.

Eicosanoids

Prostaglandins

Although they were first discovered in semen and misnamed as a product of the prostate gland, **prostaglandins** are now known to be produced by all tissues. They are derived from the essential fatty acids and are characterized by a fatty acid chain of 20 carbons containing a five-membered ring.

Prostaglandin (PG) nomenclature distinguishes the number of double bonds and the existence of keto or hydroxyl groups on the ring. The 1 series, which is derived from linoleate, has only a single double bond. The 2 series, which is derived from arachidonate, has two double bonds. The 3 series, derived from linolenate, has three double bonds. The E series of prostaglandins has a ketone group, whereas the F series has a hydroxyl group on the five-membered ring.

PGE$_2$

PGF$_2$

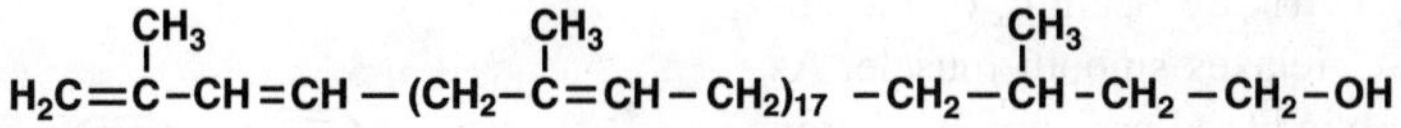

Fig. 20-6. Structure of dolichol.

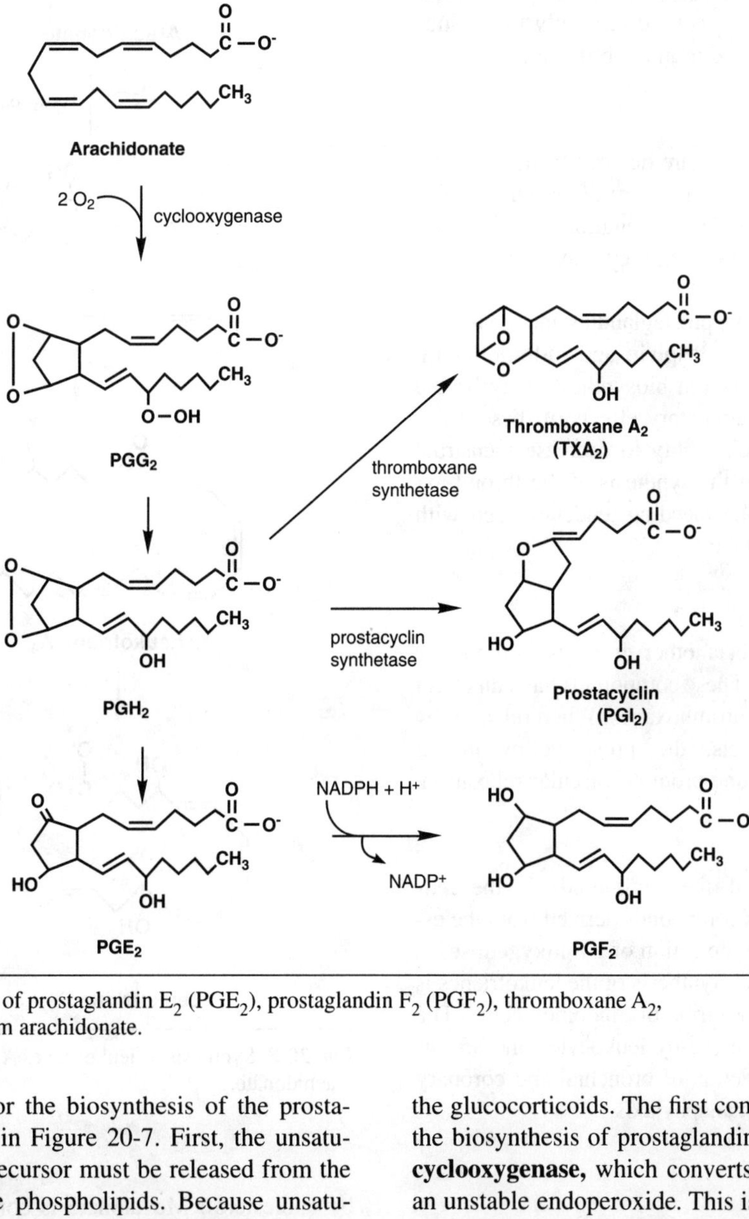

Fig. 20-7. Synthesis of prostaglandin E$_2$ (PGE$_2$), prostaglandin F$_2$ (PGF$_2$), thromboxane A$_2$, and prostacyclin from arachidonate.

The pathway for the biosynthesis of the prostaglandins is given in Figure 20-7. First, the unsaturated fatty acid precursor must be released from the plasma membrane phospholipids. Because unsaturated fatty acids are usually on carbon 2 of glycerol, this requires the activity of phospholipase A$_2$. This hydrolysis reaction is stimulated by bradykinin and

the glucocorticoids. The first committed reaction for the biosynthesis of prostaglandins is catalyzed by a **cyclooxygenase,** which converts the fatty acid into an unstable endoperoxide. This is quickly converted into a hydroxyl compound that is an intermediate for the biosynthesis of both the prostaglandins and the thromboxanes.

$PGE_{2\alpha}$ comes from PGH_2 by opening of the endoperoxide ring. $PGE_{2\alpha}$ relaxes smooth muscle. As such, it causes a decease in blood pressure, a dilation of bronchial tubes, and an increase in urine flow. $PGF_{2\alpha}$, which is a reduction product of PGE_2, has the opposite physiological effects. Because it stimulates smooth muscle, $PGF_{2\alpha}$ is used clinically to produce uterine contractions and promote birth and delivery.

Thromboxanes

The **thromboxanes** also are derived from PGH_2 by the pathway illustrated in Figure 20-7. These compounds increase platelet aggregation and smooth-muscle contraction. As such, they play an important role in blood clotting.

Synthesis of both the prostaglandins and the thromboxanes is inhibited by **aspirin** and **indomethacin.** Inhibition of prostaglandin biosynthesis may be the cause of the antiinflammatory effects of these drugs and may explain their ability to decrease menstrual cramps. Inhibition of the synthesis of the thromboxanes may produce the bleeding tendency seen with prolonged aspirin intake.

Prostacyclins

Prostacyclins are yet another group of compounds derived from PGH_2. These compounds have an effect opposite that of the thromboxanes. When released by injured blood vessels, the prostacyclins inhibit platelet aggregation and promote vascular relaxation.

Leukotrienes

The fourth group of the eicosanoids is the **leukotrienes,** noncyclic compounds derived from the essential fatty acids by the action of a **lipoxygenase,** as shown in Figure 20-8. Synthesis of the leukotrienes is not inhibited by aspirin or indomethacin. The leukotrienes are produced by leukocytes and are involved in the constriction of bronchial and coronary arteries that occurs in anaphylaxis.

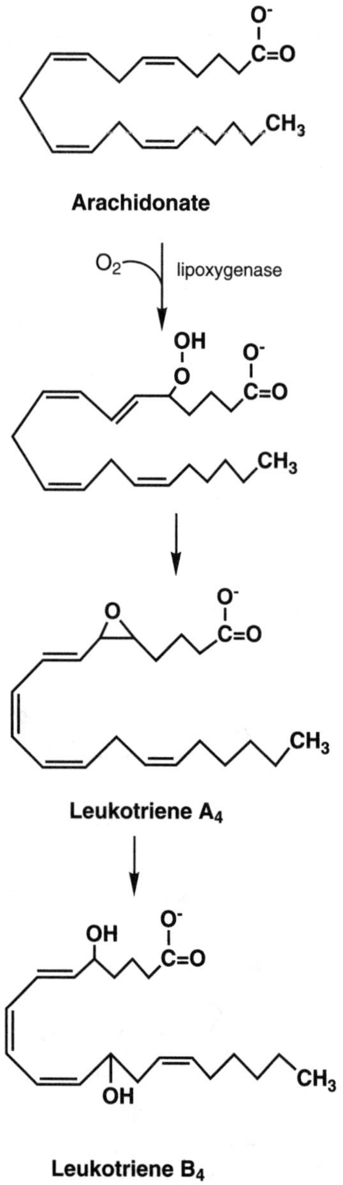

Fig. 20-8. Synthesis of leukotrienes A_4 and B_4 from arachidonate.

Summary

Cholesterol is made from acetyl-CoA. First, three units of acetyl-CoA are condensed to form mevalonate. The rate-controlling enzyme HMG-CoA reductase is regulated by hormones and diet as well as by cholesterol. Mevalonate is converted to the pyrophosphate and then decarboxylated and dehydrogenated to yield active isoprene, the unit from which all steroids are made. Two isoprene units are condensed to form geranyl pyrophosphate. The addition of one more isoprene unit gives farnesyl pyrophosphate. This is the precursor to the steroids ubiquinone and dolichol. The condensation of two farnesyl pyrophosphates makes squalene, which is cyclized to form the first steroid, lanosterol. Lanosterol is converted into cholesterol.

Cholesterol is degraded to the bile salts, which are excreted by the liver and aid in the emulsification of dietary fats. Most of the bile salts are resorbed in the intestine. Atherosclerosis is associated with the deposition of cholesterol and its esters in arteries. This process can be slowed by restriction of dietary cholesterol and by treatment with drugs that increase fatty acid oxidation or decrease the biosynthesis of cholesterol or the resorption of bile acids.

The eicosanoids include prostaglandins, thromboxanes, prostacyclins, and leukotrienes. Prostaglandins are derived from the essential fatty acids. After release from plasma membrane phospholipids, the fatty acids react with a cyclooxygenase to produce the prostaglandins, thromboxanes, and prostacyclins. The cyclooxygenase reaction is inhibited by aspirin and indomethacin. The leukotrienes also are derived from the essential fatty acids, but this pathway is initiated by the action of a lipoxygenase. Synthesis of the leukotrienes is not inhibited by aspirin or indomethacin.

21 Membranes

Membranes are composed of phospholipids, sphingolipids, cholesterol, proteins, and carbohydrates. The plasma membrane divides the inside of cells from the outside, provides the means for entry of nutrients and export of products, and allows recognition and response to hormones and growth factors. As discussed in the first chapter, there are a number of membrane-surrounded organelles inside cells. These include the mitochondria, lysosomes, peroxisomes, and nuclei. In addition, the endoplasmic reticulum and Golgi complex are composed of membranes and connect to other membranes. Membranes are necessary also for the electron transport chain, vision, and transmission of nerve impulses.

Objectives

After completing this chapter, you should be able to

Describe the plasma membrane in regard to type and orientation of lipids, proteins, and carbohydrates.

List the functions of membranes.

Define the terms *micelle, amphipathic,* and *liposome.*

Discuss the mechanisms by which compounds enter and leave a cell. Define the terms *endocytosis, pinocytosis,* and *exocytosis.*

Discuss the characteristics of passive diffusion, facilitated diffusion, and active transport.

Define the characteristics and give an example of uniport, symport, and antiport transport systems.

Structure of Membranes

The lipids found in membranes are **amphipathic** lipids, characterized by having one end that is hydrophobic and the other end hydrophilic. In an aqueous environment, the hydrophobic portion of the molecule associates with other hydrophobic molecules and away from water. The hydrophilic portion associates with other hydrophilic molecules and with water. When added to water in small amounts, amphipathic lipids form droplets or **micelles.** When added in larger amounts, they form **bilayers** in which the hydrophobic portions of the molecules are on the inside and the hydrophilic portions are on the outside. In contrast to phospholipids and sphingolipids, cholesterol esters and triglycerides do not have a hydrophilic end and do not form micelles or bilayers when in an aqueous environment.

The ability of amphipathic lipids to form micelles has been used as a means of encapsulating drugs and enzymes. To form a **liposome,** a polar lipid is mixed vigorously with a solution containing the drug or enzyme. The lipid forms a micelle with a small amount of the solution trapped inside. While inside a lipo-some, the drug or enzyme is protected from enzy-matic and chemical degradation and from dilution. Once in the circulation, the liposome will transport its contents to a cell, fuse with the plasma membrane, and release its contents. It is hoped that the use of liposomes containing specific lipids can direct drugs and enzymes to particular cells and decrease side ef-fects caused by reaction of drugs with nontarget cells.

The types of lipid found in a membrane and their overall proportions affect the physical properties and the function of the membrane. Because of this, membranes with different functions contain differ-ent lipids. For example, cholesterol is found in plasma membranes but is absent from mitochondrial membranes. Different types of cells also have differ-ent lipid proportions in similar membranes. For ex-ample, sphingolipids occur in higher concentration in the plasma membranes of cells of the nervous sys-tem than in cells from other tissues.

In addition to amphipathic lipids, membranes con-tain proteins and carbohydrates. The model dia-grammed in Figure 21-1 is the fluid mosaic model for membranes. It contains two classes of proteins—in-tegral proteins and peripheral proteins. **Integral pro-teins** are embedded in or pass completely through the membrane. They contain segments with a large pro-portion of hydrophobic amino acids. These amino acids are folded to be on the outside of the molecule in contact with the lipids of the membrane. Because they are embedded in the membrane, integral pro-teins cannot be removed from the membrane without total disruption of its structure. **Peripheral proteins** are found near the surface of the membrane. They contain a hydrophobic segment that anchors the pro-tein to the membrane and a hydrophilic segment that extends into the aqueous environment.

Like the membrane lipids, the amount and kind of protein varies among membranes. In general, the amount of protein is greater in cells with more meta-

Fig. 21-1. The fluid mosaic model of the plasma membrane showing the arrangement of phospholipids, proteins, and carbohydrates.

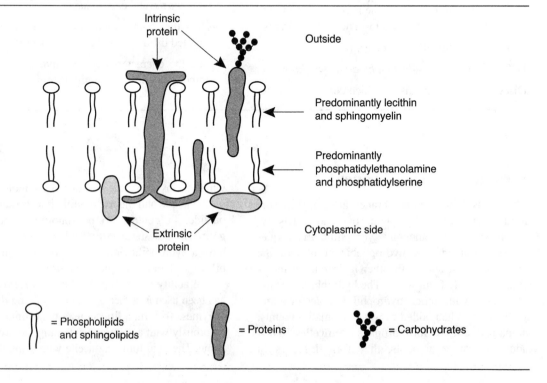

bolic activity and with a need to respond to more external stimuli. For example, there is more protein in the plasma membrane of a hepatocyte than in an erythrocyte.

The carbohydrate found in membranes is in the form of glycoproteins and glycolipids. These molecules extend beyond the lipid boundary of the membrane, with the carbohydrate portion in the aqueous phase. All the carbohydrate in plasma membranes is on the noncytoplasmic side of the membrane, where it functions as a recognition and binding site for hormones and other extracellular messengers.

Specific proteins and lipid molecules also are located either on one side of the membrane or the other. For example, in erythrocytes, there is a predominance of choline phospholipid and sphingomyelin on the outside of the plasma membrane and a preponderance of ethanolamine and serine phospholipids on the cytoplasmic side of the membrane. Although neither lipids nor proteins migrate from one side of the membrane to the other, they do move laterally within the membrane. The amount of this movement is related to the temperature and the degree of unsaturation of the fatty acids in the membrane.

Membrane Transport

Endocytosis and Exocytosis

There are many ways for substances to enter and leave cells. In **endocytosis** (Fig. 21-2), large molecules or particles are bound to specific sites on the membrane called **coated pits.** A vacuole is formed that fuses with the lysosomes. Then the lysosomal enzymes can degrade the ingested material. Endocytosis can also be triggered by contact between a compound and a specific surface receptor. However, molecules entering the cell bound to a receptor are

Fig. 21-2. Process of transport of molecules and solutions by endocytosis and exocytosis.

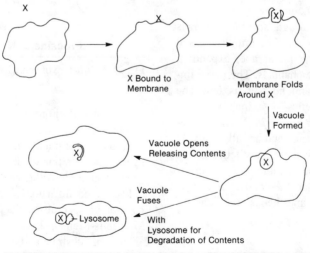

PROCESS OF ENDOCYTOSIS OF MOLECULE X — TRANSPORT INTO A CELL

PROCESS OF EXOCYTOSIS — TRANSPORT OUT OF A CELL

not usually destined for degradation by the lyso-somes. In a specialized form of endocytosis called **pinocytosis** a portion of the membrane surrounds a molecule. This is followed by a release of the mole-cule on the inside of the cell. In a reversal of this process, **exocytosis,** an internal vesicle fuses with the plasma membrane and releases its contents out-side the cell. The digestive enzymes are released from pancreatic cells by this mechanism.

Passive Diffusion

For lipid-soluble molecules, entry into cells occurs by **passive diffusion.** The molecules dissolve in the lipid core of the membrane and pass into or out of the cell. Molecules that are more lipid-soluble dif-fuse into cells more readily than molecules that are less lipid-soluble. Passive diffusion has no energy requirement and shows no saturation when the con-centration of the entering substance is increased.

Some small water-soluble molecules and ions also enter cells by diffusion. Urea, water, oxygen, carbon dioxide, and ammonia appear to pass through the membrane through aqueous pores rather than by dis-solving in the lipid of the membrane.

Sodium enters cells through specific channels during the transmission of a nerve impulse. Local **anesthetics** dissolve in the membrane, expand it, and effectively block these channels. The nerve im-pulse is not transmitted and no pain is felt. The structures of three anesthetics—cocaine, lidocaine, and procaine—are given in Figure 21-3. These com-pounds have similar structures and are all relatively lipid-soluble. The effect of these anesthetics is lost when they diffuse from the membrane or are meta-bolically degraded. Similar membrane disruption occurs with the inhalation anesthetics ether, chloro-form, and halothane.

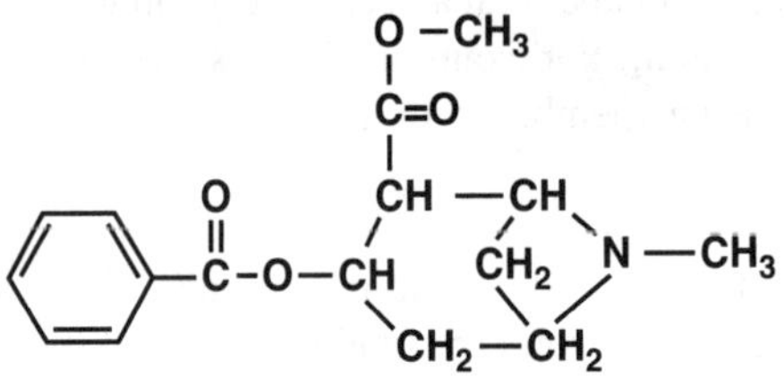

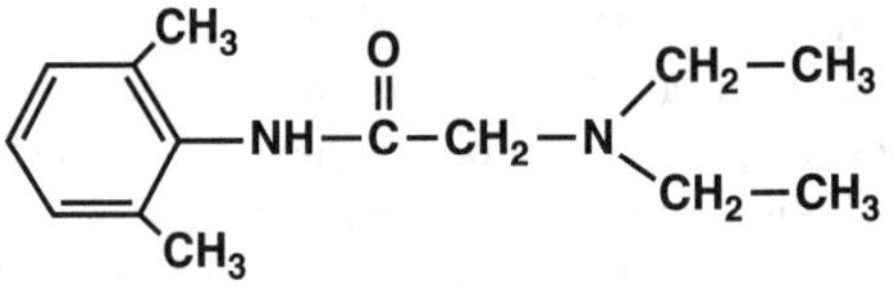

Facilitated Diffusion

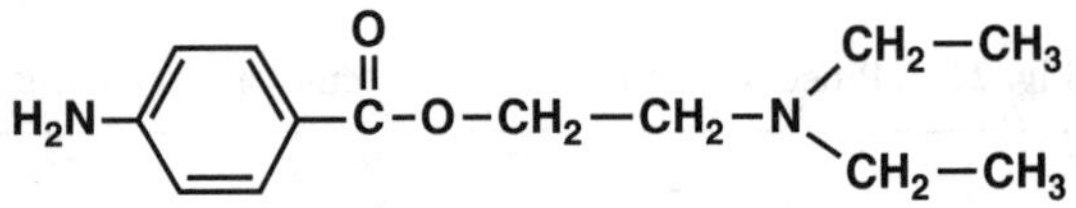

Carrier-mediated diffusion or **facilitated diffusion** allows entry of molecules that have a low solubility in lipid. Such molecules bind to specific carrier proteins on the outside of the membrane. They are transported through the membrane while bound to this carrier and then are released on the other side. Facilitated diffu-sion can function for molecules entering the cell or leaving it. However, transport must go from a more

Fig. 21-3. Structure of some local anesthetics.

concentrated environment to a less concentrated one. It is the concentration gradient that provides the driv-ing force for the transport. Because transport is driven by the concentration gradient, no ATP or other energy source is used in facilitated diffusion.

Facilitated diffusion requires binding to a fixed number of specific carriers, so the reaction shows Michaelis-Menten kinetics. There is a saturating amount of substrate and a maximum velocity neces-sary for the transport. The transport of one substance can be competitively inhibited by the presence of a structurally similar molecule.

Glucose uptake in erythrocytes, liver, and muscle takes place by facilitated diffusion. The immediate phosphorylation of glucose after entering the cell prevents its being transported out again. This phos-phorylation also maintains the concentration gradient for glucose in the proper direction for glucose entry.

Active Transport

Active transport requires the use of a carrier. Unlike facilitated diffusion, active transport can occur against a concentration gradient. The energy needed to drive active transport comes from the hydrolysis of ATP. Because there is a carrier involved in active transport, it exhibits saturation kinetics, a maximum velocity, and competitive inhibition. Another characteristic of active transport is that it is unidirectional—that is, compounds are transported across the membrane in only one direction.

Uniport, Antiport, and Symport Systems

The transport of a single molecule across a membrane is a **uniport** system. This can be either facilitated diffusion or active transport. Calcium transport into cells is an active transport uniport system.

In **symport** or coupled transport, two different molecules must be bound to the carrier before trans-

port can occur. Both molecules are transported in the same direction at the same time. Glucose transport is a symport system that requires sodium ion as the other component.

In **antiport** or exchange transport, one compound binds to a carrier, passes through the membrane, and is released on the other side. Then a different molecule binds to the same carrier, which transports it across the membrane in the opposite direction. Both molecules must be present and on opposite sides of the membrane for antiport transport to function. The exchange of ATP and ADP across a mitochondrial membrane is an example of an antiport system.

Sodium-Potassium Pump

All mammalian cells contain more potassium in the inside of the cell and more sodium on the outside. The maintenance of this concentration gradient is due to the function of the sodium-potassium pump, an active transport exchange reaction. As shown in Figure 21-4,

Fig. 21-4. Function of the sodium-potassium pump of the plasma membrane.

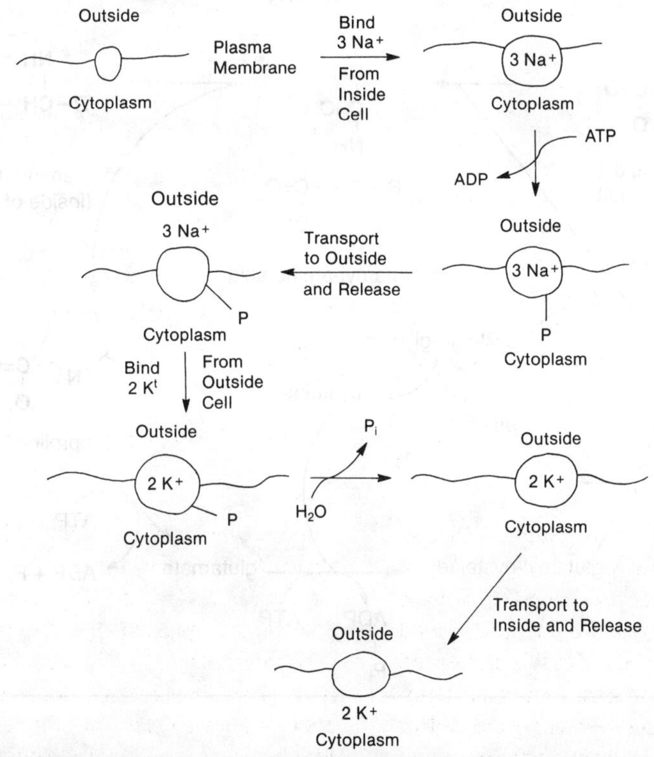

three sodium ions on the inside of the cell are bound to a carrier. The carrier is then phosphorylated by ATP. The carrier moves to the other side of the membrane and releases the sodium. Two molecules of potassium are bound. The carrier is dephosphorylated and moves to the inside, releasing the potassium.

The sodium-potassium pump is unidirectional. It works only when sodium is bound to the carrier on the inside and potassium is bound on the outside. Vanadium on the inside of a cell inhibits the pump by inhibiting the binding of sodium. Ouabain on the outside of a cell inhibits the pump by inhibiting potassium binding.

Amino Acid Transport

The entry of amino acids into cells is coupled with the entry of sodium. Amino acids enter *against* a concentration gradient at the same time sodium en-

ters *with* a concentration gradient. The sodium gradient drives the transport toward entry rather than exit of the amino acids. The transport of amino acids also appears to be involved with the attachment of the amino acid to glutathione in what is called the γ-glutamyl or **Meister cycle** (Fig. 21-5). On reaching the inside of the plasma membrane, the amino acid is attached to the glutamyl end of glutathione with the release of cysteinylglycine, a dipeptide containing the other two amino acids of glutathione. Then the amino acid is released with the formation of 5-oxoproline, a cyclized glutamate. Glutamate is formed by ring opening accompanied by hydrolysis of ATP. Next, cysteine and glycine are added to resynthesize glutathione. This requires the expenditure of two more molecules of ATP, for a total of three.

The γ-glutamyl cycle has been shown to function for the uptake of neutral amino acids in the kidney

Fig. 21-5. γ-Glutamyl or Meister cycle for the transport of amino acids.

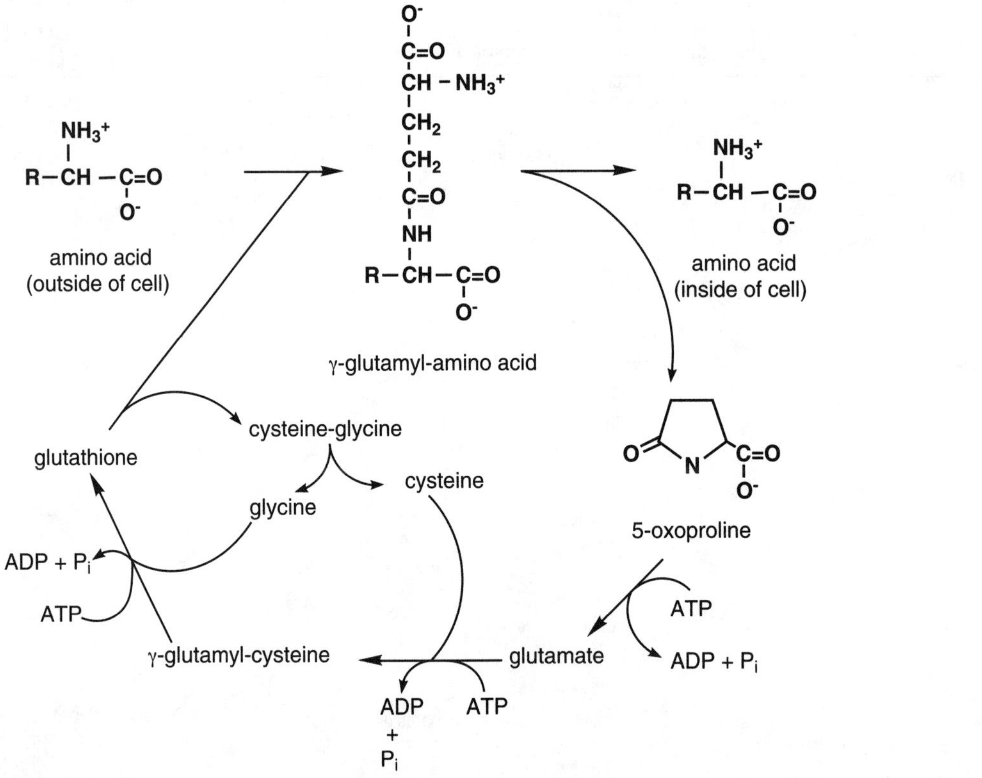

and gastrointestinal tract. However, the cycle does not explain the transport of all amino acids in all tissues. It does not appear to function for any of the acidic amino acids. Furthermore, erythrocytes transport amino acids but do not have mitochondria to produce the large amount of ATP necessary to run the transport cycle.

Summary

Membranes contain proteins, carbohydrates, and amphipathic lipids. The lipids are arranged in two layers, with the hydrophobic portion inside and the hydrophilic portion outside. Proteins can be embedded in the membrane (integral proteins) or more loosely attached (peripheral or extrinsic). All carbohydrates are in the form of glycoproteins and glycolipids and are located on the noncytoplasmic side of the plasma membrane.

Liposomes are micelles of phospholipids containing small amounts of drugs or enzymes in a solution trapped inside. They can travel to target cells with their contents protected from degradation or reaction with nontarget cells.

Compounds enter cells by a variety of mechanisms. Entry by simple diffusion occurs for compounds soluble in lipids and for very small water-soluble ones. Larger compounds and hydrophilic ones require a carrier. If no energy is involved, the process is facilitated diffusion. If energy is used, it is active transport. Only the latter can transport compounds against a concentration gradient. All transport systems that use a carrier exhibit saturation kinetics, a maximum velocity, and inhibition by compounds with similar structures. A system that transports only one compound is a uniport system. One that transports two compounds at the same time in the same direction is a symport system. A system that transports two compounds in the opposite direction is an antiport system. Calcium, sodium, and potassium are transported by active transport systems called *pumps*. They are characteristic of all mammalian cells and are required for survival. Some amino acids enter the circulation from the kidney and gastrointestinal system by reaction with glutathione. This process, called the *γ-glutamyl cycle,* requires three ATPs for each amino acid transported.

Compounds can also enter cells by endocytosis. In this process, part of the membrane folds around a compound or fluid and brings it into the cell. Exocytosis, the reversal of this process, is the mechanism by which the digestive enzymes are secreted by the pancreas.

Part V Questions: Lipid Metabolism

1. The rate-limiting reaction for fatty acid biosynthesis is the
   A. condensation of acetyl-CoA with oxaloacetate in the mitochondria.
   B. transfer of citrate from the mitochondria.
   C. synthesis of malonyl-CoA.
   D. transfer of malonyl-CoA to fatty acid synthetase.
   E. condensation of malonyl-CoA with acetyl-CoA.

2. During fatty acid biosynthesis,
   A. all but the first two carbons of the growing acyl chain come directly from malonyl-CoA.
   B. $FADH_2$ reduces the β-keto group.
   C. ATP is cleaved in the dehydration step.
   D. the growing acyl chain is attached to coenzyme A.
   E. the product is released after a chain of 24 carbons is synthesized.

3. The desaturation system
   A. produces fatty acids with *trans* double bonds.
   B. yields fatty acids with conjugated double bonds.
   C. is activated by citrate.
   D. along with elongation can convert linoleic acid into arachidonic acid.
   E. can convert palmitic acid into oleic acid.

4. The essential fatty acids are required in the mammalian diet because they
   A. are needed for the maintenance of normal membrane function.
   B. can be oxidized to carbon dioxide and water more rapidly than the nonessential fatty acids.
   C. are related metabolically to the polyamines.
   D. can be synthesized from unsaturated fatty acids.
   E. are required for the synthesis of steroids.

5. Diacylglycerol is a precursor to
   A. ceramides.
   B. sulfatides.
   C. triglycerides.
   D. sphingosine.
   E. prostaglandins.

6. All of the following are produced in the oxidation of fatty acids with an odd number of carbon atoms *except*
   A. propionyl-CoA.
   B. malonyl-CoA.
   C. succinyl-CoA.
   D. acetyl-CoA.
   E. methylmalonyl-CoA.

7. Tay-Sachs disease
   A. is due to the lack of a hexamidase.
   B. produces an accumulation of sphingomyelin.
   C. causes excessive degradation of ethanol plasmalogen.
   D. is characterized by abnormal glycogen structure.
   E. is due to a failure to remove NANA from sphingolipids.

8. Which of the following is a ketone body?
   A. Mevalonate
   B. Acetyl-CoA
   C. 3-Hydroxybutyrate
   D. Hydroxyacetone phosphate
   E. Hydroxymethylglutarate

9. What is the net ATP yield from the complete oxidation of the following compound by mitochondrial β-oxidation?

$$CH_3-(CH_2)_4-CH=CH-(CH_2)_5-COO^-$$

   A. 119
   B. 117
   C. 105

D. 102

E. 98

10. Oxidation of palmitoyl-CoA requires all the following enzymatic steps *except*

A. dehydration.

B. dehydrogenation.

C. oxidation of a hydroxyl group.

D. cleavage by coenzyme A.

E. hydration.

11. In lipid digestion and absorption,

A. small fatty acids enter the portal circulation bound to albumin.

B. cholesterol and its esters make up approximately 40% of the chylomicrons.

C. high levels of LDLs are associated with resistance to atherosclerosis.

D. triglycerides are transported from the liver to other tissues in the HDLs.

E. triglycerides are hydrolyzed by the action of amylase.

12. Carnitine plays an important role in metabolism because it

A. is an essential cofactor for the biosynthesis of fatty acids containing an odd number of carbon atoms.

B. is required for the extracellular transport of activated long-chain fatty acids.

C. accumulates in adipose tissue when there is a deficiency in essential fatty acids.

D. is required for the transport of activated long-chain fatty acids into the mitochondria.

E. is an allosteric effector for acetyl-CoA carboxylase.

13. Which compound bears the least structural relationship to cholesterol?

A. Lanosterol

B. Squalene

C. Sphingosine

D. Farnesyl pyrophosphate

E. Isopentenyl pyrophosphate

14. The rate-controlling reaction for the biosynthesis of cholesterol is

A. formation of malonyl-CoA.

B. condensation of two isoprene units to form farnesyl pyrophosphate.

C. cyclization of squalene.

D. production of mevalonate.

E. hydroxylation of glycocholate.

15. PGE_2 is characterized by all the following *except*

A. a five-membered ring.

B. two nonconjugated double bonds.

C. having arachidonate as its precursor.

D. requiring molecular oxygen for its biosynthesis.

E. serving as a precursor to the bile salts.

16. Thromboxanes

A. contain a seven-membered ring.

B. are derived from cholesterol.

C. exhibit accelerated biosynthesis in the presence of aspirin.

D. stimulate platelet aggregation during blood clotting.

E. are structurally related to carnitine.

17. All the following are true about the sodium-potassium pump *except*

A. it is an active transport system.

B. it is unidirectional.

C. it is the mechanism for transporting positive ions into the mitochondria.

D. it is an antiport system.

E. it is located in the plasma membrane of all mammalian cells.

18. Which of the following types of compounds are *most* often found associated with biological membranes?

A. Proteins and phospholipids

B. Proteins and nucleic acids

C. Nucleic acids and carbohydrates

D. Carbohydrates and amino acids

E. Fatty acids and amino acids

19. Both active transport and facilitated diffusion are characterized by

A. transport in only one direction.

B. a plateau in the rate of transport with a very high concentration of substrate.

C. hydrolysis of ATP.

D. transport only in the direction of a concentration gradient.

E. competitive inhibition when the competitor is on the opposite side of the membrane from the substrate.

20. Amino acids can be maintained inside cells against a concentration gradient because
    A. they bind to sites inside the cell.
    B. they are trapped inside peroxisomes so that they are not in equilibrium with the extracellular fluid.
    C. they are rapidly metabolized to carbon dioxide and water.
    D. they enter the cell more rapidly than they leave it because of metabolically coupled reactions.
    E. they are precipitated before being attached to tRNA for protein synthesis.

Part VI Amino Acid Metabolism

22 Amino Acid Biosynthesis

Humans do not have the ability to synthesize 8 of the 20 natural amino acids used in protein synthesis. These are the essential amino acids and must be provided in the diet. Two additional amino acids, arginine and histidine, can be synthesized but not in sufficient quantities for optimal growth in infants and children. These amino acids are sometimes called **semiessential amino acids.** The rest of the amino acids can be synthesized either from metabolic intermediates or from essential amino acids.

Biosynthesis of several of the amino acids depends on the availability of one-carbon units. These exist at levels of oxidation from carboxylate to methyl and are bound to tetrahydrofolate, biotin, and choline. One-carbon units are also important for the biosynthesis of carbohydrates, fatty acids, and nucleotides. Inherited or acquired deficiencies in one-carbon metabolism can produce serious clinical consequences.

Objectives

After completing this chapter, you should be able to

List the eight essential amino acids.

List the starting compound and the major intermediates required for the biosynthesis of the semiessential and nonessential amino acids.

Identify the sources of one-carbon units at the various levels of oxidation and diagram their interconversion.

Essential Amino Acids

There are eight essential amino acids for humans that must be supplied in the diet in sufficient amounts for protein synthesis. These are isoleucine, leucine, lysine, methionine, phenylalanine, threonine, tryptophan, and valine. Additional amounts of methionine and phenylalanine are required as these amino acids serve as precursors for the synthesis of cysteine and tyrosine, respectively.

Semiessential Amino Acids

Arginine and histidine are classified as semiessential amino acids. Arginine is synthesized as one of the

intermediates of the urea cycle. However, its production is not in sufficient quantities for maximal growth, especially in children. Histidine is classified as a semiessential amino acid on the basis of dietary studies in which adult volunteers were fed diets totally lacking in histidine and still maintained nitrogen balance. Nonetheless, histidine sometimes is listed as an essential amino acid.

Arginine

Arginine is formed as an intermediate of the urea cycle. Because most of the arginine is cleaved to form urea, not enough remains to satisfy the needs of protein synthesis. The route for the biosynthesis of arginine will be discussed in detail in Chapter 23 as part of the urea cycle. However, it is formed by cleavage of arginosuccinate to produce arginine and fumarate.

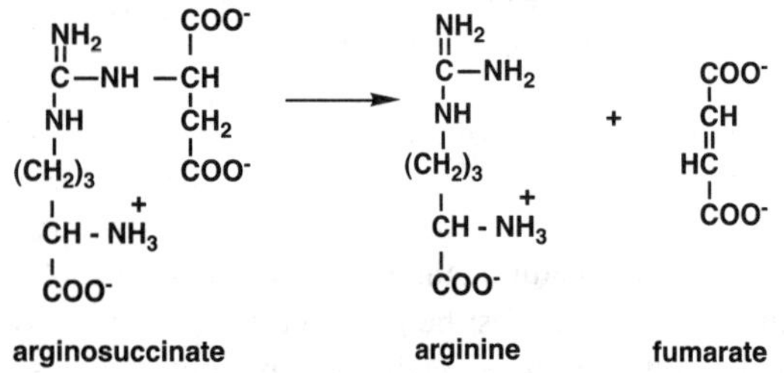

arginosuccinate   arginine   fumarate

Histidine

The major intermediates in the complex pathway for the biosynthesis of **histidine** are given in Figure 22-1. This pathway was determined in bacteria. However, the pathway in humans is probably similar. The starting material for the biosynthesis of histidine is ribose 5-phosphate, which comes from the pentose phosphate pathway. In the first reaction, ribose 5-phosphate reacts with ATP, producing **phosphoribosyl pyrophosphate** (PRPP). You will encounter this reaction again in the biosynthesis of nucleotides.

Fig. 22-1. Biosynthesis of histidine starting from ribose 5-phosphate.

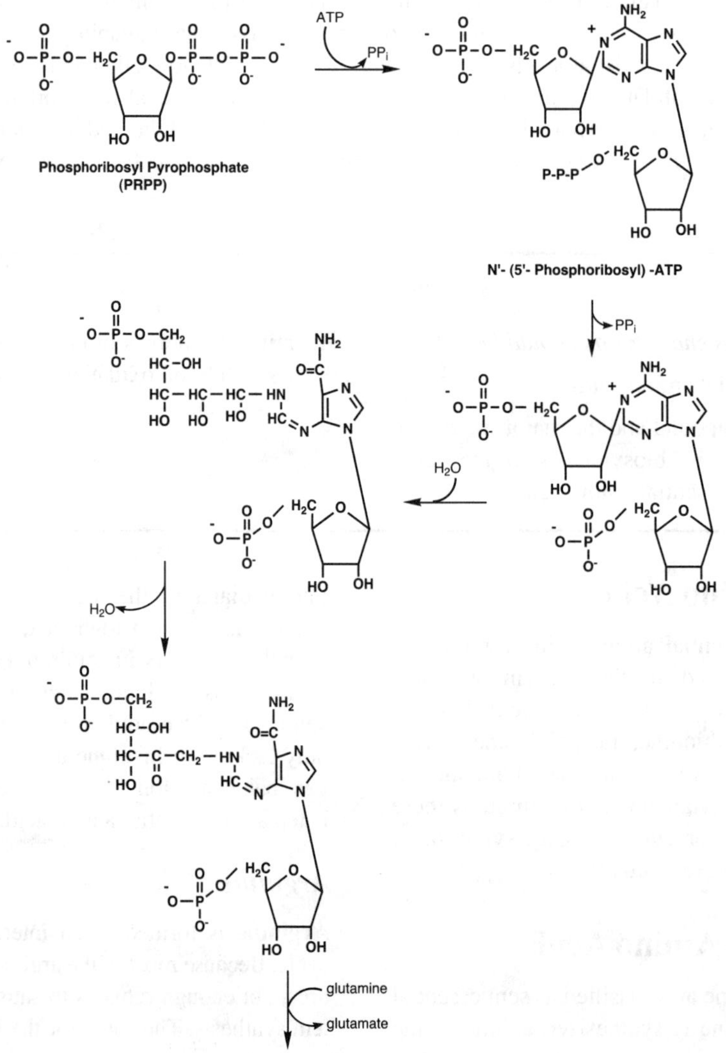

Phosphoribosyl Pyrophosphate
(PRPP)

N'- (5'- Phosphoribosyl) -ATP

PRPP reacts again with ATP, yielding phosphoribosyl-ATP, which then loses phosphate to form phosphoribosyl-AMP. The six-member adenine ring is opened, and dehydration follows. A nitrogen is transferred from glutamate and the ring recyclized and cleaved from what remains of the AMP. After dehydration, transamination, and several oxidations, histidine is produced.

Biosynthesis of Nonessential Amino Acids

There are 10 nonessential amino acids for humans: alanine, aspartate, asparagine, cysteine, glutamate, glutamine, glycine, proline, serine, and tyrosine. These amino acids can be synthesized in sufficient amounts to provide optimal growth and maintenance for both adults and children. In general, the biosynthesis of the nonessential amino acids is regulated by availability of these amino acids in the diet. If an enzyme in the biosynthesis pathway for one of the nonessential amino acids is defective, the amino acid cannot be synthesized. Then it becomes an essential amino acid for individuals with that enzyme defect.

The hydroxylated amino acids, **hydroxyproline** and **hydroxylysine,** occur almost exclusively in collagen. They are formed by oxidation of their respective amino acids after the collagen protein chain has been synthesized. These oxidations require molecular oxygen and ascorbic acid (vitamin C). The amino acid **cystine** also is produced only in a protein chain that has already been synthesized. It is formed by oxidizing two cysteine residues, forming a disulfide bond.

Fig. 22-1 (continued).

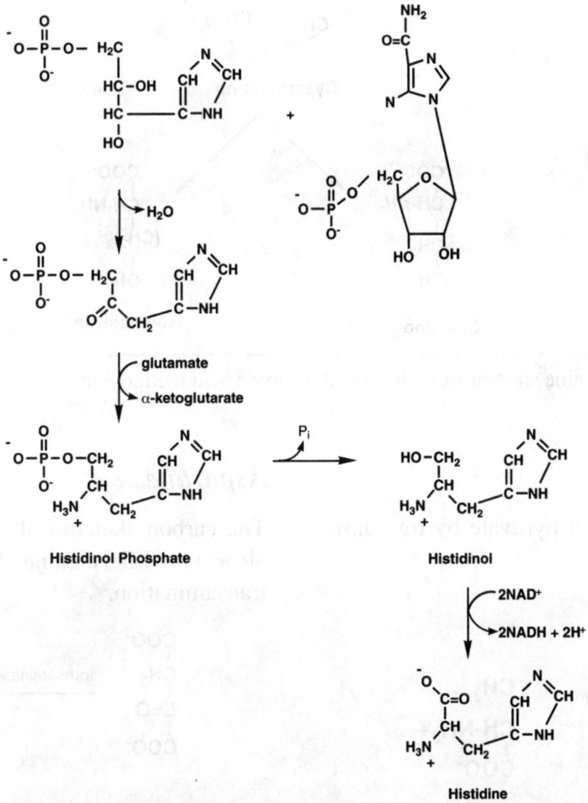

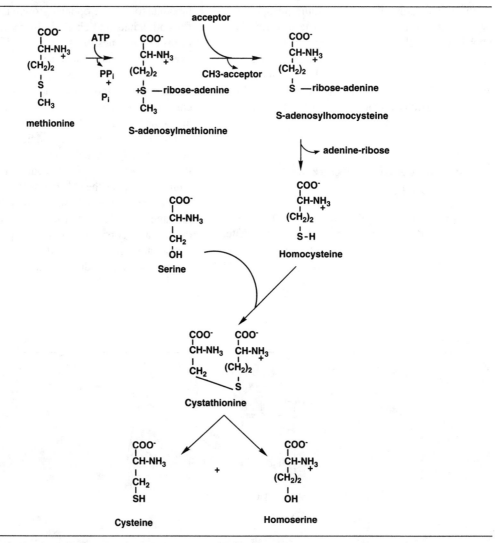

Fig. 22-2. Biosynthesis of cysteine starting from the essential amino acid methionine.

Alanine

Alanine is synthesized from pyruvate by transamination.

Aspartate

The carbon skeleton of **aspartate** comes from oxaloacetate. Like alanine, the amino acid is formed by transamination.

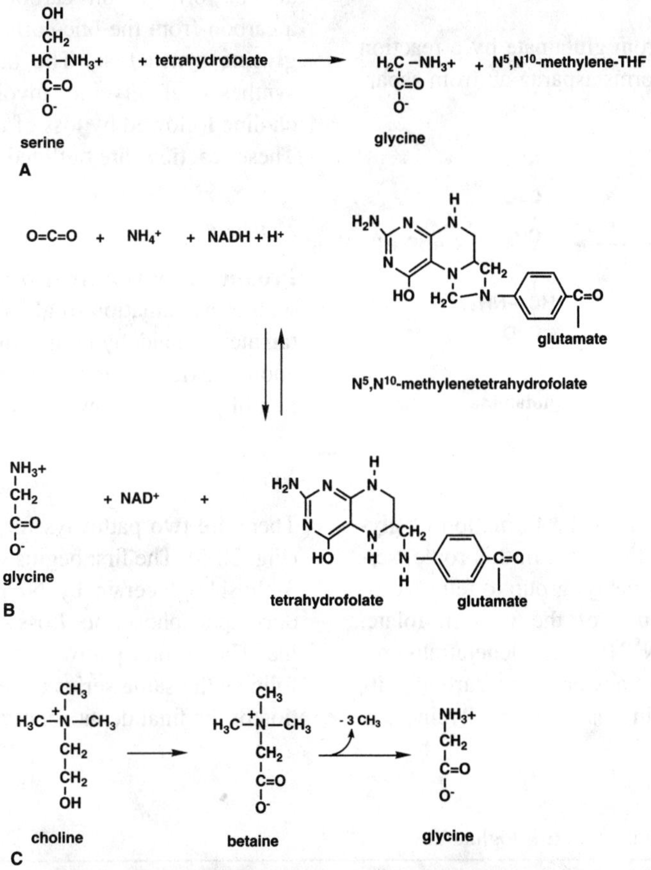

Fig. 22-3. Three pathways for biosynthesis of glycine.

Asparagine

Asparagine is formed from aspartate by the addition of ammonia in a reaction that requires the hydrolysis of ATP.

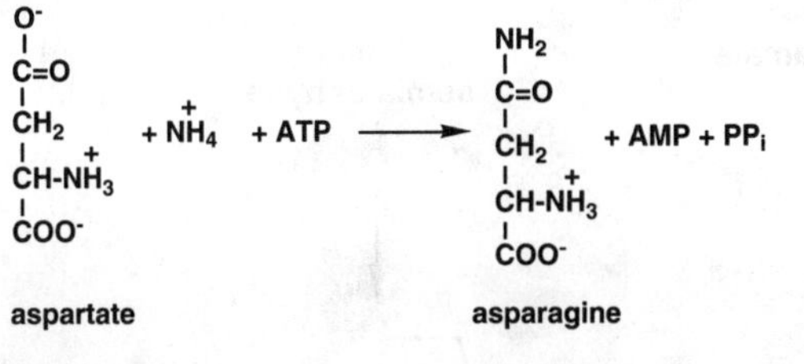

Cysteine

The carbon skeleton of **cysteine** comes from serine and the sulfur from methionine, an essential amino acid. The reaction sequence is given in Figure 22-2.

Glutamate

The carbon skeleton of **glutamate** comes from α-ketoglutarate. Like alanine and aspartate, glutamate is formed by transamination of the keto acid.

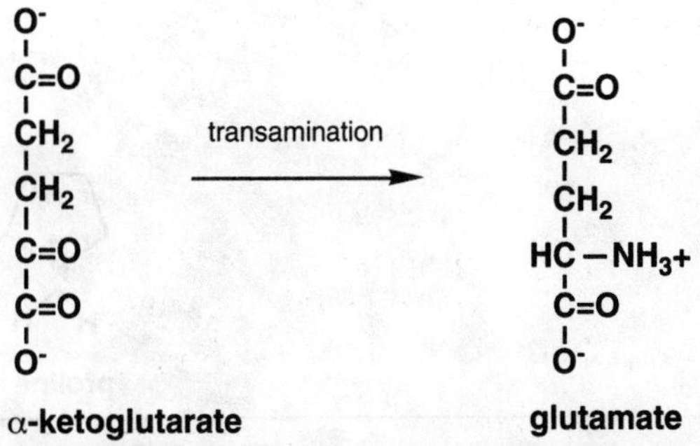

Glutamine

Glutamine is formed from glutamate by a reaction similar to the one that forms asparagine from aspartate.

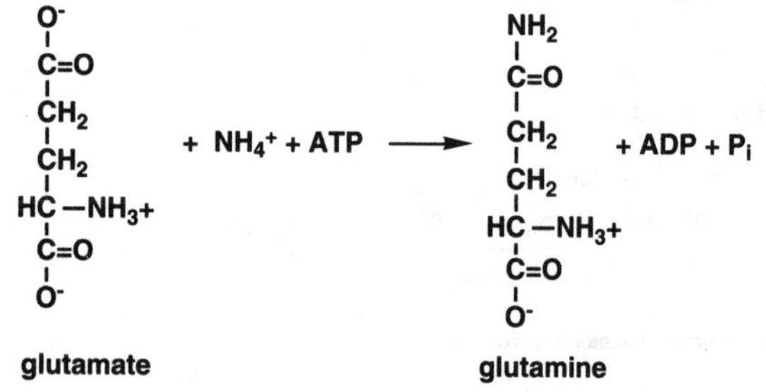

glutamate $\xrightarrow{+ NH_4^+ + ATP}$ glutamine $+ ADP + P_i$

Glycine

There are three pathways for the formation of **glycine.** The first begins with serine. In this route, serine transfers its hydroxymethyl group to tetrahydrofolate (THF), a derivative of the vitamin folate, producing glycine and N^5,N^{10}-methylenetetrahydrofolate. Its function in the transfer of one-carbon units will be discussed later in this chapter. Glycine can also be formed from carbon dioxide, ammonia, and a carbon from the one-carbon pool by the action of **glycine synthetase.** The third mechanism for the synthesis of glycine involves the oxidation of choline followed by loss of the three methyl groups. These reactions are outlined in Figure 22-3.

Proline

Proline is formed from α-ketoglutarate, beginning with transamination to glutamate. Reduction to glutamate semialdehyde and ring closure followed by another oxidation form proline (Fig. 22-4). Degradation of proline follows a reversal of these reactions.

Serine

There are two pathways for the synthesis of **serine** (Fig. 22-5). The first begins with dehydrogenation of 3-phosphoglycerate by NAD. Transamination produces phosphoserine. Loss of phosphate yields serine. The second pathway begins with glycerate and follows the same series of reactions with the exception of the final dephosphorylation step.

Fig. 22-4. Synthesis of proline from α-ketoglutarate.

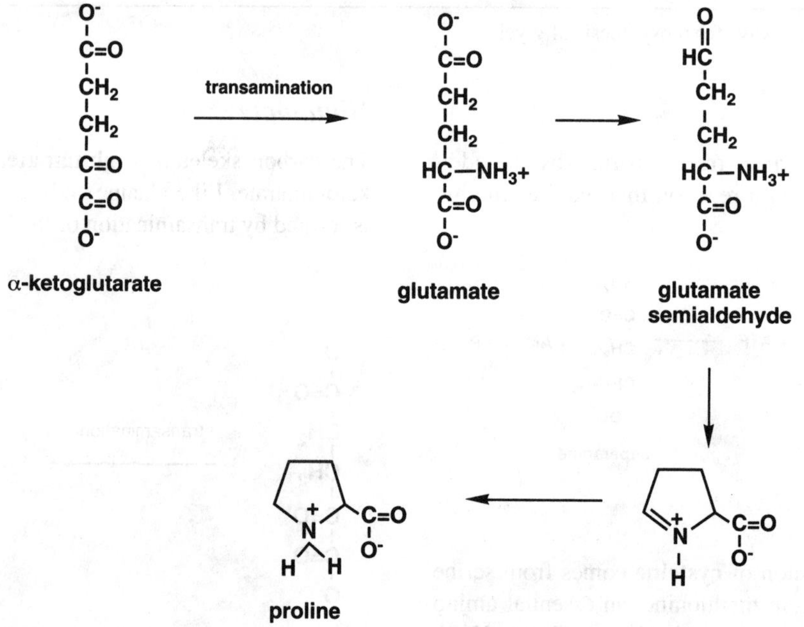

α-ketoglutarate $\xrightarrow{transamination}$ glutamate $\longrightarrow$ glutamate semialdehyde $\longrightarrow$ proline

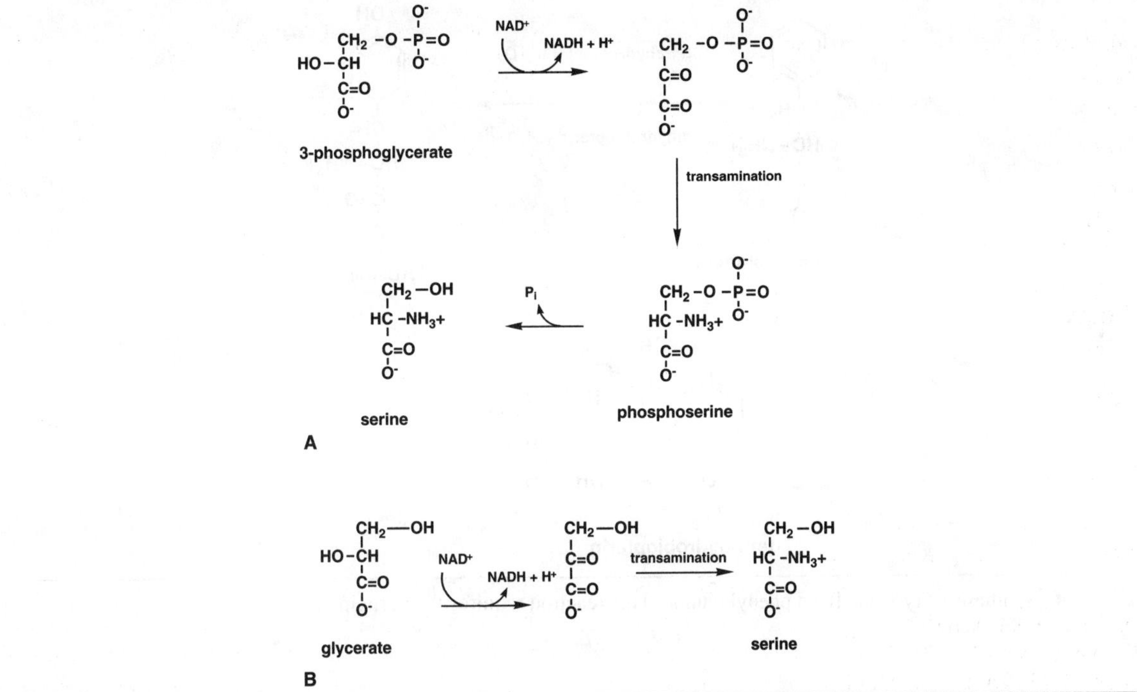

Fig. 22-5. Two pathways for the synthesis of serine. A. starting from 3-phosphoglycerate. B. Starting from glycerate.

Tyrosine

Tyrosine is formed from phenylalanine, an essential amino acid, by oxidation in a reaction that is also part of the degradative pathway. The **dihydrobiopterin** formed in the reaction is reduced to tetrahydrobiopterin by NADPH (Fig. 22-6).

Transfer of One-Carbon Units

One-carbon units are essential for the synthesis of many biologically important compounds. You have already encountered methyl transfers from S-adenosylmethionine in the synthesis of phospholipids and maturation of RNAs. Another source of methyl groups is choline. As shown in the reaction sequence in Figure 22-7, the hydroxymethyl group of choline is oxidized to a carboxylic acid, producing **betaine.** Transfer of one of the methyl groups to homocysteine gives methionine.

The majority of one-carbon transfers in biosynthetic reactions come from **tetrahydrofolate** (THF).

THF is formed by reduction of the vitamin folate first to dihydrofolate and then to tetrahydrofolate.

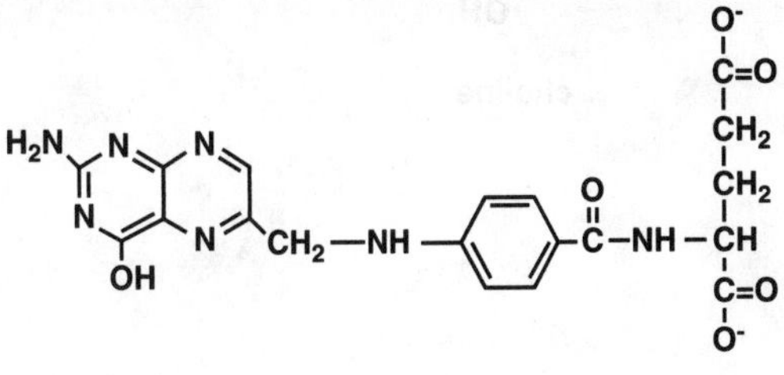

folate

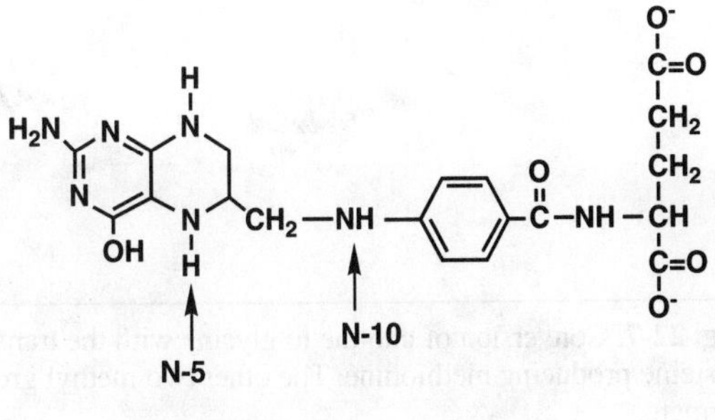

tetrahydrofolate

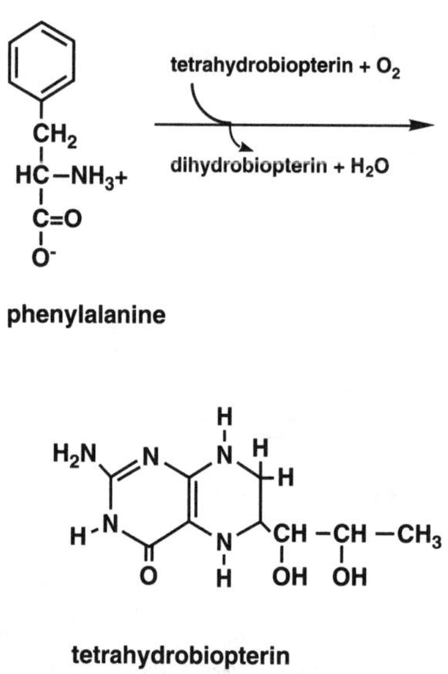

Fig. 22-6. Synthesis of tyrosine from phenylalanine. This reaction requires the participation of tetrahydrobiopterin.

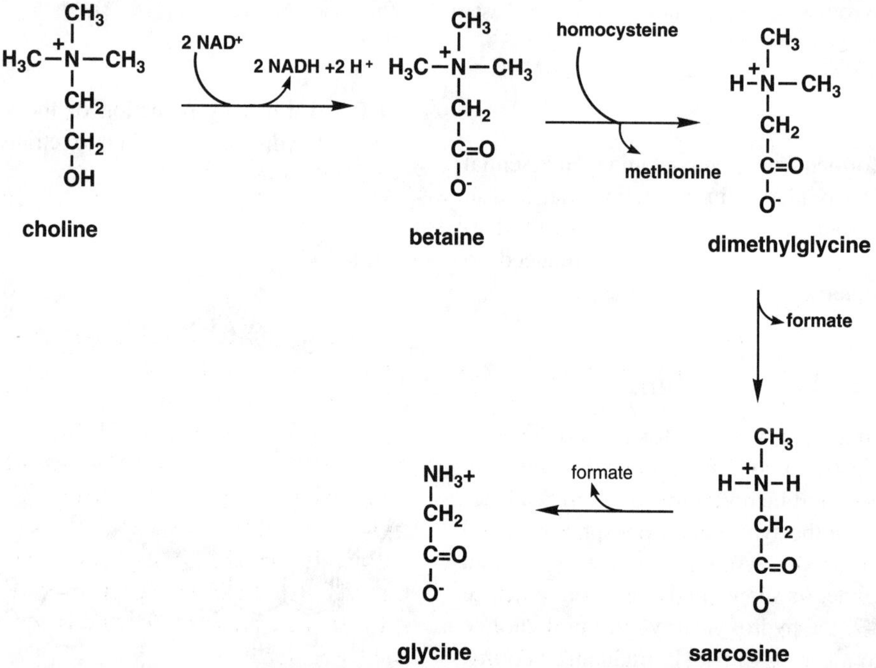

Fig. 22-7. Conversion of choline to glycine with the transfer of one methyl group to homocysteine producing methionine. The other two methyl groups are oxidized and lost as formate.

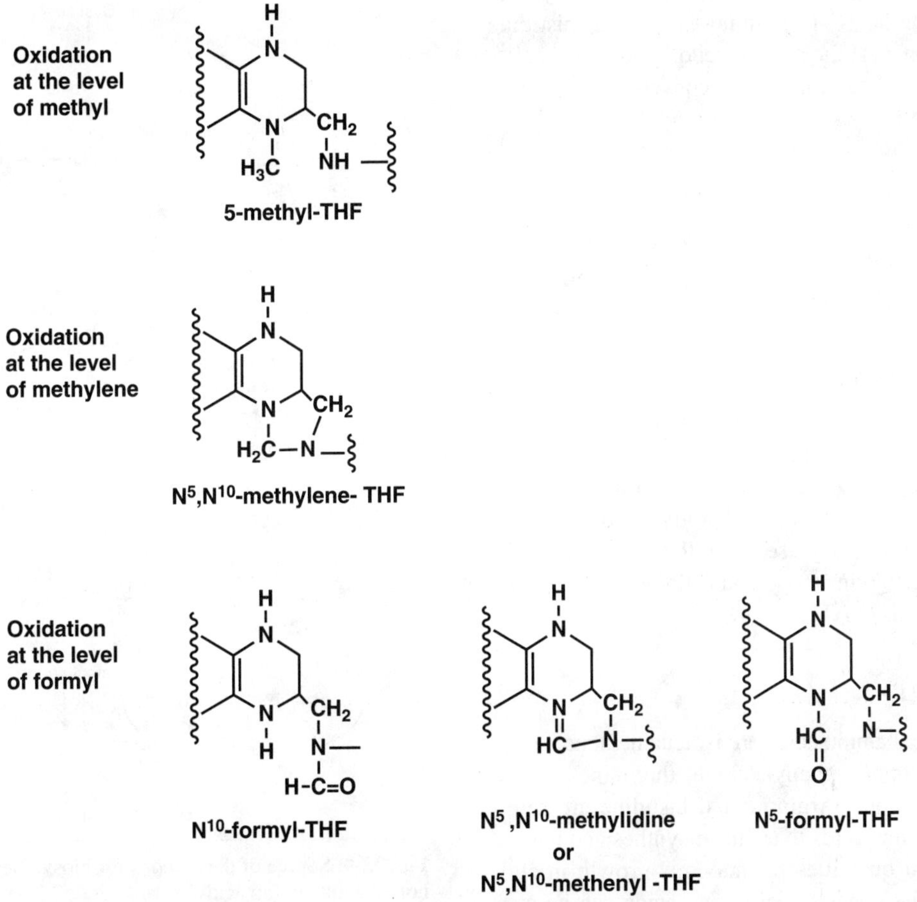

Fig. 22-8. Structure of one-carbon derivatives of tetrahydrofolate (THF) at different levels of oxidation.

One-carbon units are transferred from glycine or serine and attached to the N^5 or N^{10} positions of THF. All levels of oxidation of carbon can be bound to THF except for carbon dioxide. The various levels of oxidation are interconvertible by oxidation or reduction while the carbon remains attached to THF. The structures of the major one-carbon THF intermediates are shown in Figure 22-8.

N^5,N^{10}-**methylene-THF** is formed when serine is converted into glycine. This is the form of THF required in the biosynthesis of the nucleotide thymidine. N^5-**methyl-THF** is formed from N^5,N^{10}-methylene-THF in an irreversible reduction by NADH. N^5-methyl-THF can transfer its methyl group to ho-

mocysteine, producing methionine. This transfer requires the participation of vitamin B_{12}. If the vitamin is deficient, N^5-methyl-THF accumulates. Because of the irreversibility of the formation of N^5-methyl-THF, under these conditions, most of the THF will be trapped as N^5-methyl-THF. This will stop all one-carbon transfers because there will be no one-carbon THF at any oxidation level except methyl. **Megaloblastic** or **pernicious anemia** results from the effect of this inhibition on nucleotide and DNA synthesis.

Oxidation of N^5,N^{10}-methylene-THF gives N^{10}-**formyl-THF**, a compound used in the biosynthesis of purines and in the formation of formylmethionine, which is required in the initiation of protein synthesis in bacteria. N^5-formimino-THF, from histidine degradation, is at the same level of oxidation as N^{10}-formyl-THF. They can be interconverted by the gain or loss of NH_4.

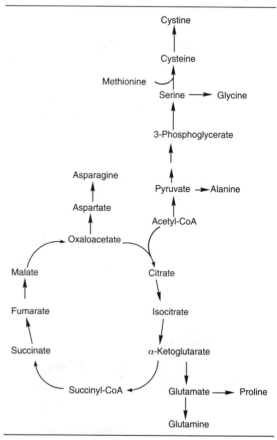

N^5-formimino-THF   **N^5,N^{10}-methenyl-THF**

You have already encountered one-carbon transfers at the carboxylate level in gluconeogenesis (pyruvate carboxylase), fatty acid oxidation (propionyl-CoA carboxylase) and fatty acid synthesis (acetyl-CoA carboxylase). All these reactions use the vitamin biotin and carbon dioxide and require the hydrolysis of ATP.

Summary

The essential amino acids are isoleucine, leucine, lysine, methionine, phenylalanine, threonine, tryptophan, and valine. Arginine and histidine are semiessential amino acids that can be synthesized but not in sufficient quantities for maximal growth in children. All the nonessential amino acids can be synthesized by humans. They include alanine, aspartate, asparagine, cysteine, glutamate, glutamine, glycine, proline, serine, and tyrosine. The relationship among these amino acids and the mainstream of metabolism is summarized in Figure 22-9.

One-carbon units at the level of methyl can be transferred from N^5-methyl-THF, S-adenosylmethionine, or betaine. Transfer at the methylene level occurs from N^5,N^{10}-methylene-THF. N^{10}-formyl- and N^5-formimino-THF are at the next higher oxidation level.

Fig. 22-9. Source of the carbons for biosynthesis of the nonessential amino acids.

One-carbon units attached to THF can be interconverted by oxidation or reduction. All these reactions are reversible except the formation of N^5-methyl-THF. Lack of vitamin B_{12} prevents transfer of this methyl group, which results in inhibition of one-carbon metabolism and megaloblastic anemia. Transfer of carboxylate groups occurs using the vitamin biotin as a cofactor.

23 Protein Degradation and Disposal of Amino Acid Nitrogen

Cellular proteins and those ingested as part of the diet are degraded to amino acids by specific proteases. To prevent self-digestion of the proteases before they can function, the digestive enzymes are secreted as zymogens and then activated. The cellular proteases are protected by storage in the lysosomes. If the amino acids produced from protein degradation are not needed for protein synthesis, the α-nitrogen is removed and the carbon skeleton oxidized or converted to glucose or ketone bodies. The nitrogen is collected as the α-amino group of glutamate. From there, the nitrogen can be released in the liver or kidney as ammonia. Mammals convert 80% of the ammonia to urea. Because ammonia is toxic, inherited or acquired defects in the transport of ammonia or its conversion to urea produce serious and frequently fatal consequences.

Objectives

After completing this chapter, you should be able to

Outline the steps in the digestion of protein.

List the enzymes involved in protein digestion and their activation process.

Describe the process of HCl secretion in the stomach.

Describe the removal of nitrogen by transamination and oxidative deamination.

Explain the concept of nitrogen balance.

List the reactions producing ammonia and describe how ammonia is transported from the tissues to the liver.

Name the intermediates of the urea cycle and the enzymes involved in the cycle and identify the points at which energy is required.

For each of the inherited diseases in nitrogen disposal, list the disease, the enzyme deficiency, the major symptoms, and the rationale for treatment.

Digestion of Proteins

In the Stomach

Digestion of proteins begins in the stomach. Here the enzyme **pepsin** digests proteins by breaking peptide bonds on the amino side of aromatic (phe, tyr, trp), hydrophobic (leu, ile, met), or dicarboxylic (glu, asp) amino acids, as shown in Figure 23-1. Pepsin has a pH optimum of 2, so it has high activity in the acidic environment of the stomach and ceases activity when proteins move into the

more alkaline environment of the small intestine.

Pepsin is secreted as the zymogen **pepsinogen.** It is activated by the acid in the stomach and autocatalytically by the action of pepsin itself. In the activation

Fig. 23-1. Peptide bonds hydrolyzed by pepsin.

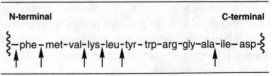

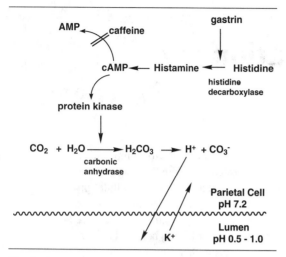

Fig. 23-2. Production of H⁺ by the parietal cells of the stomach.

process, 42 amino acids are removed from the N-terminal end of the zymogen. This cleavage alters the conformation of the protein and exposes the active site.

Because the low pH of the stomach is essential for the action of pepsin and, in fact, is capable of hydrolyzing some peptide bonds, let us digress for a short discussion of how this unusual pH is attained. The H⁺ for HCl comes from the carbonic anhydrase reaction outlined in Figure 23-2. The parietal cells of the stomach secrete the H⁺ against a concentration gradient of 10^{-7} M inside the cells to 10^{-1} M outside the cells. The secretion is driven by hydrolysis of ATP in a reaction that exchanges H⁺ for K⁺. **Histamine** increases H⁺ secretion by increasing the intracellular concentration of cAMP. This, in turn, activates cAMP-dependent protein kinase, which activates carbonic anhydrase by phosphorylation. Any agent, such as caffeine, that inhibits cAMP breakdown increases acid secretion. The hormone **gastrin** has a similar effect on acidity by activating histidine decarboxylase, which produces histamine. This is followed by the rest of the cAMP cascade.

In approximately 4% of the total adult population and 40% of those older than 60 years HCl secretion by the stomach is absent, a condition known as **achlorhydria.** In these persons, pepsin is not active and digestion of proteins does not begin until food passes into the intestine. Because there are a number of proteases found in the intestine, no major clinical symptoms result.

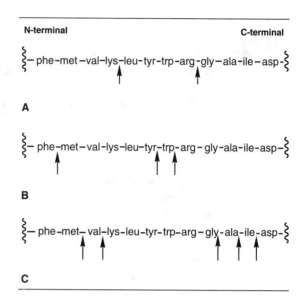

Fig. 23-3. Peptide bonds hydrolyzed by trypsin (A), chymotrypsin (B), and elastase (C).

In addition to pepsin, the enzyme **rennin** (chymosin) also is found in the stomach. Rennin coagulates milk. This is an important reaction in infants because it prevents milk from passing too quickly into the intestine. In the presence of calcium, rennin converts the casein in milk to paracasein which, unlike casein, is a substrate for hydrolysis by pepsin.

In the Intestine

In the intestine, proteins are acted on by a number of pancreatic enzymes including **trypsin, chymotrypsin, carboxypeptidase,** and **elastase.** All these enzymes are secreted as zymogens. Enteropeptidase or enterokinase is capable of converting trypsinogen to trypsin. Trypsin, in turn, can activate the zymogens for the other peptidases—chymotrypsinogen, procarboxypeptidase, and proelastase. Trypsin breaks peptide bonds next to basic amino acids. Chymotrypsin acts on peptide bonds adjacent to amino acids with aromatic side chains. Elastase acts on peptide bonds next to amino acids with aliphatic side chains. The combined action of these three enzymes produces free amino acids, dipeptides, and tripeptides, as shown in Figure 23-3. Further action by carboxypeptidase, which cleaves amino acids from the carboxyl end of a protein or peptide, and by aminopeptidase, which cleaves from the N-terminal end, completes the digestion.

Trypsin, chymotrypsin, and elastase have serine in the active site and are called **serine proteases.** We encountered serine proteases before in our discussion of the blood-clotting factors. During the hydrolysis of proteins by these enzymes, the carboxyl end of one amino acid is covalently bound to a serine residue in the active site of the enzyme. The remainder of the protein or peptide is released. After water hydrolyzes the serine ester, the reaction is complete.

Fig. 23-4. Mechanism of transamination by enzymes containing pyridoxal phosphate.

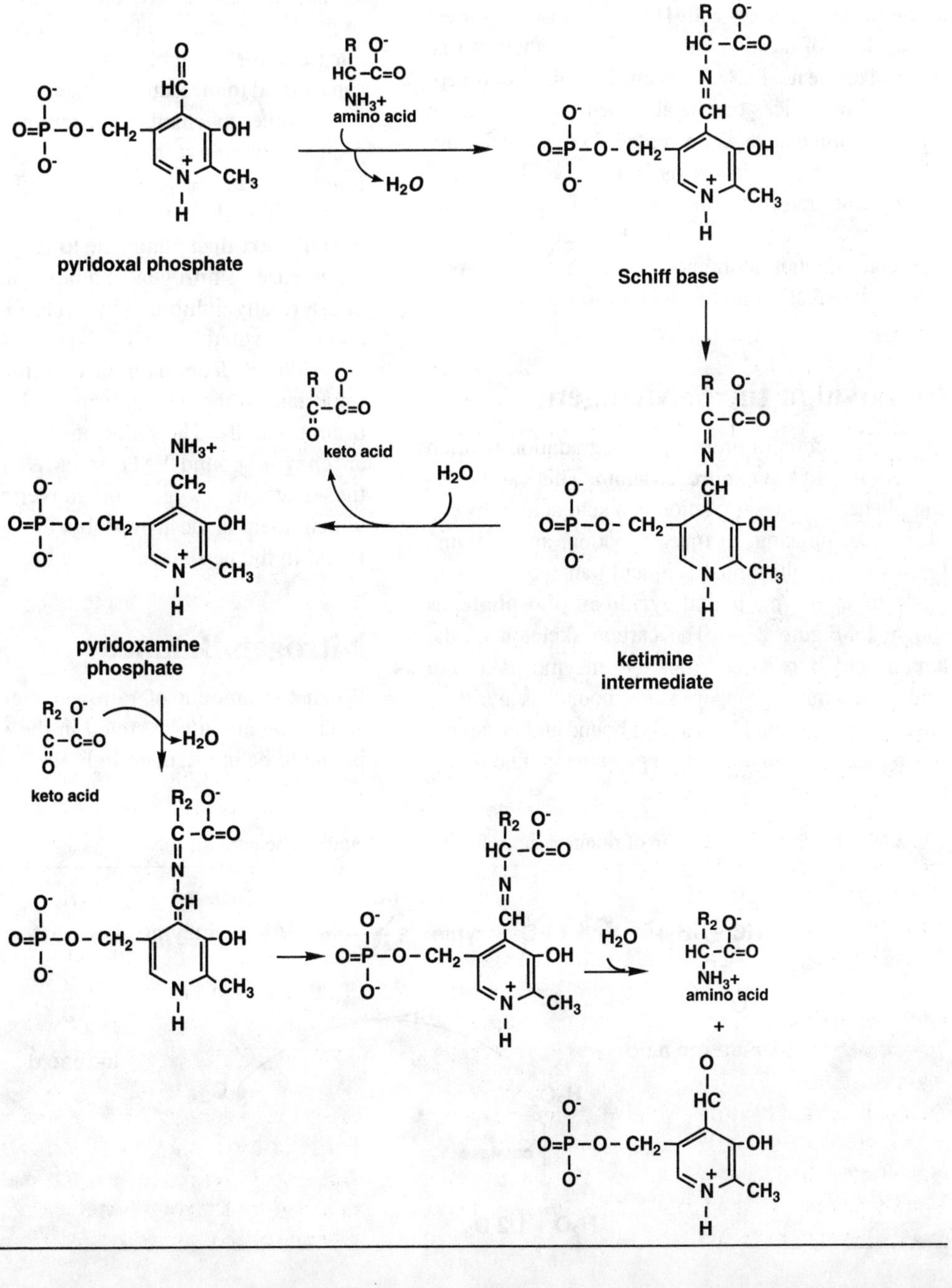

Absorption of Amino Acids

After protein digestion is complete, the amino acids must be absorbed, transported to the tissues, and taken up by cells. At least two inherited syndromes result from the inability to complete these processes. In **Hartnup's disease,** there is impaired uptake of neutral amino acids. The unabsorbed amino acids are excreted in the urine. A person with Hartnup's disease suffers from a lack of these amino acids unless they can be synthesized de novo. Because small amounts of dipeptides and tripeptides can be absorbed, total starvation for these amino acids is prevented. In **cystinuria,** uptake of the basic amino acids is defective. Defects in amino acid transport affect intestinal, cellular, and renal uptake. The effect in the kidney accounts for the high concentration of amino acids in the urine in persons with defects in amino acid transport.

Removal of the α-Nitrogen

The first reaction of amino acid degradation is often the removal of the α-nitrogen atom. This can be accomplished by transamination to a keto acid or by oxidative deamination. In transamination, amino transferases bind to the amino acid and transfer the amino group to an enzyme-bound **pyridoxal phosphate,** as shown in Figure 23-4. The carbon skeleton of the amino acid is released from the enzyme as a keto acid. The amino group remains bound as pyridoxamine. Then a second keto acid is bound and the amino group transferred to it, producing another amino acid.

Pyruvate and α-ketoglutarate are the most common keto acid recipients in transamination reactions. The transaminases are specific for one keto acid and its corresponding amino acid but have a wide specificity for the other amino and keto acids. Alanine or **alanine-pyruvate transaminase** and glutamate or **glutamate-α-ketoglutarate transaminase** are the major transamination enzymes and are present in the cytoplasm of most mammalian cells. When cell damage causes a release of these soluble enzymes, their levels in serum can be used to measure the extent of damage.

The other mechanism for removal of the α-amino group is oxidative deamination. This is especially important for the degradation of glutamate (Fig. 23-5). Liver glutamate dehydrogenase uses NAD or NADP to oxidize glutamate to α-ketoglutarate, with the release of nitrogen as ammonia. The reaction is allosterically inhibited by ATP, GTP, and NADH and is activated by ADP.

Oxidative deamination by amino acid oxidases is important in the metabolism by the peroxisomes of D-amino acids. The reductant for these reactions is an enzyme-bound FAD or FMN. In the course of the reaction, oxygen is converted to peroxide, which then is destroyed by the enzyme catalase found in the peroxisomes (see Fig. 23-5).

Nitrogen Balance

When the amount of nitrogen ingested in the diet equals the amount excreted in the urine, the person is said to be in **nitrogen balance.** For a 150-lb adult,

Fig. 23-5. Oxidative deamination of D-amino acids by enzymes found in the peroxisomes.

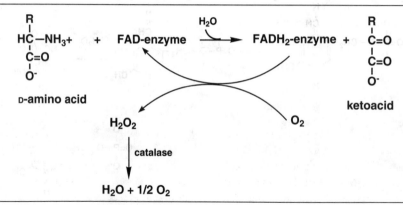

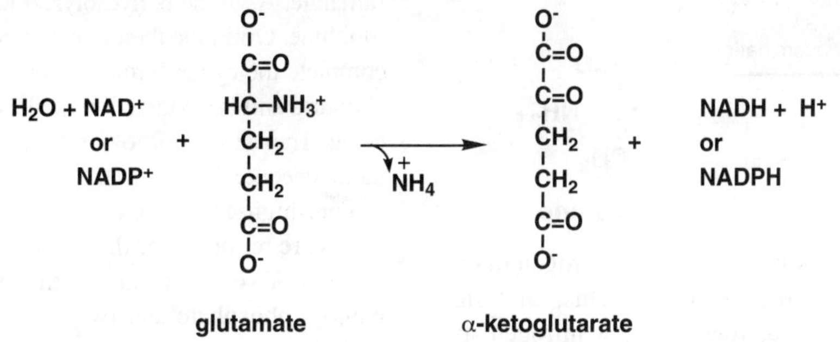

Fig. 23-6. Oxidative deamination of glutamate by glutamate dehydrogenase.

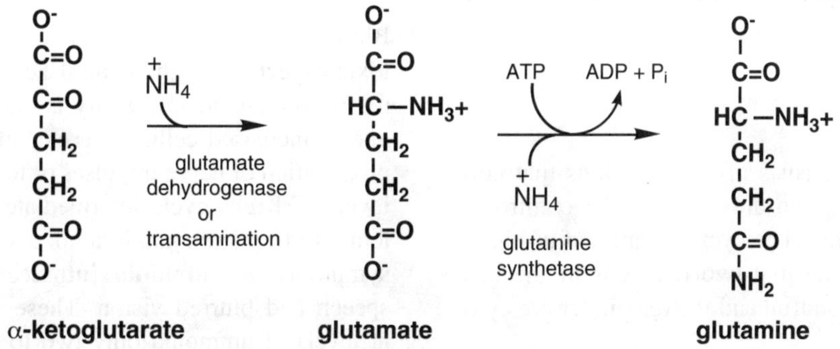

Fig. 23-7. Trapping of two molecules of ammonia by converting α-ketoglutarate first to glutamate and then to glutamine.

the protein requirement necessary to maintain nitrogen balance is only 30 g, or less than 2 oz of protein per day. A **positive nitrogen balance** occurs when there is a net gain in the amount of nitrogen in the body. This occurs in growth, healing, and pregnancy.

In **negative nitrogen balance,** there is a net loss of body nitrogen. Because there is no mechanism for storing nitrogen as amino acids or protein, a negative nitrogen balance is detrimental to health. A negative nitrogen balance can occur in malnutrition, after surgery, and in the elderly, especially if they are confined to bed.

Sources of Ammonia

Ammonia is produced by the cellular degradation of amino acids and nucleic acids as well as by the action of intestinal bacteria. This degradation occurs in persons with positive, negative, and normal nitrogen balance. Glutaminase, asparaginase, and the amino acid oxidases all produce ammonia. Histidase, serine dehydratase, and cysteine dehydratase are other sources of cellular ammonia. However, the major source, especially in the liver, is a product of the reaction catalyzed by the mitochondrial enzyme **glutamate dehydrogenase,** shown in Figure 23-6. This enzyme releases ammonia from the nitrogen collected in glutamate from amino acid transamination reactions. Because it is a mitochondrial enzyme, the ammonia can be used immediately for urea synthesis.

Tissues other than liver and kidney do not synthesize urea. Consequently, tissues must be able to dispose of the ammonia produced in cellular reactions and transport it to the liver in some nontoxic form. Starting from α-ketoglutarate, two molecules of ammonia can be safely transported as glutamine (Fig. 23-7). This ammonia-trapping reaction is especially important for the central nervous system.

Muscle uses glutamine and also alanine to transport excess nitrogen. Alanine is produced from pyruvate, a readily available substrate in tissues metabolizing glucose or glycogen.

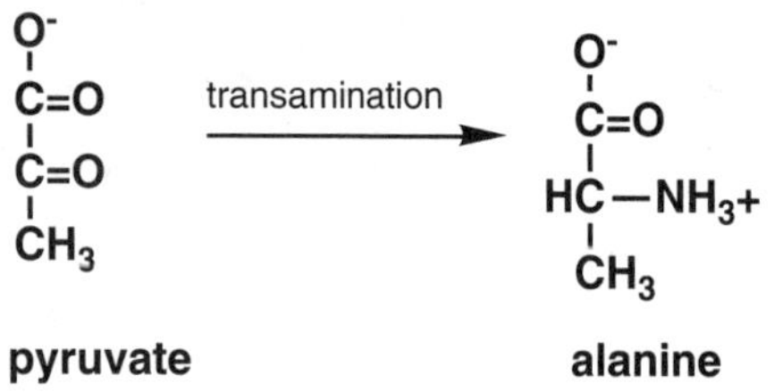

pyruvate   **alanine**

Once in the liver, alanine transfers its nitrogen to α-ketoglutarate, producing glutamate. This, in turn, can be dehydrogenated, releasing the nitrogen as ammonia. Glutamine arriving in the liver can release two molecules of ammonia by the combined actions of glutaminase and glutamate dehydrogenase (Fig. 23-8).

Urea Cycle

The urea cycle consists of five reactions that convert ammonia, carbon dioxide, and the α-nitrogen of aspartate to urea. The cycle is outlined in Figure 23-9. Note that the first two reactions of the urea cycle are mitochondrial and the remainder are cytoplasmic.

In the first reaction, mitochondrial carbon dioxide is phosphorylated by ATP and then condensed with ammonia, using the energy produced by hydrolysis of another ATP. The final product is **carbamoyl phosphate. Carbamoyl phosphate synthetase,** the enzyme catalyzing this rate-limiting reacting, requires N-acetylglutamate as a cofactor.

In the second reaction, carbamoyl phosphate condenses with ornithine, producing citrulline and releasing phosphate. This is the second rate-controlling reaction of the urea cycle. The citrulline leaves the mitochondria and condenses with aspartate in the third reaction of the cycle. This condensation requires the hydrolysis of ATP to AMP. The product, arginosuccinate, is converted in the fourth reaction to arginine and

fumarate. Arginine is hydrolyzed to produce urea and ornithine. Ornithine then re-enters the mitochondria to complete the cycle. Fumarate also can enter the mitochondria and be oxidized to oxaloacetate in the Krebs cycle. Transamination of oxaloacetate regenerates aspartate needed for reaction 3.

Four high-energy phosphate bonds or ATP equivalents are required for the synthesis of one molecule of urea. Two are used for the production of carbamoyl phosphate and two more for the condensation of aspartate and citrulline.

Defects in Nitrogen Disposal

Blood ammonia levels in excess of 5×10^{-5} M are toxic especially to the central nervous system. The reason for this neurotoxicity is unclear but may be due to increased cellular permeability affecting the propagation of nerve impulses or to a decrease in the levels of Krebs cycle intermediates used in the attempt to trap ammonia in a nontoxic form. The first symptoms of **ammonia intoxication** are slurred speech and blurred vision. These symptoms occur at levels of ammonia only two to three times normal. At higher ammonia concentrations, vomiting, lethargy, and anorexia occur. If the ammonia concentration is not decreased, seizures, coma, and death can follow.

Ammonia intoxication can be caused by inherited or acquired defects in ammonia trapping or the urea cycle. These inherited defects, which are summarized in Table 23-1 occur at a rate of 1 in every 30,000 births. **Hyperammonemia type I** is due to a defect in carbamoyl phosphate synthetase and **type II** to a defect in ornithine transcarbamoylase. Both of these defects in mitochondrial enzymes are maternally inherited and lead to an increase of ammonia and glutamine levels in the blood. Early death results if the defect produces a total enzyme deficiency.

Fig. 23-8. Release of two molecules of ammonia in the liver by conversion of glutamine to glutamate and then to α-ketoglutarate.

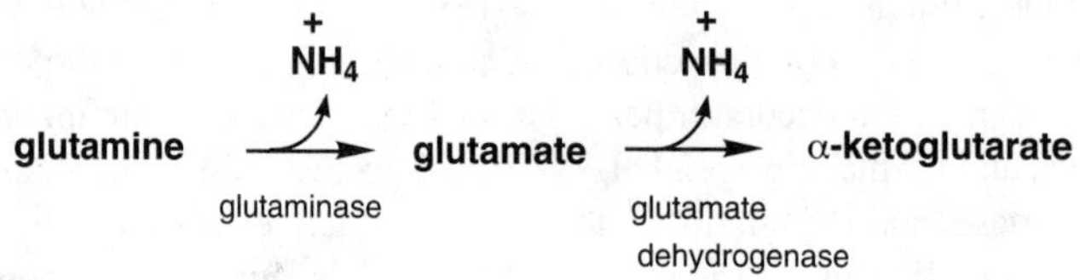

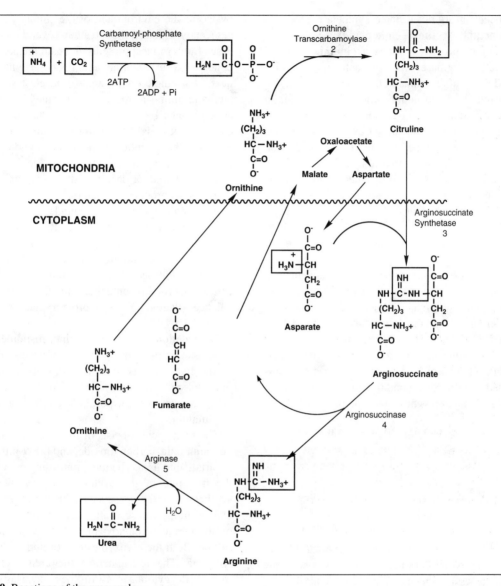

Fig. 23-9. Reactions of the urea cycle.

Table 23-1. Inherited defects in the urea cycle

Disease	Defective enzyme	Substance produced in excess
Hyperammonemia		
Type I	Carbamoyl phosphate synthetase	Ammonia
Type II	Ornithine transcarbamoylase	Ammonia
Citrullinemia	Arginosuccinate synthetase	Citrulline
Arginosuccinic aciduria	Arginosuccinate	Arginosuccinate
Arginemia	Arginase	Arginine

A defect in arginosuccinate synthetase produces **citrullinemia. Arginosuccinic aciduria** is due to a lack of arginosuccinase. Finally, **arginemia** is caused by a lack of arginase. All these defects lead to increased levels of ammonia in the blood and to severe mental retardation in infants who survive the initial episodes of ammonia intoxication.

Ammonia intoxication caused by inherited defects in the urea cycle enzymes after reaction 3, production of arginosuccinate, can be treated by a diet low in protein and amino acids and supplemented by arginine and citrulline. These amino acids increase the urinary excretion of arginosuccinate by competing with its resorption in the kidney. Treatment with sodium benzoate can produce additional disposal of nonurea nitrogen by combining with glycine. The product, **hippuric acid,** is excreted in the urine. Treatment with phenylacetate is even more effective as it condenses with glutamine, the major carrier of excess nitrogen. The resulting compound, **phenylacetylglutamine,** is excreted, carrying two nitrogens with it. Another mechanism of treatment of defects in the urea cycle is the administration of keto acids. These can be transaminated to their corresponding amino acids and excreted, carrying excess nitrogen with them.

hippurate
sodium benzoate + glycine

phenylacetylglutamine

A defect in glutamine synthetase also can lead to ammonia intoxication as the defect blocks one of the normal routes for the trapping of ammonia and its transport from peripheral tissues, especially the brain. Any disease or condition that adversely affects liver mitochondria also is capable of producing an increased level of ammonia in the blood. Such conditions include liver cirrhosis, alcoholism, hepatitis, and Reye's syndrome.

Summary

Pepsin hydrolyzes proteins in the stomach. It is secreted as a zymogen—pepsinogen—and activated by the acidic environment of the stomach and also autocatalytically. In the intestine, several endopeptidases further digest proteins. Trypsin breaks peptide bonds next to arginine and lysine, chymotrypsin next to aromatic amino acids, and elastase next to aliphatic amino acids. These enzymes are secreted as zymogens by the pancreas. Further breakdown occurs at the ends of the resulting peptides by the action of carboxypeptidase and aminopeptidase. A defect in the transport of neutral amino acids produces Hartnup disease; a defect in the transport of basic amino acids produces cystinuria.

Nitrogen-containing compounds are continually being synthesized and degraded. Excess nitrogen is converted to urea and excreted. This occurs in healthy persons who are in nitrogen balance, those in positive nitrogen balance (synthesizing a net increase in protein), and those in negative nitrogen balance (degrading more protein than is synthesized).

Degradation of serine, cysteine, histidine, glutamate, glutamine, and asparagine produces ammonia. In the liver, this is converted directly to urea. In other tissues, ammonia is added to pyruvate or α-ketoglutarate and transported to the liver as alanine or glutamine.

The urea cycle consists of five enzymes that convert ammonia, carbon dioxide, and the α-nitrogen of aspartate into urea. The rate-controlling enzymes are the first two of the cycle—carbamoyl phosphate synthetase, which produces carbamoyl phosphate, and ornithine transcarbamoylase, which condenses ornithine and carbamoyl phosphate to produce citrulline. Both these enzymes are located in the mitochondria. The remainder of the urea cycle takes place in the cytoplasm. Citrulline condenses with aspartate to make arginosuccinate, using ATP as the energy source. Arginosuccinate is split to form fumarate and arginine. Finally, arginine is hydrolyzed to urea and ornithine.

Defects in the urea cycle produce ammonia intoxication, which is characterized by central nervous system abnormalities followed, at higher levels, by coma and death. Some of the defects are treatable by a low-protein diet supplemented with sodium phenylacetate, sodium benzoate or keto acids to produce alternate routes of nitrogen disposal.

24 Amino Acid Degradation: Metabolism of the Carbon Skeleton

After the removal of the alpha nitrogen, the carbon skeletons of the amino acids either are oxidized directly to carbon dioxide and water or are converted to glucose (glucogenic amino acids) or ketone bodies (ketogenic amino acids). If metabolism of an amino acid yields products that can be converted into glucose, the amino acid is called **glucogenic.** If metabolism produces ketone bodies, the amino acid is **ketogenic.** The metabolism of leucine produces only ketone bodies. Five amino acids—phenylalanine, tyrosine, tryptophan, isoleucine, and lysine—produce both ketone bodies and intermediates for the synthesis of glucose. The remaining 15 amino acids are purely glucogenic.

Objectives

After completing this chapter, you should be able to

Name the glucogenic and ketogenic amino acids.

List the points at which each amino acid enters the mainstream of metabolism and the major intermediates involved.

Match the inherited defects of amino acid degradation with the disease name and the affected enzyme.

Amino Acids Converted to Pyruvate

Parts of seven amino acids enter the mainstream of metabolism as pyruvate. These are alanine, cysteine, serine, glycine, hydroxyproline, threonine, and tryptophan.

Alanine

Alanine is converted to pyruvate by transamination with α-ketoglutarate.

Cystine and Cysteine

The amino acids **cystine** and **cysteine** are both converted into pyruvate. First cystine is reduced by

269

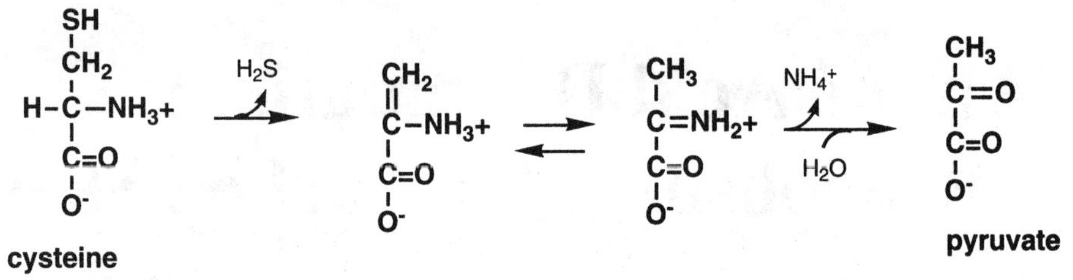

Fig. 24-1. Conversion of cysteine to pyruvate by cysteine dehydratase.

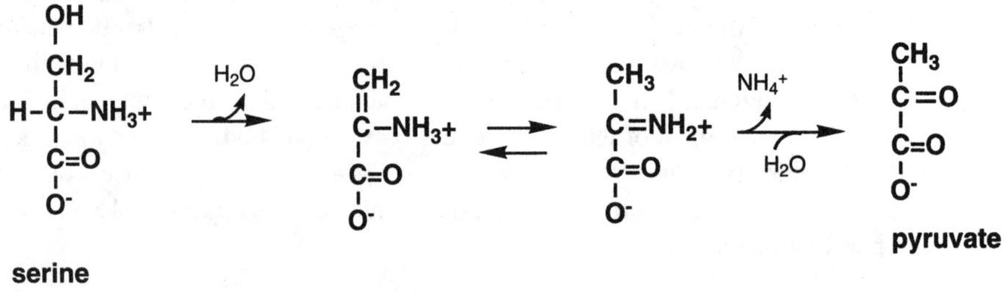

Fig. 24-2. Conversion of serine to pyruvate by serine dehydratase.

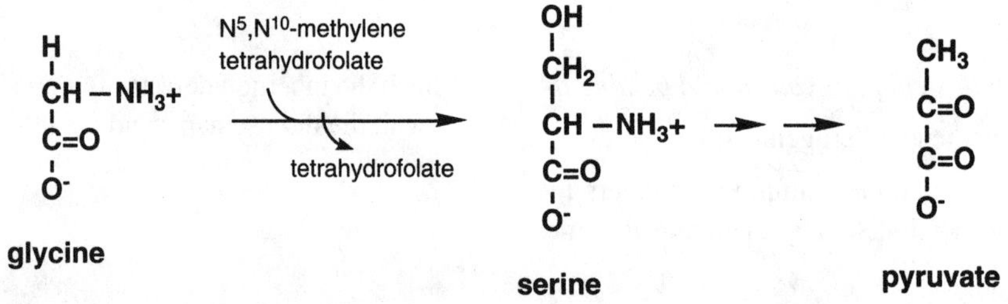

Fig. 24-3. Conversion of glycine to serine and then to pyruvate. The extra carbon is supplied from the one-carbon pool in the form of N^5,N^{10}-methylenetetrahydrofolate.

NADH to two cysteines. Cysteine can be converted into pyruvate by two different pathways. In the first one, transamination produces thiopyruvate. Loss of H_2S yields pyruvate.

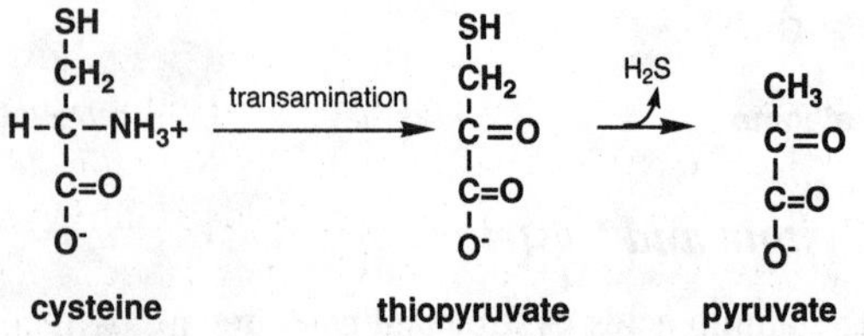

In the second pathway, H_2S is lost and then **cysteine dehydratase** gives pyruvate and ammonia (Fig. 24-1).

Serine

Serine is dehydrated by **serine dehydratase,** followed by hydrolysis, to give pyruvate and ammonia (Fig. 24-2). These reactions are similar to those for the second route of cysteine degradation.

Glycine

Glycine is converted into serine and then into pyruvate. The extra carbon atom comes from the one-carbon pool in the form of N^5,N^{10}- methylene-tetrahydrofolate (Fig. 24-3).

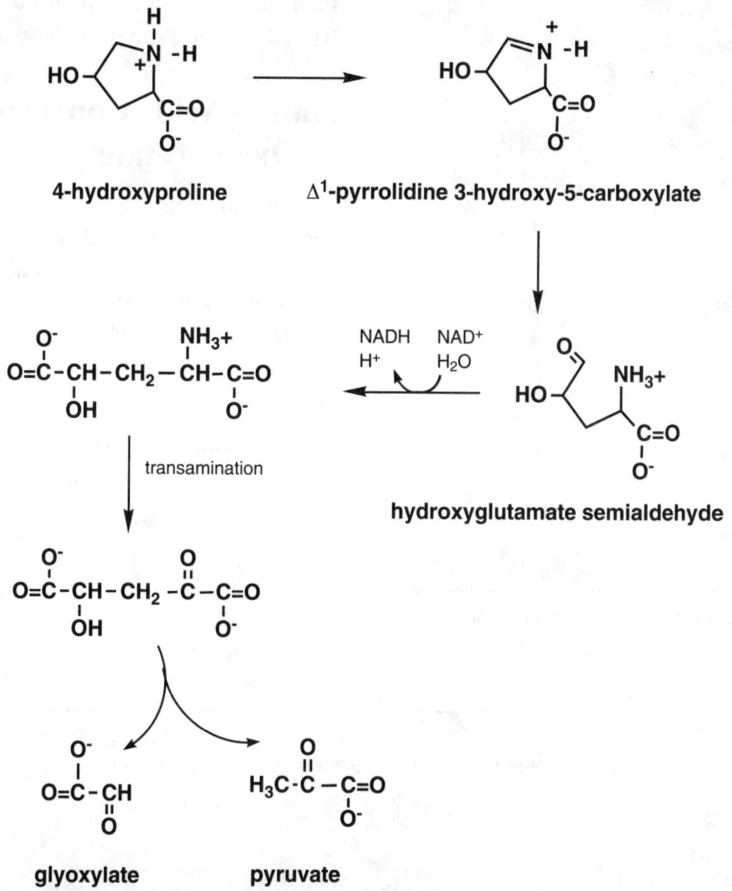

Fig. 24-4. Catabolism of hydroxyproline to glyoxylate and pyruvate.

Hydroxyproline

Hydroxyproline is found exclusively in collagen. It is degraded to pyruvate and glyoxylate by the series of reactions shown in Figure 24-4. In this pathway, transamination is one of the last reactions rather than one of the first.

Threonine

The initial reaction of **threonine** degradation produces glycine and acetaldehyde. Glycine is converted to pyruvate. Acetaldehyde is oxidized to ac-

etate and then condensed with coenzyme A to yield acetyl-CoA, as shown in Figure 24-5.

Tryptophan

During degradation, the side chain of **tryptophan** is converted to pyruvate and the rest of the molecule into acetoacetate. The pathway is outlined in Figure 24-6. The first and rate-controlling reaction is the ring opening, which is catalyzed by **tryptophan oxygenase.** The synthesis of this enzyme is induced by the presence of tryptophan. Note that the vitamin **nicotinamide** can be produced from one of the in-

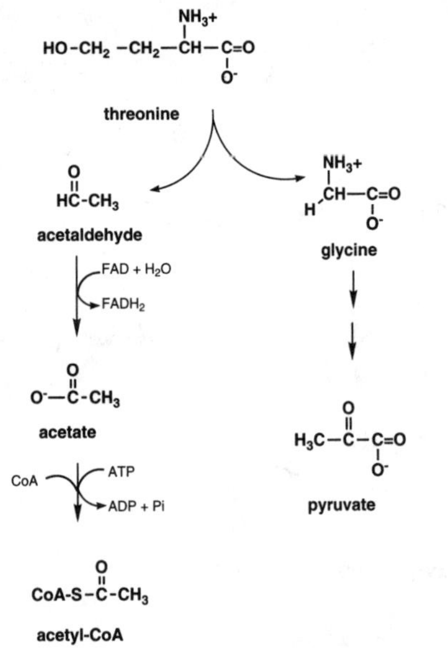

Fig. 24-5. Conversion of threonine to pyruvate and acetyl-CoA.

nicotinamide can be produced from one of the intermediates of tryptophan degradation.

Amino Acids Converted to Oxaloacetate

Two amino acids, asparagine and aspartate, are converted directly into oxaloacetate. However, the carbons from many other amino acids pass through oxaloacetate during gluconeogenesis or during oxidation through the Krebs cycle.

Aspartate

Transamination of aspartate with α-ketoglutarate gives glutamate and oxaloacetate.

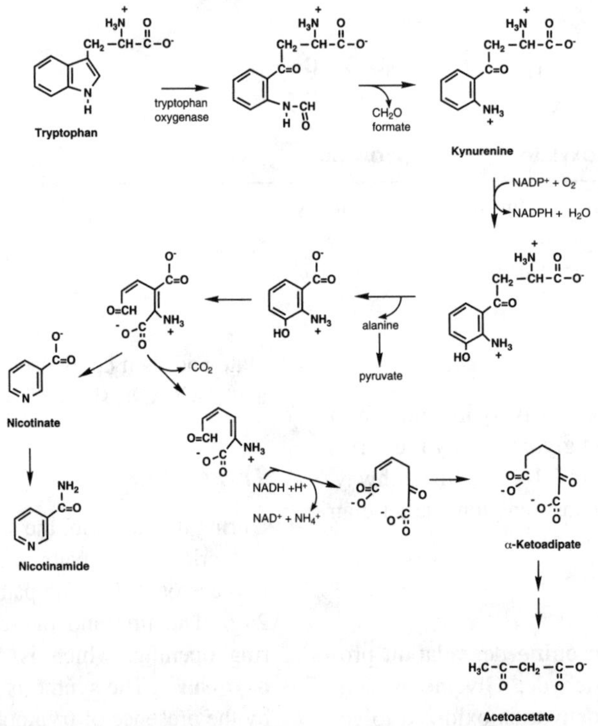

Fig. 24-6. Degradation of tryptophan to pyruvate and acetoacetate and the synthesis of nicotinamide.

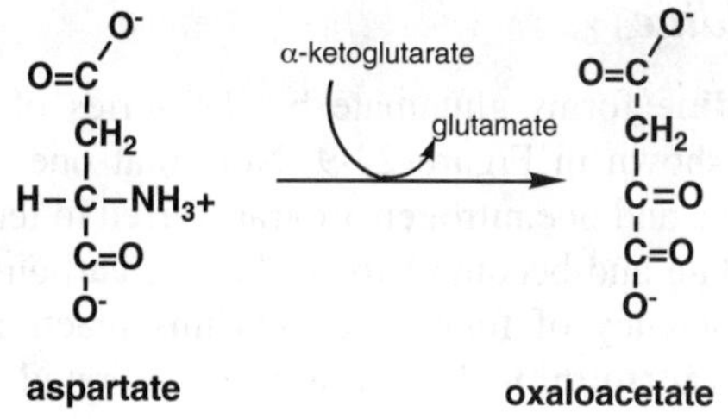

aspartate   oxaloacetate

Asparagine

Asparagine is deaminated by **asparaginase** to produce aspartate, which can then be transaminated to produce oxaloacetate (Fig. 24-7).

Amino Acids Converted to α-Ketoglutarate

Four amino acids are converted into glutamate and, by oxidative deamination, into α-ketoglutarate. These are glutamine, asparagine, histidine, and proline.

Glutamate

Glutamate is converted into α-ketoglutarate by oxidative deamination, as we saw in the previous chapter. In another route, glutamate can be transaminated by pyruvate, yielding alanine and α-ketoglutarate.

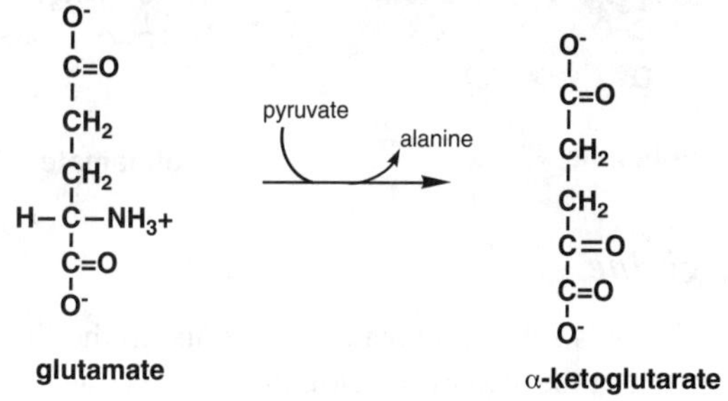

glutamate   α-ketoglutarate

Glutamine

Glutamine is converted into glutamate by **glutaminase.** This reaction, which we saw previously as a source of ammonia, is similar to the one producing aspartate from asparagine.

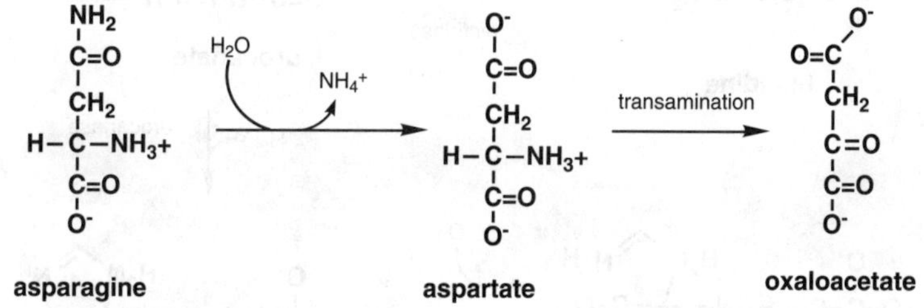

asparagine   aspartate   oxaloacetate

Fig. 24-7. Conversion of asparagine to aspartate by asparaginase. Aspartate can then be transaminated to produce oxaloacetate.

arginine   ornithine   glutamate semialdehyde   glutamate

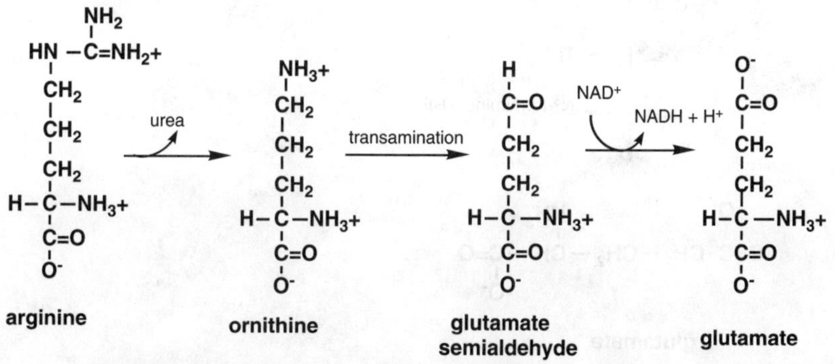

Fig. 24-8. Conversion of arginine to glutamate. The first reaction produces urea and ornithine, which can then be oxidized to glutamate.

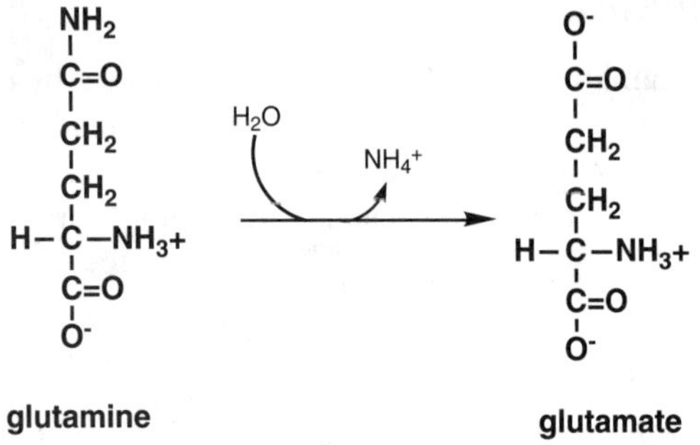

glutamine   glutamate

Arginine

Arginine is split into urea and ornithine in the final reaction of the urea cycle. Ornithine then is transaminated to glutamate semialdehyde and is subsequently oxidized to glutamate (Fig. 24-8).

Histidine

Histidine forms glutamate by the series of reactions shown in Figure 24-9. Note that one of the carbons and one nitrogen are transferred to tetrahydrofolate and become part of the one-carbon pool. A deficiency of folate prevents this reaction and causes **formiminoglutamate** to be excreted in the urine.

Proline

Oxidation of **proline** by NAD produces an unstable intermediate that is readily hydrolyzed to glutamate semialdehyde (Fig. 24-10). As we saw with the metabolism of arginine, another oxidation gives glutamate. Reversal of this reaction sequence allows proline to be synthesized from glutamate.

Fig. 24-9. Conversion of histidine to glutamate.

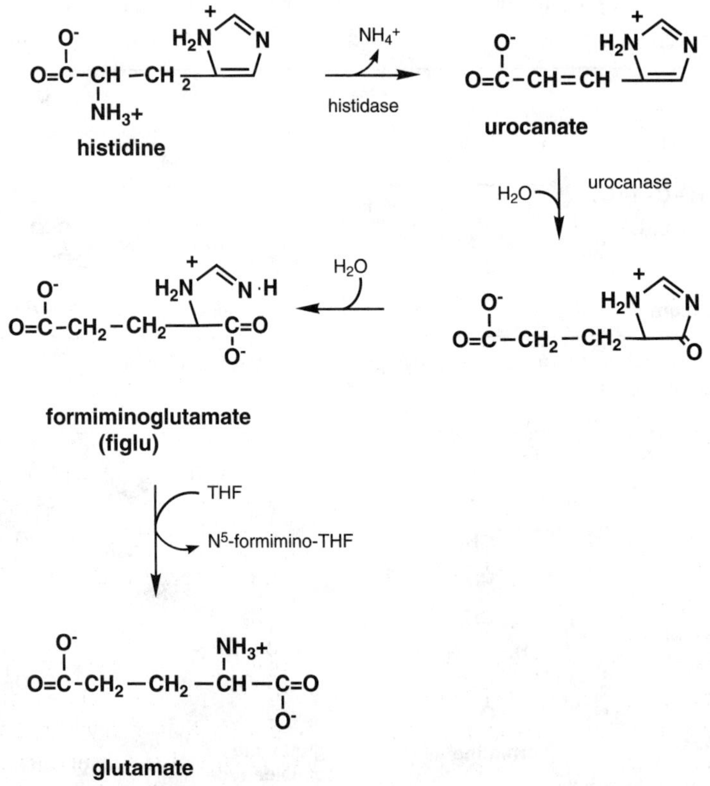

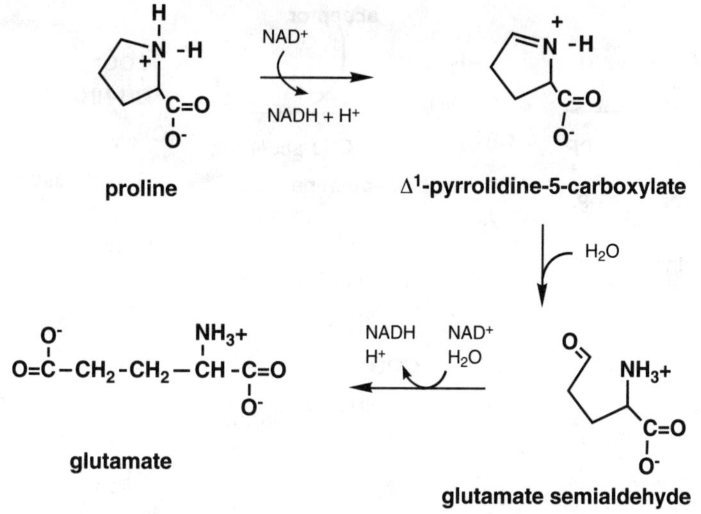

Fig. 24-10. Oxidation of proline to produce glutamate semialdehyde and then glutamate.

Amino Acids Converted to Succinyl-CoA

Parts of three amino acids are converted to the Krebs cycle intermediate succinyl-CoA. These are methionine, isoleucine, and valine. These amino acids all form either propionyl-CoA or methylmalonyl-CoA and yield succinyl-CoA by the same reactions used in the oxidation of fatty acids with an odd number of carbon atoms.

Methionine

The first step in the metabolism of methionine is production of **S-adenosylmethionine** (SAM). You have encountered SAM before as a donor of methyl groups in phospholipid biosynthesis and in the methylation of nucleotides in tRNA and rRNA. When SAM loses its methyl group, the product is **S-adenosylhomocysteine** (SAH). Loss of adenosine and the addition of serine yields **cystathionine.** This can be cleaved to cysteine and homoserine.

Cysteine can be converted to pyruvate. Oxidative deamination of homoserine gives α-ketobutyrate, which can be decarboxylated and attached to CoA to yield propionyl-CoA. Carboxylation to methyl-malonyl-CoA and rearrangement gives succinyl-CoA, as outlined in Figure 24-11.

Valine and Isoleucine

As you can see in Figure 24-12, the initial reactions of the degradation of **valine** and **isoleucine** are similar and are catalyzed by the same enzymes. The amino acids are first transaminated and then decarboxylated and attached to CoA. Reduction produces a double bond at the β-carbon. The product from valine is methacryl-CoA and from isoleucine, tigyl-CoA.

Hydration of the double bond, loss of CoA, and oxidation of methacryl-CoA yields methylmalonate semialdehyde, which can be converted to methyl-malonate and then to methylmalonyl-CoA. Alternatively, the methylmalonate semialdehyde can be transaminated to β-aminoisobutyrate, which is excreted in the urine.

Hydration of the double bond of tigyl-CoA gives α-methyl-β-hydroxybutyryl-CoA. Oxidation of the hydroxyl group to a ketone, followed by cleavage of the β-keto compound by CoA, yields acetyl-CoA and propionyl-CoA. These reactions are the same as those involved in the β-oxidation pathway for fatty acids.

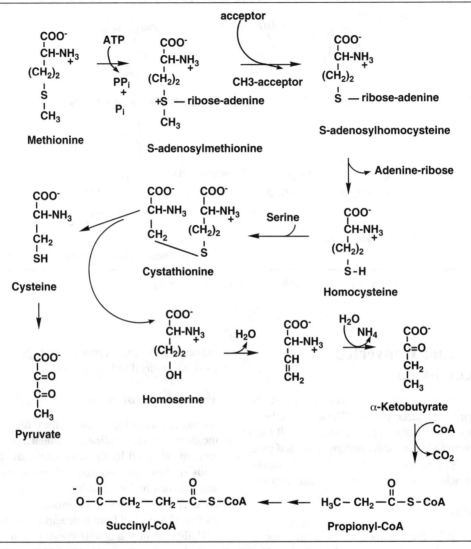

Fig. 24-11. Degradation of methionine to pyruvate and succinyl-CoA.

Amino Acids Converted to Fumarate

Phenylalanine and **tyrosine** are converted to fumarate, an intermediate of the Krebs cycle that can be converted into glucose. Because conversion of phenylalanine to tyrosine is the first step in the degradation of phenylalanine, metabolism of these two amino acids will be discussed together.

The **reaction sequence** for the degradation of phenylalanine and tyrosine is outlined in Figure 24-13. First, phenylalanine is oxidized to tyrosine by **phenylalanine hydroxylase.** This reaction requires **tetrahydrobiopterin,** which is oxidized to the dihydro compound in the course of the reaction. Dihydrobiopterin is reduced to tetrahydrobiopterin by NADPH (Fig. 24-14).

In the next reaction, tyrosine is transaminated

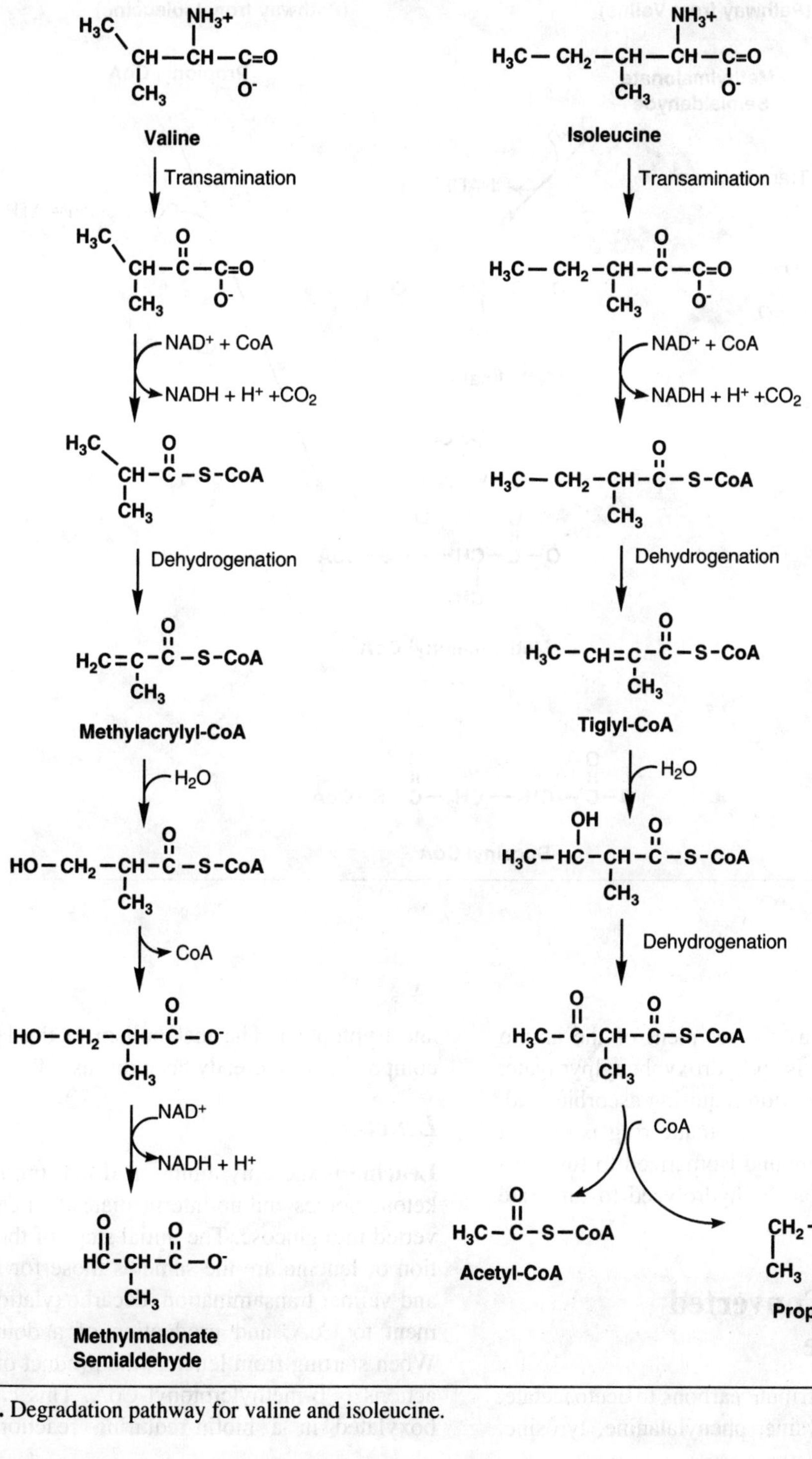

Fig. 24-12. Degradation pathway for valine and isoleucine.

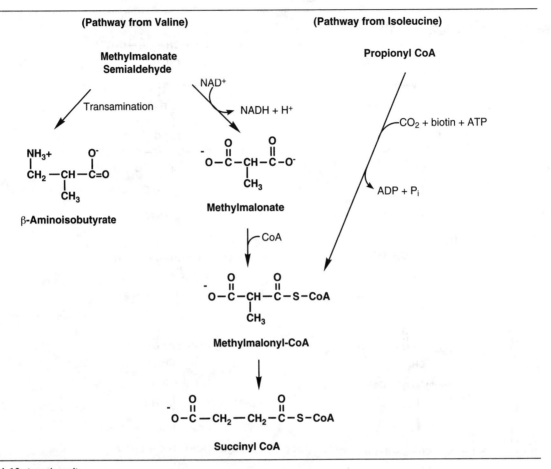

Fig. 24-12 (continued).

with α-ketoglutarate as the recipient of the amino group. The product is p-hydroxyphenylpyruvate. Decarboxylation, a reaction requiring ascorbic acid, gives **homogentisate.** The aromatic ring is opened and the resulting compound isomerized to fumaryl-acetoacetate, which can be hydrolyzed to fumarate and acetoacetate.

Amino Acids Converted to Acetoacetate

Five amino acids contribute carbons to acetoacetate. These are leucine, lysine, phenylalanine, tyrosine,

and tryptophan. The metabolism of the latter three compounds has already been discussed.

Leucine

Leucine is the only amino acid that produces only ketone bodies and no intermediates that can be converted into glucose. The initial steps of the degradation of leucine are the same as those for isoleucine and valine: transamination, decarboxylation, attachment to CoA, and production of a double bond. When starting from leucine, the product of these reactions is β-methylcrotonyl-CoA. This can be carboxylated in a biotin-requiring reaction to give

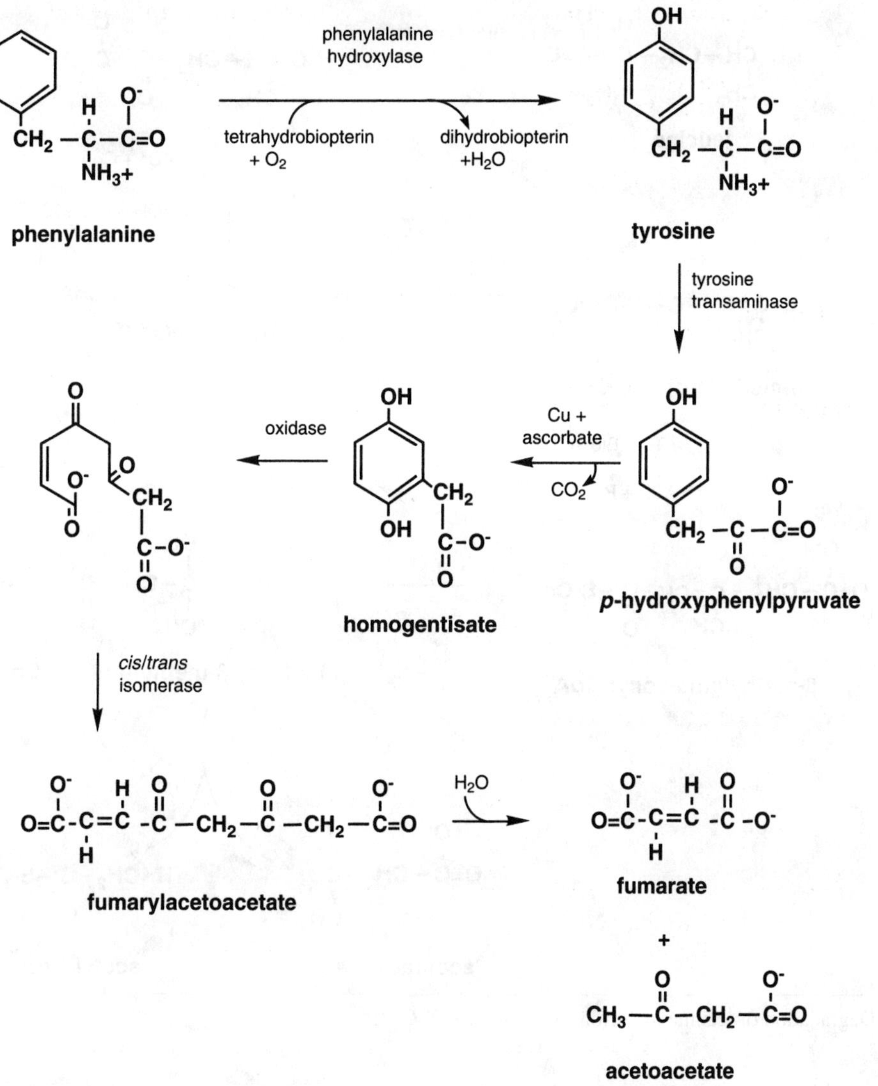

Fig. 24-13. Metabolism of phenylalanine and tyrosine to fumarate and acetoacetate.

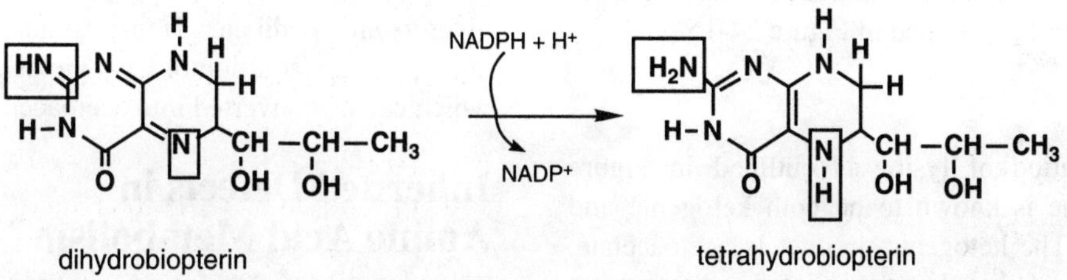

Fig. 24-14. Reduction of dihydrobiopterin to tetrahydrobiopterin by NADPH.

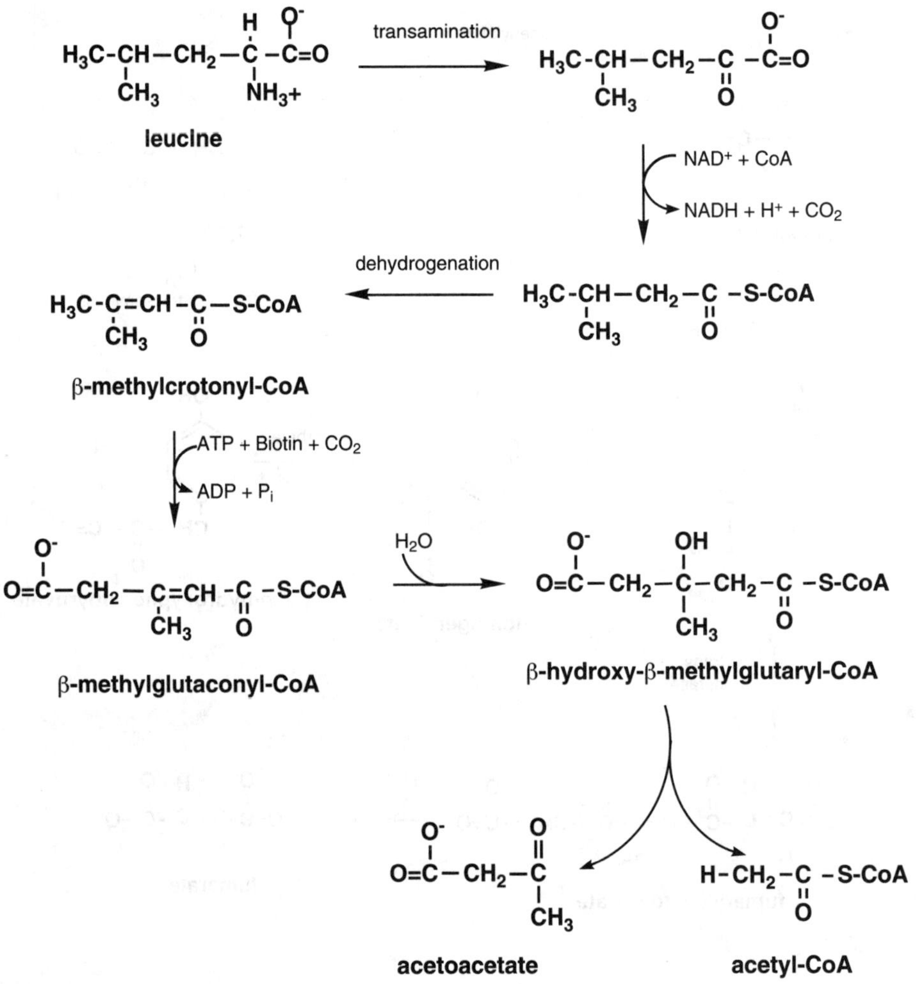

Fig. 24-15. Degradation of leucine to acetoacetate and acetyl-CoA.

β-methylglutaryl-CoA. The addition of water produces β-hydroxy-β-methylglutaryl-CoA, which can be cleaved to acetoacetate and acetyl-CoA. The reaction sequence is outlined in Figure 24-15.

Lysine

The degradation of **lysine** is outlined in Figure 24-16. Lysine is known to be both ketogenic and glucogenic. The ketogenic product is acetoacetate. The identity of the glucogenic intermediate is not known. The first reaction of lysine metabolism is condensation with α-ketoglutarate, followed by re-

duction. Reoxidation and hydrolysis result in the loss of glutamate, which carries with it the ε-amino group. Further oxidation of the remaining compound gives α-aminoadipate. Transamination, decarboxylation, and the addition of CoA yields glutaryl-CoA, which can be converted into acetoacetate.

Inherited Defects in Amino Acid Metabolism

There are a large number of inherited defects in the metabolism of amino acids, some of which produce severe mental retardation. Fortunately, most are rare

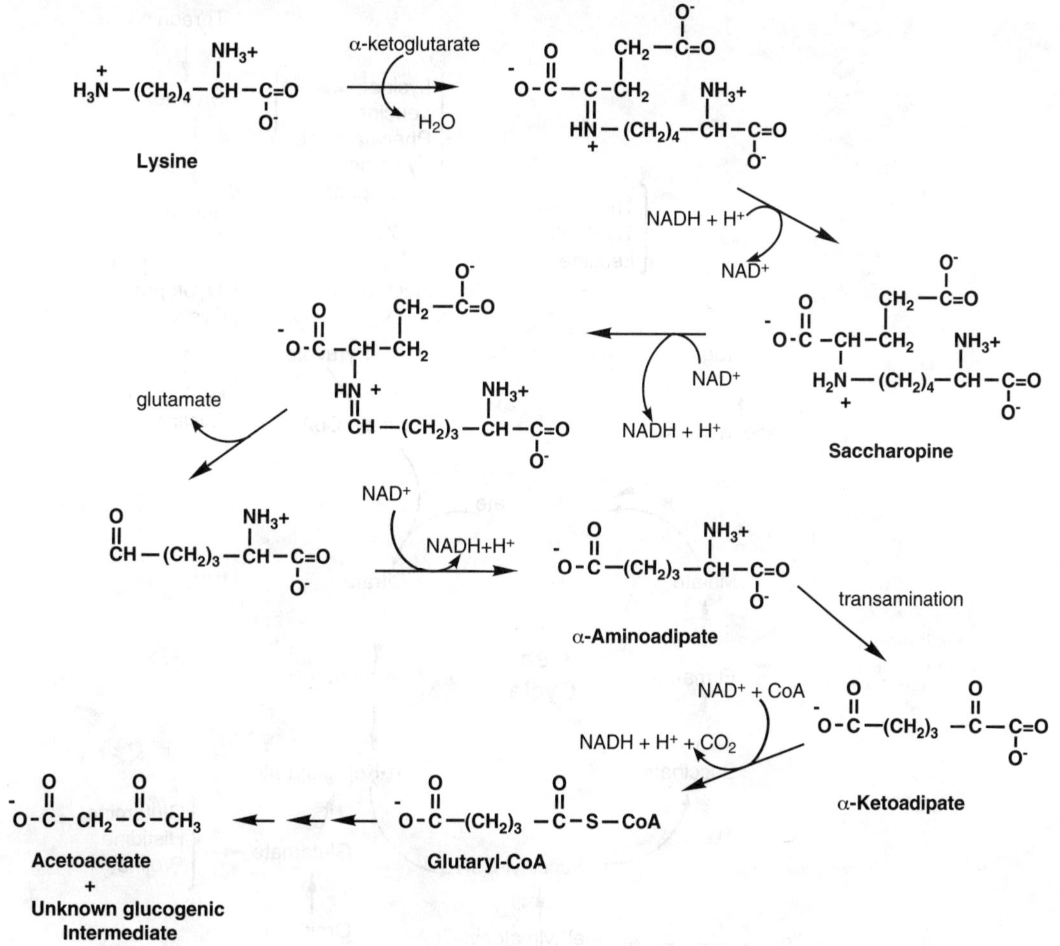

Fig. 24-16. Degradation of lysine.

recessive characteristics. The most common diseases, the enzymes affected, and the products excreted in excess are listed in Table 24-1.

Phenylketonuria is an inherited defect in phenylalanine metabolism in which phenylalanine hydroxylase is the defective enzyme. Because no tyrosine is produced from phenylalanine, tyrosine becomes an essential amino acid. In the absence of phenylalanine hydroxylase, phenylalanine is transaminated to phenylpyruvate and then decarboxylated to phenylacetate, which can in turn be reduced to phenyllactate. All these are excreted in the urine. Unless the amount of phenylalanine in the diet is controlled

from birth, severe mental retardation will result. Because this defect is easily detectable and highly treatable, testing for phenylketonuria in all newborn infants is required by law.

In **alkaptonuria,** the enzymatic defect is also in the degradation pathway of phenylalanine and tyrosine. Lack of homogentisate oxidase causes homogentisate to be excreted. On oxidation in air, such as on a baby's diaper, it turns black. There are no major clinical effects of this defect.

Cystinuria is due to faulty renal absorption of cysteine, lysine, arginine, and ornithine. Problems occur when cysteine, the least soluble of these

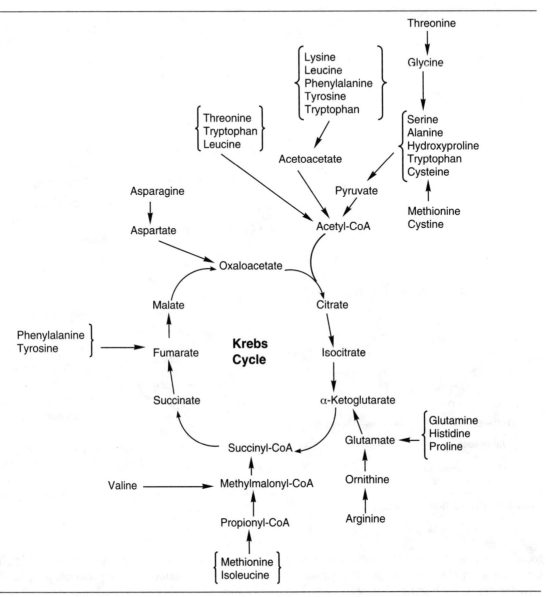

Fig. 24-17. Summary of the points at which the carbon skeletons of the amino acids enter the mainstream of metabolism.

Table 24-1. Inherited defects in amino acid degradation

Disease	Affected enzyme	Excreted product
Alkaptonuria	Homogentisate oxidase	Homogentisate
Cystinuria	Renal absorption of cys, lys, arg	Cysteine, lysine, arginine
Histidinemia	Histidase	Histidine
Homocystinuria	Cystathionine synthetase	Homocysteine
Hyperlysinemia	Lysine α-ketoglutarate reductase	Lysine
Hyperprolinemia		
Type I	Proline oxidase	Proline
Type II	Pyrroline 5-carboxylate decarboxylase	Proline
Maple syrup urine disease	Keto acid decarboxylase	Valine, isoleucine, leucine, and their keto acids
Methylmalonic aciduria	Methylmalonyl-CoA carboxymutase	Methylmalonate
Phenylketonuria	Phenylalanine hydroxylase	Phenylpyruvate, phenylacetate, phenyllactate
Tyrosinemia	Tyrosine aminotransferase	Tyrosine

amino acids, precipitates in the kidney, forming insoluble stones.

Homocysteinuria is caused by a defect in cystathionine synthetase, an enzyme in the degradative pathway for methionine. In the absence of this enzyme, methionine cannot be converted to cysteine. Then cysteine becomes an essential amino acid. Besides urinary excretion of homocysteine, individuals with homocysteinuria suffer osteoporosis, dislocated lenses, and mental retardation.

Maple syrup urine disease or **branched-chain ketonuria** causes the urine to have the smell of burnt sugar. Leucine, isoleucine, valine, and their keto acids are excreted in excess. The enzymatic defect is in the β–keto acid decarboxylase, which decarboxylates these acids and attaches them to CoA.

Without a severe reduction in the amount of these amino acids in the diet, death will occur before the age of 1 year.

Summary

A summary of the points at which the various amino acids enter the mainstream of metabolism is given in Figure 24-17. Degradation of the carbon skeletons of five amino acids—phenylalanine, tyrosine, tryptophan, isoleucine, and lysine—give both ketogenic and glucogenic products. Leucine yields only ketogenic products. All the rest of the amino acids are glucogenic. Defects in amino acid degradation produce an accumulation of precursors and sometimes severe clinical effects.

25 Specialized Compounds from Amino Acids

Amino acids serve not only as the subunits from which proteins are made but also as precursors for many other nitrogen-containing compounds necessary for cellular function. These include heme, creatine, polyamines, the pigment melanin, and a number of neurotransmitters and hormones.

Objectives

After completing this chapter, you should be able to

Recognize the structures of and list the amino acid precursors, rate-controlling reactions, and clinical conditions associated with deficiency of the following compounds: creatine, carnitine, heme, acetylcholine, dopamine, epinephrine, norepinephrine, γ-aminobutyric acid, serotonin, histamine, spermidine, and spermine.

Identify the degradation products of heme and the catecholamines.

Describe the enzyme affected and the mechanism of action of neostigmine, parathion, pyridine aldoxime, and succinylcholine.

Heme

The heme portion of hemoglobin is essential for the reversible transport of oxygen from the lungs to the peripheral tissues. Because erythrocytes turn over with a lifetime of 120 days, new erythrocytes and new hemoglobin must be made continuously. Heme is also a necessary part of myoglobin, cytochrome c, and the enzyme catalase.

Synthesis of Heme

Heme is synthesized from glycine and succinyl-CoA by the pathway outlined in Figure 25-1. Glycine and succinyl-CoA condense, with the loss of CoA, to form α-amino-β-ketoadipate. This is decarboxylated to form γ-aminolevulinate. The same enzyme, **γ-aminolevulinate synthetase,** catalyzes both the condensation and decarboxylation reac-

tions. The enzyme is located in the outer mitochondrial membrane and is the rate-controlling enzyme for heme biosynthesis. The activity of aminolevulinate synthetase is subject to feedback inhibition by heme, hemin, and hemoglobin, the final products of the biosynthesis pathway. As an additional control point, high concentrations of heme inhibit the synthesis of the enzyme.

In the next stage of heme synthesis, two molecules of γ-aminolevulinate condense to form **porphobilinogen.** Condensation of two porphobilinogens with the loss of NH_2 yields a dipyrolle. This condensation can produce two products with different configurations (Fig. 25-2). In A, the side chains are alternating acetate and propionate. In B, the order of the side chains is acetate, propionate, propionate, acetate. Combination of two molecules of A into a tetramer and then into a cyclic compound

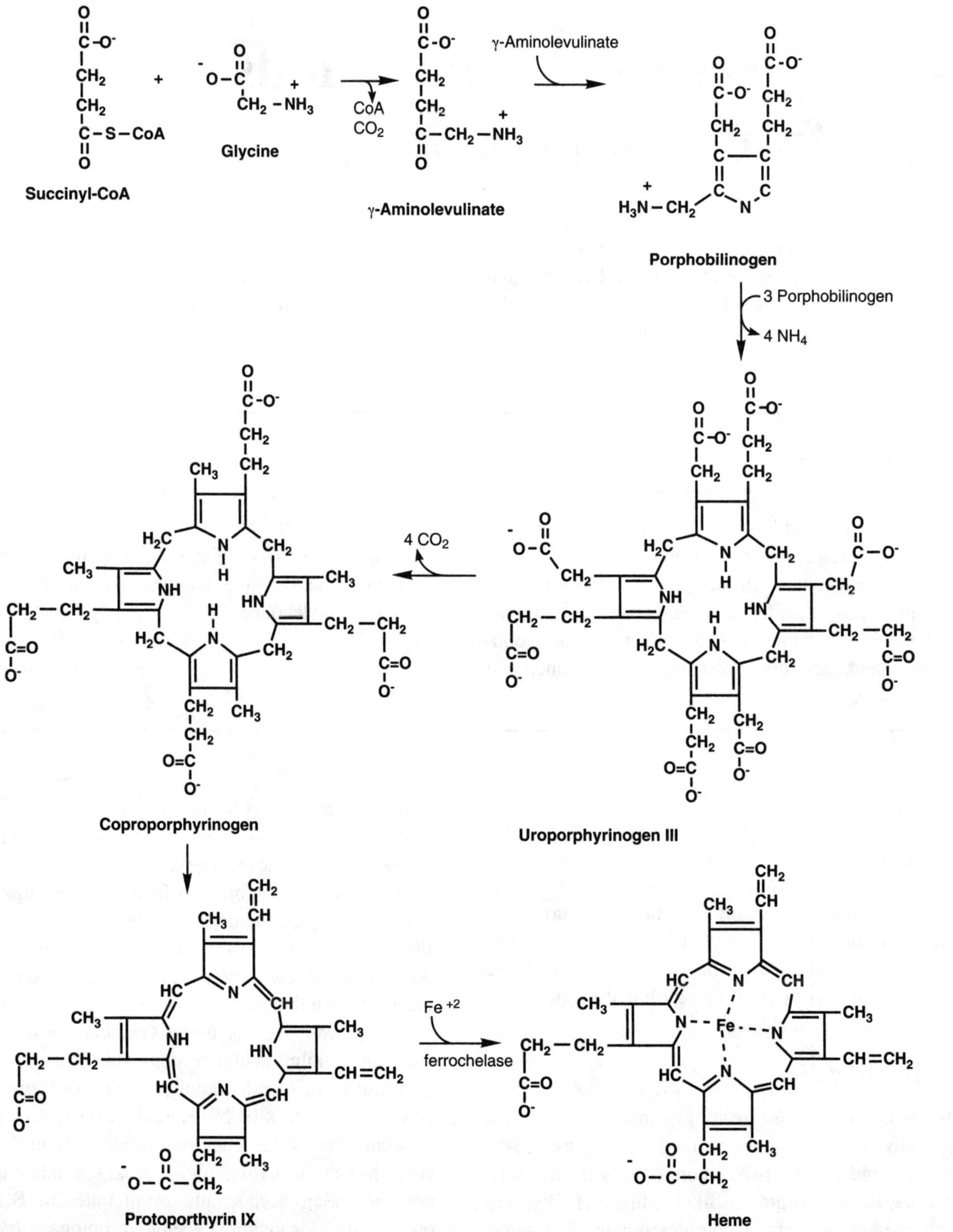

Fig. 25-1. Pathway for the synthesis of heme starting from succinyl-CoA and glycine.

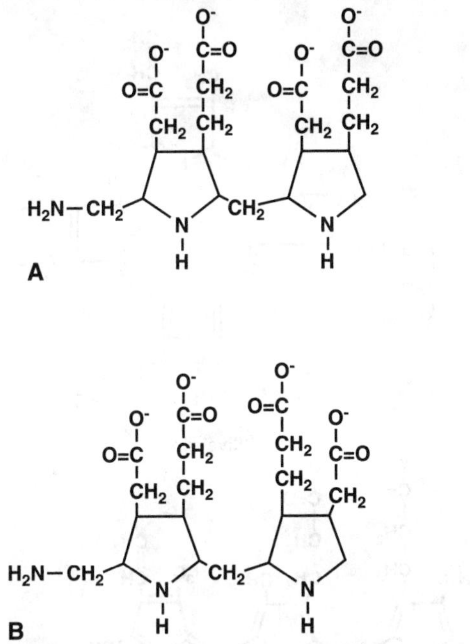

Fig. 25-2. Condensation of two porphobilinogens can produce two different dipyrroles. In A, the acetate and propionate side chains are alternating. In B, the side chain order is acetate, propionate, propionate, acetate.

gives the symmetrical uroporphyrinogen I. A combination of A and B forms the nonsymmetrical compound **uroporphyrinogen III.** Only this nonsymmetrical molecule serves as a precursor to heme.

In the final reactions, the acetate side chains of uroporphyrinogen III are decarboxylated to methyl groups producing coproporphyrinogen III. Then, two of the propionic acid side chains are changed into vinyl and the molecule dehydrogenated to yield **protoporphyrin IX.** Of the multiple isomers of protoporphyrin, only the IX isomer is a precursor to heme. In the final reaction, ferrochelatase adds the iron atom, forming heme.

A defect in a cosynthetase required to produce the nonsymmetrical uroporphyrinogen III instead of the symmetrical I isomer occurs in **congenital erythropoietic porphyria.** In this inherited disease, predominantly the I isomer is produced. Porphyrins derived from the I isomer cannot be used for heme synthesis, so they are excreted in the urine, which then turns red.

In addition, they are deposited in the skin, which becomes light-sensitive, necrotic, and deformed. This condition does not lead to mental retardation. The appearance of persons with congenital erythropoietic porphyria who have deformed faces and venture out only at night to avoid further skin damage may have been the beginning of the werewolf legend.

Hepatic porphyrias are due to increased activity of liver aminolevulinate synthetase, probably because of an increased synthesis of the enzyme. Affected persons excrete prophyrins and aminolevulinate in the urine but do not deposit them in the skin. Therefore, they are not light-sensitive. This kind of porphyria can be triggered by ingestion of heavy metals such as lead, by drugs such as barbiturates and sulfonamides, or by damage to the liver by alcoholism, hepatitis, or cirrhosis.

Heme Breakdown

When erythrocytes are degraded, the hemoglobin also is degraded. The globin portion is broken down to amino acids, the iron is salvaged, and the heme degraded by the pathway diagrammed in Figure 25-3. First, the cyclic structure is linearized, producing a green pigment, **biliverdin.** This is reduced by NADPH to form **bilirubin,** a reddish-yellow pigment. The color of these pigments gives healing bruises their characteristic tones. Bilirubin is transported to the liver, conjugated with glucuronic acid, and then excreted in the bile. In the intestine, bilirubin is reduced further to urobilinogen and urobilin.

Biliverdin and bilirubin are the **bile pigments.** If excess bile pigments are produced or the bile duct is blocked, the bile pigments are found in the circulation and tissues. This produces the yellow skin color characteristic of **jaundice.** The immature liver of newborns and premature infants is unable to conjugate bilirubin to the glucuronide. This leads to an increase in levels of circulating bilirubin and jaundice. Because the blood-brain barrier is not fully functional at birth, the excess bilirubin can enter the brain and produce toxic encephalopathy. Jaundice in newborns is treated by exposing the baby to blue light. This facilitates the breakdown of bilirubin to products that can be excreted in the urine.

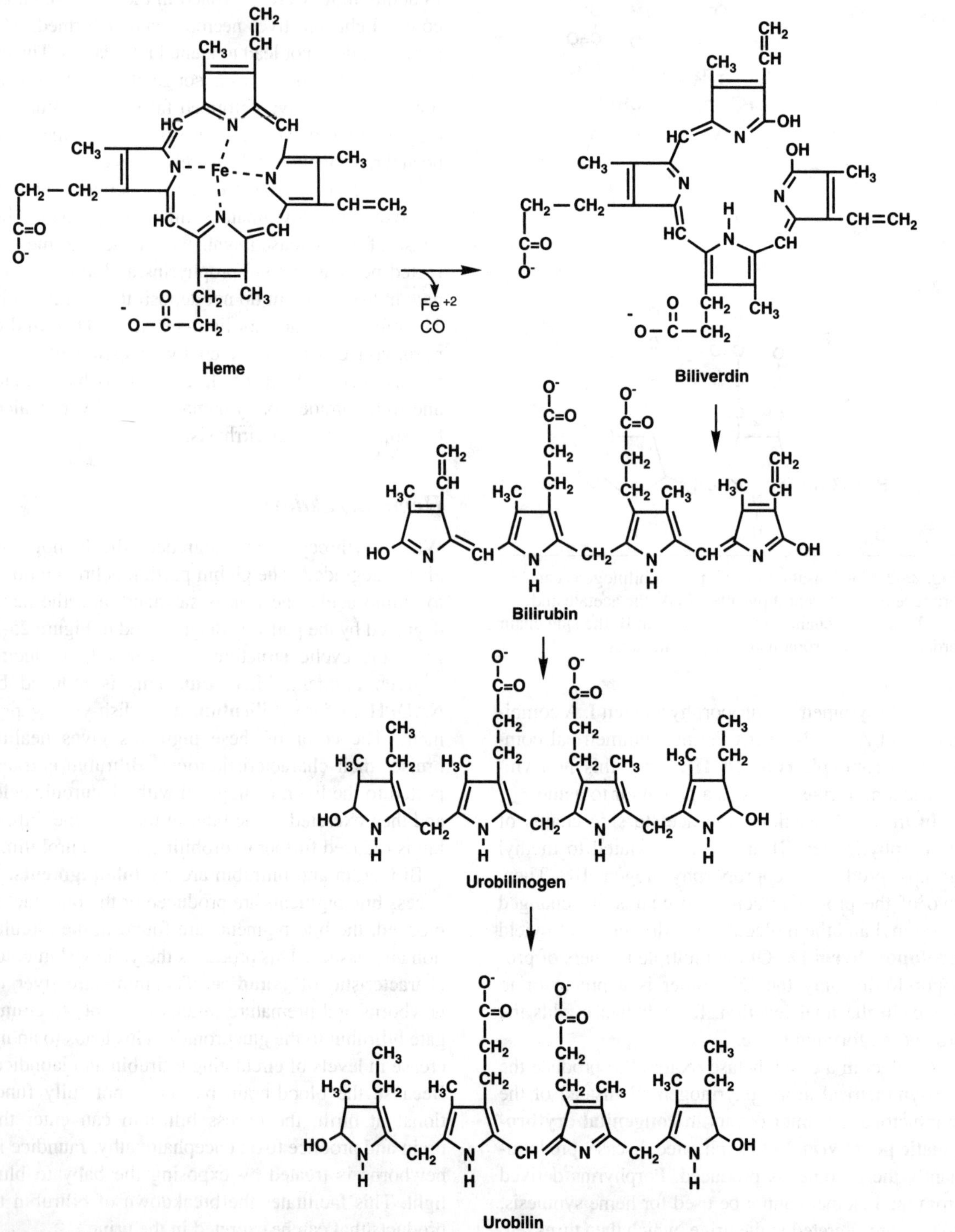

Fig. 25-3. Pathway for the degradation of heme and production of the bile pigments biliverdin and bilirubin.

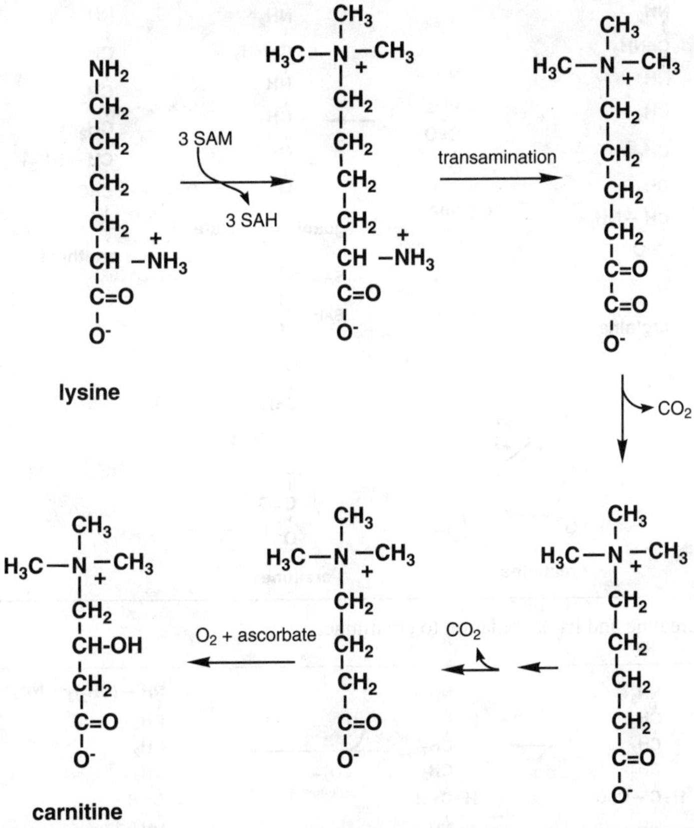

Fig. 25-4. Synthesis of carnitine from lysine.

Carnitine

You have encountered **carnitine** before in the study of fatty acid degradation in Chapter 19. Carnitine serves to accept fatty acids from CoA and carry them across the inner mitochondrial membrane. Carnitine is made from lysine by the pathway outlined in Figure 25-4. First, S-adenosylmethionine adds three methyl groups to the ε-amino group of lysine. This is followed by transamination and decarboxylation. A second carbon is lost and, finally, the β-carbon is hydroxylated in a reaction that requires ascorbic acid.

Creatine

In its phosphorylated form, **creatine** serves as a storage form of energy in muscle. Creatine can be synthesized from glycine and arginine by the pathway

shown in Figure 25-5. First, arginine donates a guanidino group (NH_2 -C $=NH_2$) to glycine, producing guanidinoacetate. Creatine is formed by methylation by S-adenosylmethionine.

Creatine is degraded to **creatinine** by a nonenzymatic reaction, also shown in Figure 25-5. The rate of production of creatinine is constant, so measurement of creatinine in the urine can be used to determine the rate of urine formation and what proportion of the daily urine a particular sample represents.

Polyamines

Spermine and **spermidine** are the major polyamines in human cells. They are found in the nucleus associated with DNA. Because the polyamines are positively charged, they are able to neutralize some of the negative charges on the phosphate groups of DNA

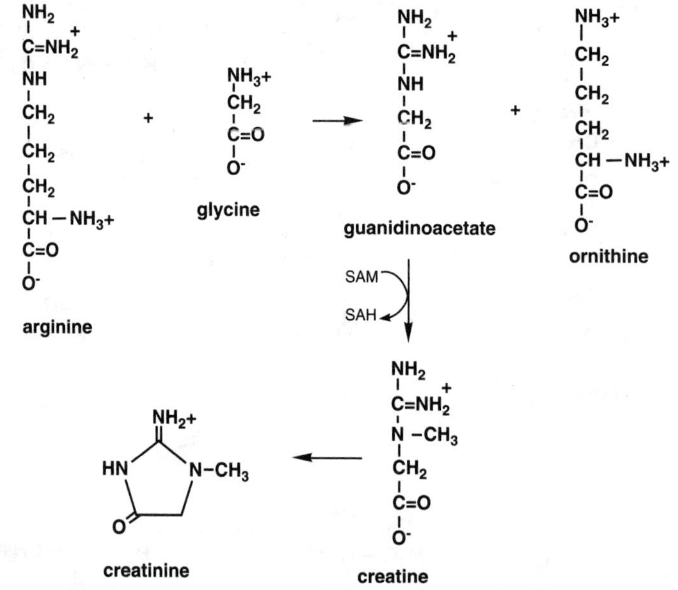

Fig. 25-5. Synthesis of creatine and its degradation to creatinine.

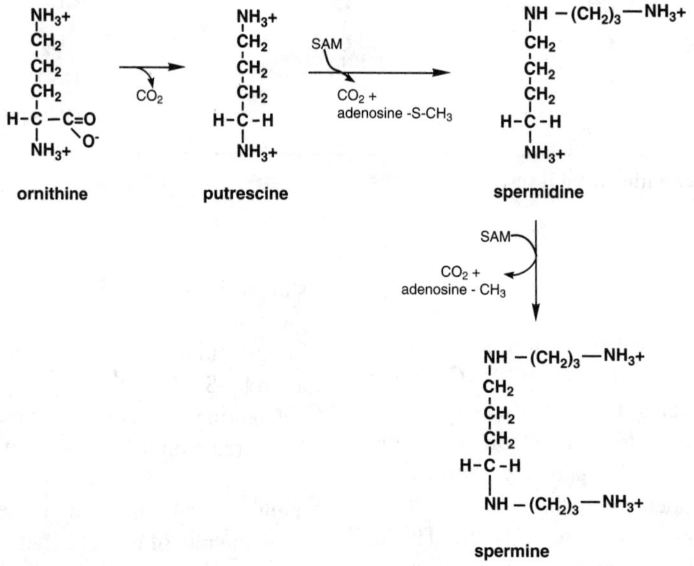

Fig. 25-6. Biosynthesis of the polyamines spermidine and spermine.

and to play a role in maintaining the stability of the nucleus. The starting material for polyamine synthesis is ornithine, as shown in Figure 25-6. First, ornithine is decarboxylated to form **putrescine.** In an unusual reaction, S-adenosylmethionine is decarboxylated and donates a propylamine group to or-

nithine, forming spermidine. A second addition of propylamine yields spermine.

The polyamines are degraded by oxidation of spermine to spermidine and then to putrescine. Some putrescine is oxidized to ammonia and carbon dioxide. The rest is excreted as an acetyl derivative.

Neurotransmitters

In the transmission of a nerve impulse between two nerve cells or between a nerve and a muscle cell, the nerve releases a chemical transmitter. The transmitter diffuses across the gap junction between the cells and binds to a receptor on the receiving cell. In response to the binding of the transmitter, the cell becomes permeable to sodium ions. This transmits the impulse along the cell. If the receiving cell is a nerve cell, the impulse causes it to release a neurotransmitter. This propagates the signal to another cell. If the receiving cell is a muscle cell, it contracts. After binding to the receptor, the transmitter molecule is degraded or sequestered. This stops transmission of the signal. Membrane permeability is restored to the resting state and muscle contraction ceases. Then the cycle of transmission can begin again.

Acetylcholine

Acetylcholine serves as the neurotransmitter between parasympathetic nerves as well as between these nerves and muscle. It is synthesized from acetyl-CoA and choline (Fig. 25-7). Following release from the nerve, acetylcholine diffuses to the target cell and binds to its receptor. Then it is quickly degraded by the enzyme **acetylcholinesterase.** In the course of the reaction, the acetyl group is bound to a serine residue of the enzyme. The activity of acetylcholinesterase can be blocked by substances that also bind to the ser-

ine residue but are hydrolyzed much more slowly than acetyl groups. **Physostigmine,** a poison from the Calavan bean, the antiglaucoma drug **neostigmine,** and the organophosphorus pesticide **parathion** act in this way. In the structures given here, the group transferred to serine is indicated with a box.

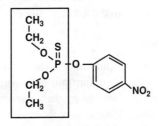

physostigmine

neostigmine

parathion

In the presence of these compounds, acetylcholine is not hydrolyzed and continues to act on its receptor. This keeps the receptor cell depolarized and prevents response to further nerve impulses. The binding of these compounds can be reversed by treatment with pyridine aldoxime methiodine, a compound that has been used clinically to treat organophosphorus pesticide poisoning.

pyridine aldoxime methiodide
(PAM)

Fig. 25-7. Synthesis of acetylcholine from acetyl-CoA and choline.

$$\underset{\text{acetyl-CoA}}{H_3C-\overset{\overset{\textstyle O}{\|}}{C}-S-CoA} \quad + \quad \underset{\text{choline}}{HO-CH_2-CH_2-\overset{\overset{\textstyle CH_3}{|}}{\underset{|}{N^+}}-CH_3}$$

CoA

$$\underset{\text{acetylcholine}}{H_3C-\overset{\overset{\textstyle O}{\|}}{C}-O-CH_2-CH_2-\overset{\overset{\textstyle CH_3}{|}}{\underset{|}{N^+}}-CH_3}$$

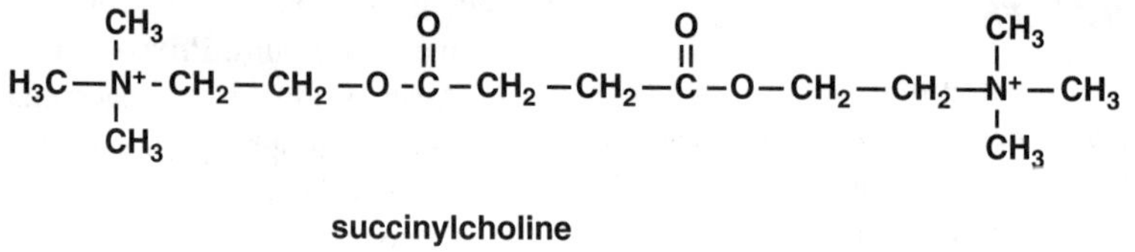

succinylcholine

Fig. 25-8. Structure of succinylcholine.

Analogs of acetylcholine also inhibit the transmission of nerve impulses. These compounds are degraded by acetylcholinesterase but at a much slower rate than acetylcholine. **Succinylcholine** (Fig. 25-8) is such an analog and is used clinically to relax muscles during surgery.

Compounds that mimic the activity of the acetylcholine receptor also interfere with the transmission of nerve impulses. **Curare,** the active component of which is is *d*-tubocurarine, and **bungarotoxin** are two such compounds. **Myasthenia gravis** is an autoimmune disease in which antibodies to the acetylcholine receptor are produced. This blocks the receptor and prevents binding of acetylcholine and the subsequent response to the nerve impulse.

Catecholamines

The catecholamines **dopamine, norepinephrine,** and **ephinephrine** act as neurotransmitters in sympathetic nerves. They are synthesized by these nerves and by the adrenal medulla. The pathway for their biosynthesis is shown in Figure 25-9. First, tyrosine is oxidized by **tyrosine hydroxylase** in a reaction similar to the one producing tyrosine from phenylalanine. The product is **dihydroxylphenylalanine** or **dopa.** This is then decarboxylated, yielding dopamine, the first of the catecholamine neurotransmitters. Dopamine can be oxidized again to give norepinephrine (noradrenaline). Norepinephrine serves as the transmitter between the sympathetic nerves and smooth-muscle cells. Transfer of a methyl group by S-adenosylmethionine yields epinephrine (adrenaline).

Loss of brain tyrosine hydroxylase is associated with **Parkinson's disease.** This produces a deficiency of dopamine, epinephrine, norepinephrine, and serotonin. The decrease of dopamine in the corpus striatum appears to be the crucial defect. The disease responds to treatment with dopa or dopa analogs that can cross the blood-brain barrier.

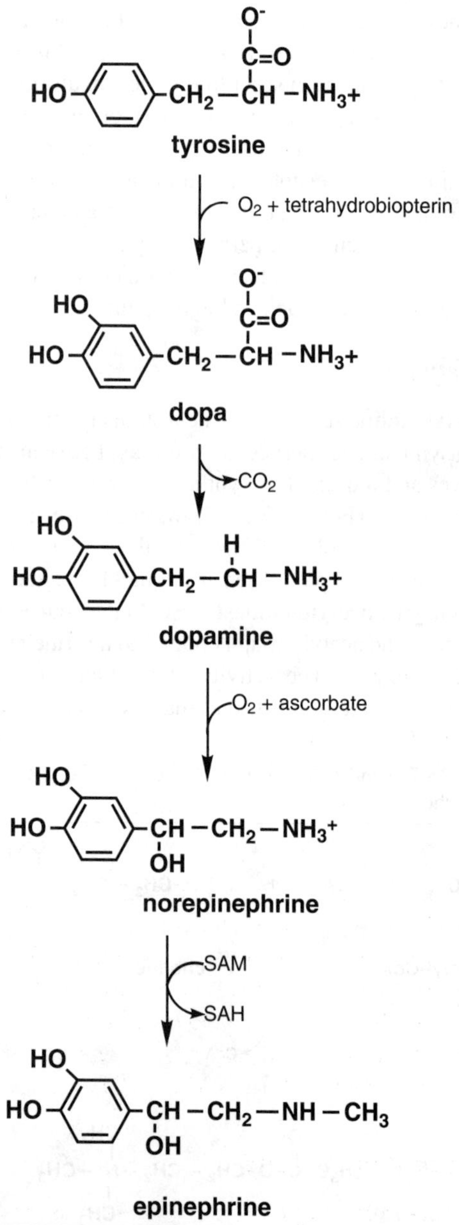

Fig. 25-9. Synthesis of the catecholamine neurotransmitters from tyrosine.

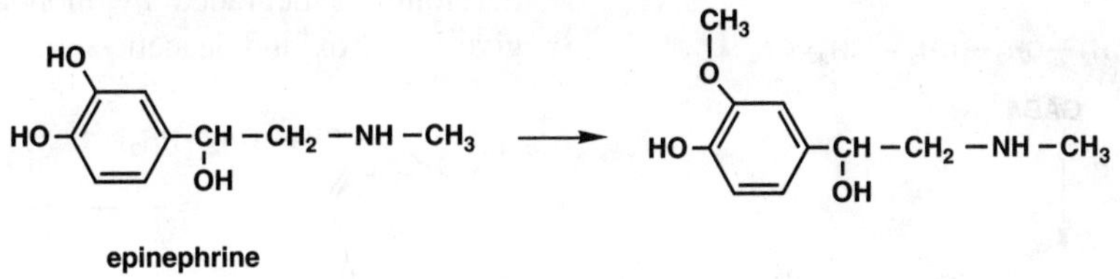

Fig. 25-10. Inactivation of epinephrine by methylation of the hydroxyl group at position 3 on the phenol ring.

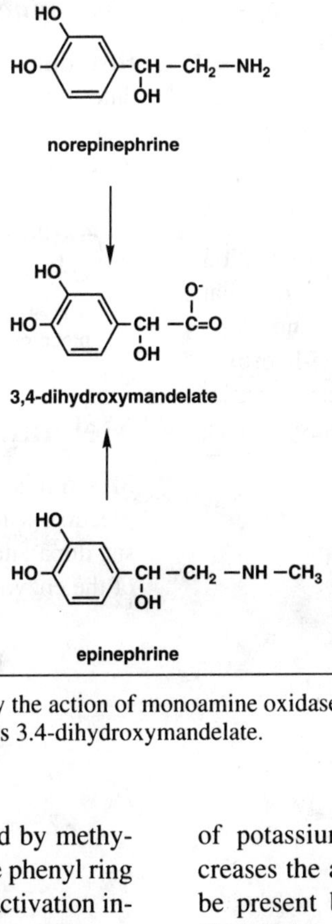

Fig. 25-11. Inactivation of catecholamines by the action of monoamine oxidase. The product from both norephinephrine and epinephrine is 3.4-dihydroxymandelate.

The catecholamines can be inactivated by methylation of the hydroxyl at position 3 on the phenyl ring (Fig. 25-10). A second mechanism of inactivation involves the action of **monoamine oxidase.** This enzyme oxidizes the terminal carbon to a carboxylic acid, with the loss of the nitrogen as ammonia. The product from both epinephrine and norepinephrine is 3,4-dihydroxymandelic acid (Fig. 25-11).

γ-*Aminobutyric Acid*

Gamma-aminobutyric acid (GABA) is an inhibitory transmitter that increases the permeability of the nerve cell membrane to potassium. The efflux of potassium depolarizes the membrane and increases the amount of another transmitter that must be present before a nerve impulse is transmitted. GABA is synthesized from glutamate by decarboxylation.

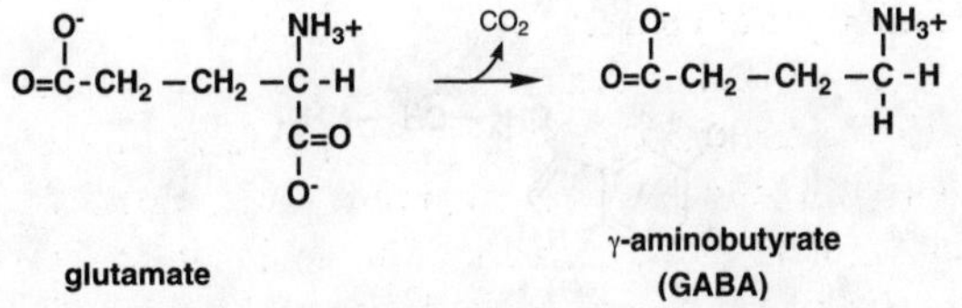

GABA is degraded by transamination, followed by oxidation to succinate.

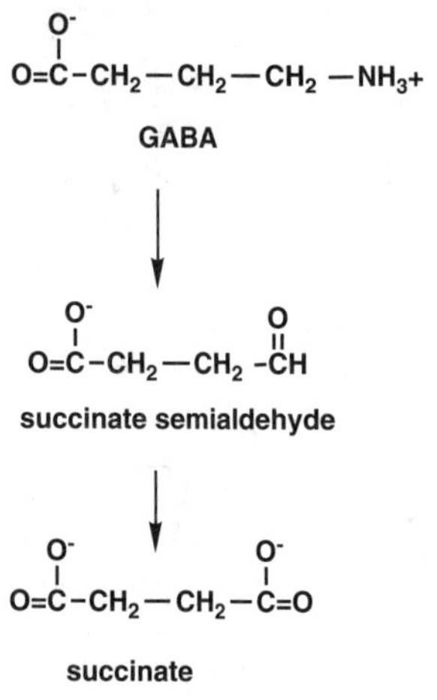

O=C-CH$_2$—CH$_2$—CH$_2$ —NH$_3$+

GABA

↓

O=C-CH$_2$—CH$_2$ -CH

succinate semialdehyde

↓

O=C-CH$_2$—CH$_2$—C=O

succinate

Serotonin

Serotonin comes from tryptophan by hydroxylation at position 5. This reaction is similar to the one that produces tyrosine from phenylalanine and dopa from tyrosine. Decarboxylation gives 5-hydroxytryptophan or serotonin. Serotonin is a vasoconstrictor and a neurotransmitter for smooth-muscle cells, especially those in the digestive tract.

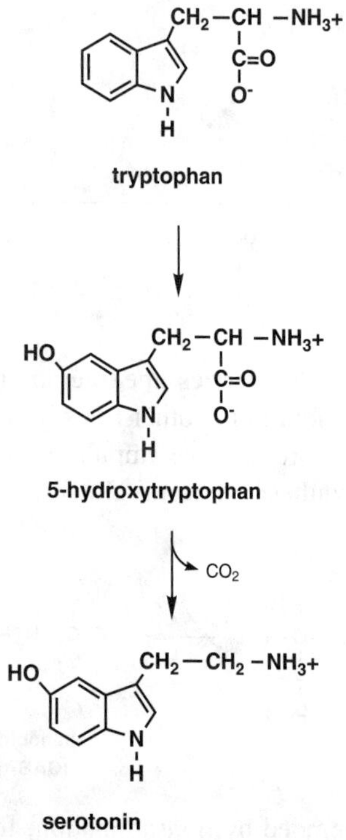

CH$_2$—CH —NH$_3$+

tryptophan

↓

HO   CH$_2$—CH —NH$_3$+

5-hydroxytryptophan

↓ CO$_2$

HO   CH$_2$—CH$_2$ —NH$_3$+

serotonin

Serotonin is degraded by monoamine oxidase to give 5-hydroxyindoleacetic acid.

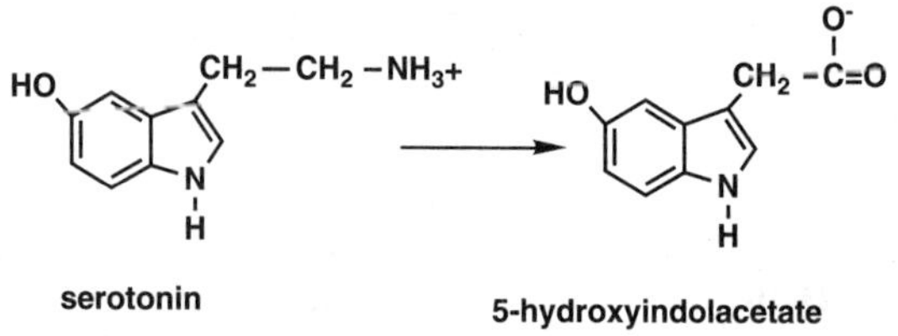

HO   CH$_2$—CH$_2$ —NH$_3$+   →   HO   CH$_2$ -C=O

serotonin   **5-hydroxyindolacetate**

Histamine

Histamine is produced by decarboxylation of histidine. It is a vasoconstrictor that also stimulates HCl secretion in the stomach. In addition, mast cells produce histamine during allergic reactions.

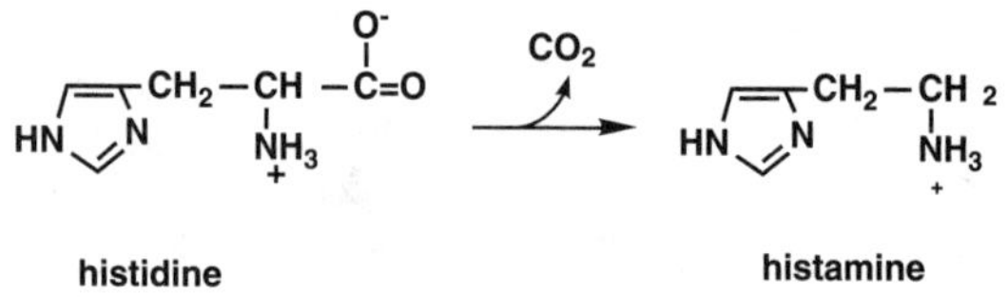

CH$_2$—CH —C=O $\xrightarrow{CO_2}$ CH$_2$—CH$_2$

histidine   **histamine**

Melanin

Melanin is a dark pigment found in skin and hair. Its precursors are tyrosine and dopa. In melanin synthesis, dopa and dopaquinone are formed by the action of the enzyme **tyrosinase.**

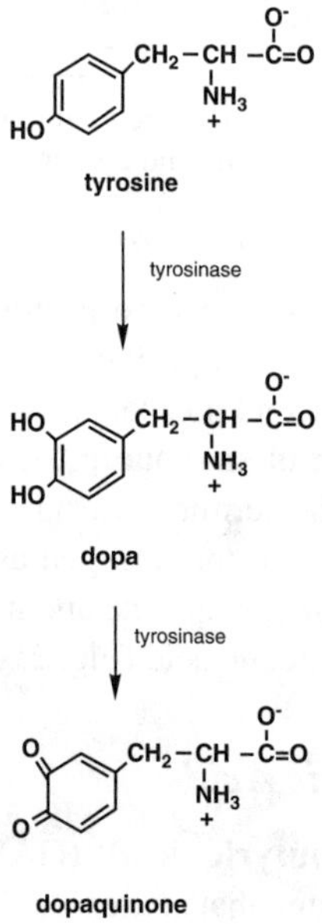

CH$_2$—CH —C=O

tyrosine

↓ tyrosinase

HO   CH$_2$—CH —C=O

dopa

↓ tyrosinase

CH$_2$-CH —C=O

dopaquinone

Details of conversion of these compounds into melanin are not completely known. However, some of the reactions many be nonenzymatic. An inherited lack of tyrosinase produces classic **albinism,** which is characterized by a lack of any skin or eye pigmentation.

Summary

Heme is synthesized from glycine and succinyl-CoA. The production of γ-aminolevulinate is the rate-controlling reaction. Both the activity of the enzyme and its synthesis are regulated by heme and heme-containing compounds. Only the asymmetrical type III prophyrins are precursors to heme. Heme is degraded to the bile pigments bilirubin and biliverdin in the liver and to urobilinogen and urobilin in the intestines.

Tyrosine is the precursor to the catecholamine neurotransmitters dopamine, norepinephrine, and epinephrine, and to the skin pigment melanin. Similar reactions yield serotonin from tryptophan. The neurotransmitter GABA comes from the decarboxylation of glutamate and histamine from decarboxylation of histidine.

Acetylcholine, another neurotransmitter, is synthesized from acetyl-CoA and choline. Inhibition of nerve impulses transmitted by acetylcholine can occur by binding of an inhibitor to the serine residue

Table 25-1. Diseases associated with defects in the metabolism of compounds derived from amino acids

Disease	Defect
Congenital erythro-poietic porphyria	Decrease in uroporphyrinogen cosynthetase, excess production of type I uroporphyrinogen
Hepatic porphyria	Increase in liver γ-aminolevulinate synthetase
Myasthenia gravis	Antibodies produced to acetylcholine receptor
Parkinson's disease	Decrease in brain tyrosine hydroxylase
Albinism	Lack of tyrosinase

of acetylcholinesterase, slow hydrolysis of analogs of acetylcholine, or the presence of receptor analogs or receptor antibodies.

Other important nitrogen-containing compounds made from amino acids include carnitine from lysine, creatine from glycine and arginine, and the polyamines from ornithine and methionine. The diseases associated with defects in synthesis and degradation of these amino acid–derived compounds are summarized in Table 25-1.

Part VI Questions: Amino Acid Metabolism

1. All the following levels of oxidation can be accommodated by the folate coenzymes *except*
   A. carbon dioxide.
   B. formate.
   C. -CH$_2$-OH.
   D. methyl.
   E. aldehyde.
2. The carbon skeleton of α-ketoglutarate is the starting material for the biosynthesis of
   A. methionine.
   B. asparagine.
   C. proline.
   D. glycine.
   E. leucine.
3. The essential amino acids include
   A. serine and aspartate.
   B. lysine and valine.
   C. alanine and glutamine.
   D. phenylalanine and tyrosine.
   E. threonine and glycine.
4. Which of the following amino acids are formed by direct transamination of the keto acid?
   A. Glycine
   B. Histidine
   C. Arginine
   D. Aspartate
   E. Methionine
5. The precursor for the carbon skeleton of cysteine is
   A. glycine.
   B. methionine.
   C. α-ketoglutarate.
   D. oxaloacetate.
   E. serine.
6. Free ammonia can be produced by a reaction catalyzed by
   A. α-ketoglutarate dehydrogenase.
   B. glutamate dehydrogenase.
   C. arginase.

D. trypsin.
E. ornithine transcarbamoylase.
7. Ammonia can be transported from the peripheral tissues in the form of
   A. free ammonia bound to albumin.
   B. citrulline.
   C. α-ketoglutarate.
   D. glutamine.
   E. arginine.
8. One of the nitrogen atoms of urea comes directly from the α-amino group of
   A. citrulline.
   B. asparagine.
   C. N-acetylglutamate.
   D. aspartate.
   E. arginine.
9. The energy-requiring reactions of the urea cycle include the reaction catalyzed by
   A. carbamoyl phosphate synthetase.
   B. ornithine transcarbamoylase.
   C. arginosuccinase.
   D. arginase.
   E. ornithine decarboxylase.
10. All the following are true about ammonia intoxication *except*
    A. it can be caused by a defect in glutamine synthetase.
    B. the symptoms can include blurred vision, slurred speech, coma, and death.
    C. it can occur in a person with alcoholic liver cirrhosis after a high-protein meal.
    D. some cases can be treated by administration of sodium benzoate.
    E. it can be produced by an excess of arginase.
11. In transamination, which amino acid is converted into α-ketoglutarate?
    A. Lysine
    B. Aspartate
    C. Glutamate

D. Threonine

E. Methionine

12. All of the following amino acids are degraded into Krebs cycle intermediates *except*

A. aspartate.

B. glutamine.

C. tyrosine.

D. leucine.

E. proline.

13. Methionine

A. is metabolized to homogentisate.

B. produces pyruvate and succinyl-CoA upon degradation.

C. is both a ketogenic and a glucogenic amino acid.

D. degradation is defective in cystinuria.

E. is deaminated to produce cystine.

14. Phenylalanine

A. is degraded to α-ketoglutarate.

B. degradation is defective in maple syrup urine disease.

C. hydroxylation to tyrosine requires tetrahydrobiopterin.

D. degradation parallels the reactions of the β-oxidation pathway.

E. is transaminated to produce oxaloacetate.

15. Ketogenic amino acids include

A. tryptophan.

B. proline.

C. cysteine.

D. glycine.

E. threonine.

16. All the following are true about the rate-controlling reaction for the synthesis of heme *except* it

A. is subject to feedback inhibition by the end products of the reaction pathway.

B. is repressed in its synthesis by heme.

C. is γ-aminolevulinate synthetase.

D. is increased in activity in hepatic porphyria.

E. catalyzes the condensation of two molecules of porphobilinogen.

17. Degradation of heme produces

A. porphobilinogen.

B. biliverdin.

C. bile salts.

D. carnitine.

E. γ-aminolevulinate.

18. The catecholamine neurotransmitters include

A. dopamine.

B. GABA.

C. histamine.

D. acetylcholine.

E. serotonin.

19. Methionine is necessary for the biosynthesis of

A. lysine.

B. dopa.

C. spermidine.

D. serotonin.

E. GABA.

20. Creatine

A. is necessary for the transport of fatty acids into the mitochondria.

B. in its phosphorylated form serves as a storage form of energy in muscle.

C. is synthesized from alanine and ornithine.

D. has a degradation rate that is dependent on the intake of amino acids in the diet.

E. is a transport form for ammonia in the blood.

Part VII Nucleotide Metabolism

26 Nucleotide Biosynthesis and Salvage

Nucleotides are the subunits from which DNA and RNA are composed. In addition, they serve as carriers of amino acids, lipids, and carbohydrates and as parts of the coenzymes FAD, NAD, and CoA. Nucleotides can be synthesized from small molecules readily available in mammalian cells or salvaged intact from dietary sources or from the breakdown of polynucleotides. In nondividing cells, there is no requirement for a net accumulation of either DNA or RNA. For these cells, the biosynthesis of individual nucleotides is not extensive because the need for nucleotides can be satisfied by salvage or from dietary sources. However, a net synthesis of nucleotides is required for proliferating cells such as the bone marrow and intestinal epithelial cells of adults and in most cells of infants and young children.

Objectives

After completing this chapter, you should be able to

Identify the source of all the atoms of the purine and pyrimidine rings.

List the rate-controlling reactions for purine and pyrimidine biosynthesis and the conditions that control them.

Outline the biosynthesis of thymidine and the deoxynucleotides.

Diagram the salvage pathways.

Recognize the structures and describe the mechanism of action of 5-fluorouracil, 5-iodouridine, 5-azacytidine, 6-mercaptopurine, cytosine arabinoside, adenosine arabinoside, azidothymidine (AZT), acyclovir, N-phosphonoacetyl-aspartate (PALA), methotrexate, sulfonamide, and azaserine.

Match the inborn errors of nucleotide biosynthesis with the affected enzymes.

Purine Synthesis

The biosynthesis of purines involves the construction of the purine ring system by adding functional groups one by one onto a preexisting ribose phosphate. This produces a nucleotide as the final product. There is no separate pathway for the synthesis of the free base or the nucleoside. The atoms of the purine ring come from the sources shown in Figure 26-1.

Synthesis of PRPP

As you can see in Figure 26-2, the first step in purine biosynthesis involves the reaction of ribose 5-phosphate with ATP, producing **5-phosphoribosyl-1-pyrophosphate** (PRPP). The ribose 5-phosphate comes from glucose by way of the pentose phosphate pathway. PRPP is involved in the biosynthesis of both purines and pyrimidines as well as purine salvage. You have already encountered PRPP in the biosynthesis of histidine.

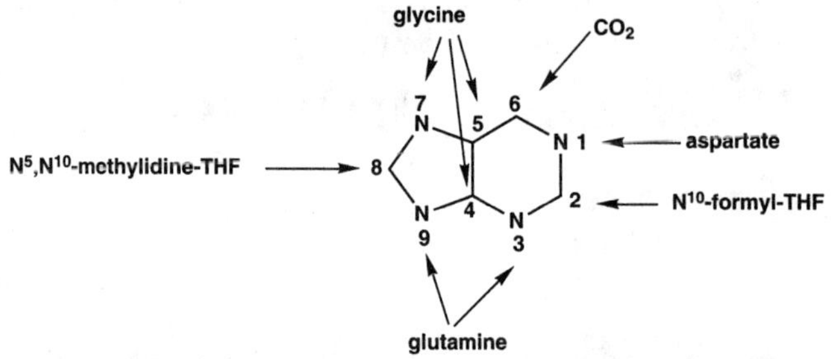

Fig. 26-1. Source of the atoms of the purine ring.

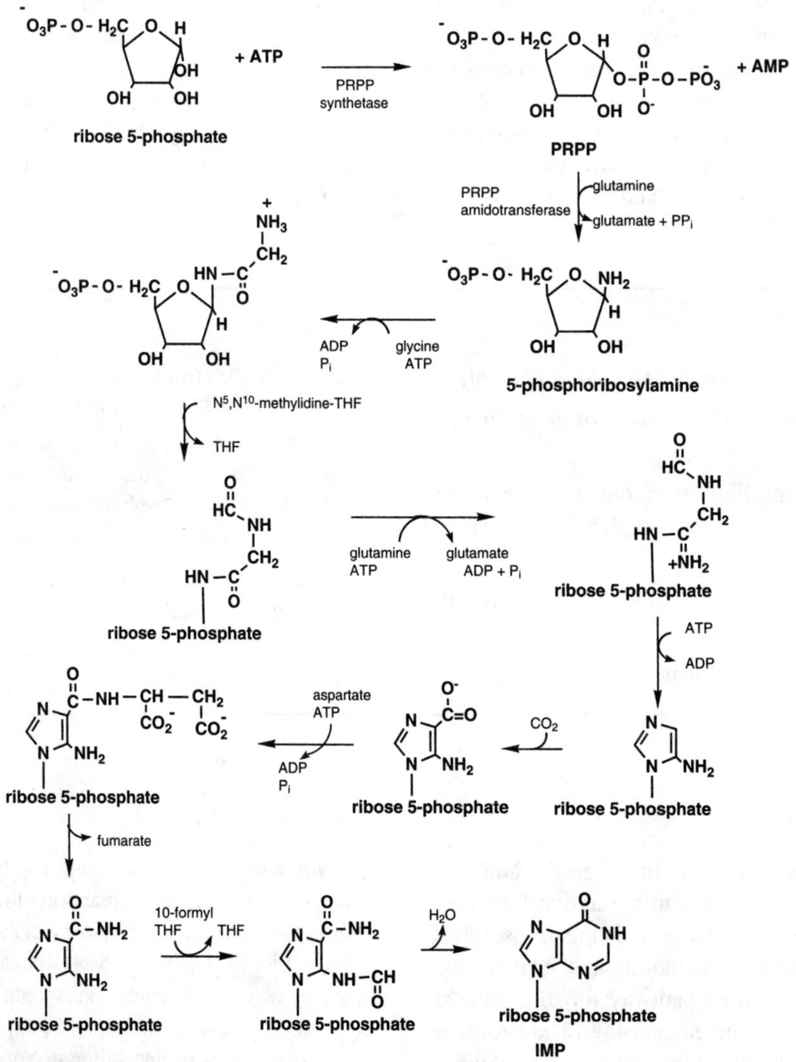

Fig. 26-2. Biosynthesis of the purine nucleotide IMP.

Because of its importance as a major intermediate, PRPP synthesis is tightly regulated. The synthetic enzyme **PRPP synthetase** is inhibited by high concentrations of nucleotides, especially deoxythymidine triphosphate (dTTP), ADP, and AMP. It is activated by high concentrations of inorganic phosphate. The enzyme responds to the availability of nucleotides by decreasing activity when they are plentiful and increasing activity when, as indicated by increasing levels of phosphate, they are scarce.

Synthesis of Inosine Monophosphate

The first reaction directed solely toward purine biosynthesis is the reaction of PRPP with glutamine, forming 5-phosphoribosylamine and releasing pyrophosphate. This reaction, catalyzed by **PRPP amidotransferase,** is irreversible under cellular conditions because of the immediate hydrolysis of pyrophosphate to inorganic phosphate. All guanosine and adenosine nucleotides act as allosteric inhibitors of PRPP amidotransferase. The addition of the amino group involves an inversion at carbon 1 from the α-configuration in PRPP to β in 5-phosphoribosylamine. The β-configuration persists in all purine nucleosides and nucleotides.

The next step in purine biosynthesis involves the addition of an entire glycine molecule. This is coupled with the hydrolysis of ATP, making it another irreversible reaction. Next, a formyl group is donated by N^5,N^{10}-methylidine tetrahydrofolate. This is followed by a transfer of the amide group of glutamine to what was the carboxyl carbon of glycine. This reaction also uses the energy from the hydrolysis of ATP.

Another molecule of ATP is used in the succeeding reaction, a ring closure. Carbon dioxide then is added. Unlike the carboxylation reactions you have seen before, this one does not use biotin, thiamine pyrophosphate, or vitamin B_{12} as a carbon dioxide carrier.

In the next reaction, the amino group of aspartate forms an amide with the newly added carboxyl group, again using ATP as the energy source. The compound then is cleaved, releasing fumarate. The addition of aspartate and the release of fumarate is similar to the reaction in the urea cycle in which aspartate was added to citrulline yielding arginosuccinate, which then was cleaved to produce fumarate and arginine.

10-Formyltetrahydrofolate provides the last atom of the ring. Dehydration causes ring closure and the formation of the first purine nucleotide, **inosine monophosphate** (IMP).

Synthesis of Adenosine Monophosphate

It is at IMP that the pathways to AMP and GMP diverge. For the synthesis of AMP, shown in Figure 26-3, IMP first reacts with aspartate, forming adenylsuccinate. This requires the hydrolysis of GTP to GDP. Fumarate then is split off, leaving AMP. The reaction is similar to the one in the urea cycle and the one discussed earlier in the synthesis of IMP.

Synthesis of Guanosine Monophosphate

For the synthesis of GMP, also shown in Figure 26-3, IMP is oxidized by NAD, producing xanthine 5′-phosphate. Next, an amino group is donated by glutamine, giving GMP. This last reaction is a familiar one that occurs twice in the synthesis of IMP. However, here ATP is hydrolyzed to AMP instead of ADP.

Pyrimidine Synthesis

In contrast to the synthesis of purines, in pyrimidine biosynthesis the ring structure is synthesized first and then attached to PRPP. The atoms of the pyrimidine ring come from the following sources:

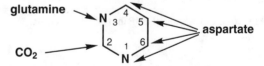

Synthesis of Carbamoyl Phosphate

The pathway for the biosynthesis of pyrimidines is shown in Figure 26-4. The first and rate-controlling reaction in humans is the condensation of a phosphate from ATP, with carbon dioxide and an amino group donated by glutamine to produce **carbamoyl phosphate.** Like the similar reaction in the urea cycle, this reaction requires the hydrolysis of two ATPs. It is inhibited allosterically by uridine triphosphate (UTP), one of the final products of the reaction sequence.

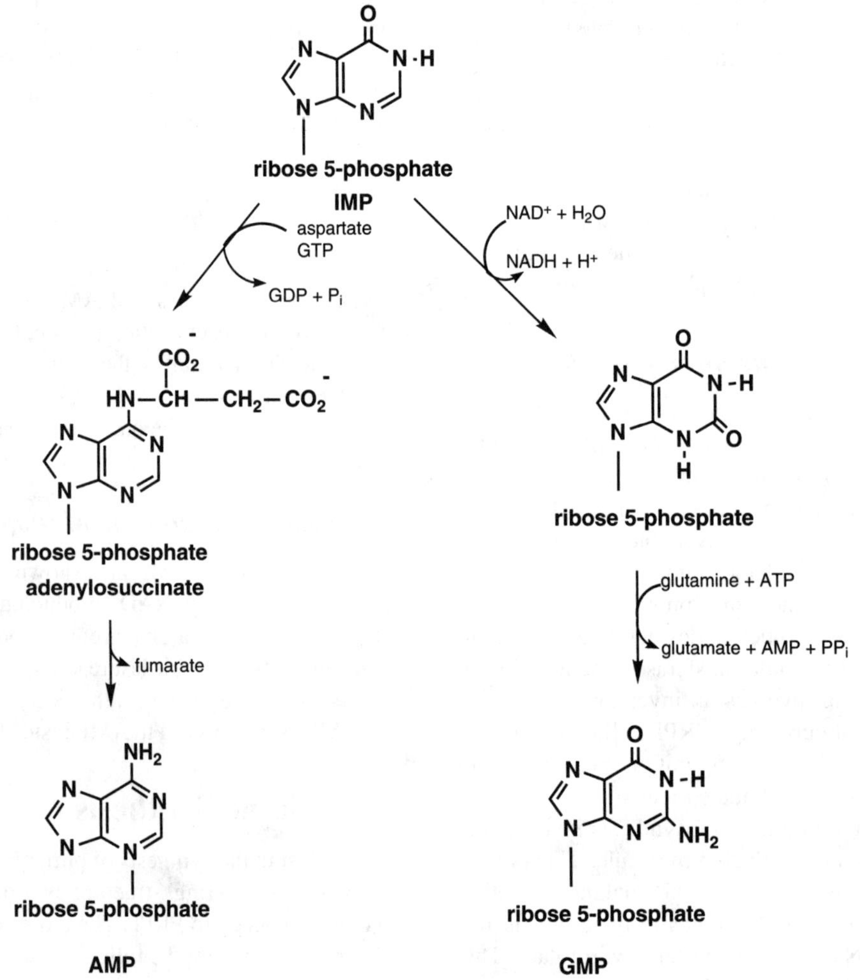

Fig. 26-3. Conversion of IMP into AMP and GMP.

Although both the urea cycle and pyrimidine biosynthesis involve the synthesis of carbamoyl phosphate, there are several major differences between the two reactions. The carbamoyl phosphate for the urea cycle is produced in the mitochondria, whereas that for pyrimidine biosynthesis is made in the cytosol. Glutamine is used as the donor of the amino group for pyrimidine biosynthesis and ammonia for the urea cycle. The enzyme catalyzing carbamoyl phosphate biosynthesis for the urea cycle requires N-acetylglutamate as a cofactor; the enzyme

for pyrimidine biosynthesis does not. In the liver and kidney, which contain both the cytosolic and mitochondrial enzymes, the amount of mitochondrial carbamoyl phosphate synthetase is 10 times greater than the cytosolic enzyme. This reflects a much greater need to synthesize urea than pyrimidines.

Synthesis of Uridine Monophosphate

Aspartate transcarbamoylase catalyzes the condensation of carbamoyl phosphate with aspartate, with

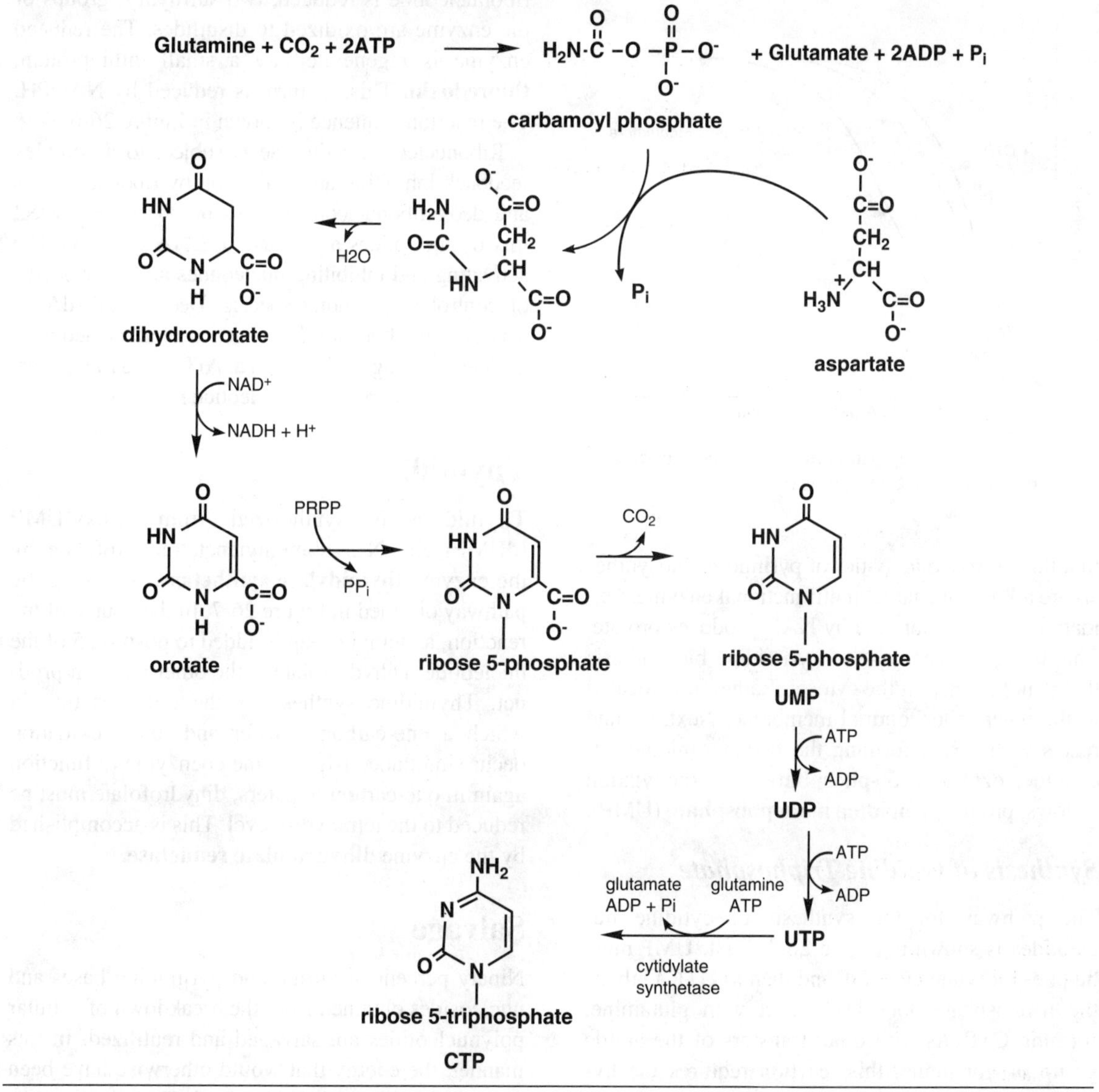

Fig. 26-4. Synthesis of the pyrimidine nucleotides UMP and CTP.

the loss of phosphate. This is the rate-controlling reaction for pyrimidine biosynthesis in bacteria. As shown in Figure 26-5, the bacterial aspartate transcarbamoylase is inhibited allosterically by cytidine triphosphate (CTP), the final product of the reaction sequence. CTP decreases the affinity of the enzyme for aspartate. The inhibition by CTP can be overcome by high concentrations of ATP, as both ATP and CTP compete for the same allosteric site on the enzyme.

The next reaction is dehydration, which causes ring closure, forming dihydroorotate. In humans, the

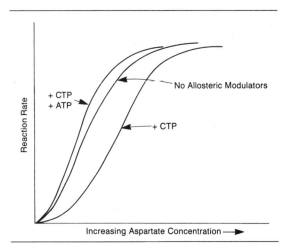

Fig. 26-5. Allosteric regulation of prokaryotic aspartate transcarbamoylase by CTP and ATP.

first three enzyme activities of pyrimidine biosynthesis are all part of a large multifunctional enzyme. Oxidation of dihydroorotate by NAD produces orotate. This is the only enzyme of pyrimidine biosynthesis that is not located in the cytosol. Rather, it is located in the outer mitochondrial membrane. Next, orotate reacts with PRPP, forming the first pyrimidine nucleotide, **orotidine 5′-phosphate.** Decarboxylation follows, producing **uridine monophosphate** (UMP).

Synthesis of Cytidine Triphosphate

The pathway for the synthesis of cytidine nucleotides is shown is Figure 26-4. First, UMP must be phosphorylated to UDP and then to UTP. Only as the triphosphate does UTP react with glutamine, forming CTP. As with other transfers of the amide group of glutamine, this reaction requires the hydrolysis of ATP. There is no separate pathway for the synthesis of cytosine, CMP, or CDP.

Deoxynucleotides

Ribonucleotides must be reduced to deoxynucleotides in order to serve as substrates for the synthesis of DNA. The reduction takes place only at the level of ribonucleoside diphosphate. NADPH is the ultimate reducing agent, and **ribonucleotide reductase** is the enzyme catalyzing the reaction. When the ribonucleotide is reduced, two sulfhydryl groups of the enzyme are oxidized to disulfides. The reduced enzyme is regenerated by a small dithioprotein, **thioredoxin.** This, in turn, is reduced by NADPH. The reaction sequence is shown in Figure 26-6.

Ribonucleotide reductase is subject to a complex feedback inhibition and activation by ribonucleotides and deoxyribonucleotides. This produces the correct mix of nucleotides necessary for DNA synthesis. The activating and inhibiting nucleotides and their degree of control vary among species. Deoxy-ATP (dATP) acts as an inhibitor for the reduction of all nucleotides, as does the drug **hydroxyurea.** ATP and dTTP generally are activators of ribonucleotide reductase.

Thymidine

Thymidine is synthesized from deoxy-UMP (dUMP) and N^5,N^{10}-methylenetetrahydrofolate by the enzyme **thymidylate synthetase,** following the pathway outlined in Figure 26-7. In the course of the reaction, a methyl group is added to position 5 of the nucleotide. Dihydrofolate is the other reaction product. Thymidine synthesis is the only reaction in which a one-carbon transfer and folate oxidation occur simultaneously. For the coenzyme to function again in one-carbon transfers, dihydrofolate must be reduced to the tetrahydro level. This is accomplished by the enzyme **dihydrofolate reductase.**

Salvage

Ninety percent of purine and pyrimidine bases and nucleosides obtained from the breakdown of cellular polynucleotides are salvaged and reutilized. In this manner, the energy that would otherwise have been expended in biosynthesis is saved. The salvage pathway is important also for the activation of purine and pyrimidine analogs used in chemotherapy. The salvage pathways for purines and pyrimidines are summarized in Figure 26-8.

Nucleotides, like other phosphorylated compounds, do not cross the plasma membrane. Extracellular nucleotides are converted into nucleosides by an enzyme located on the cell membrane, **5′-nucleotidase.** This is the first enzyme of the salvage pathway. Once inside the cell, nucleosides are either

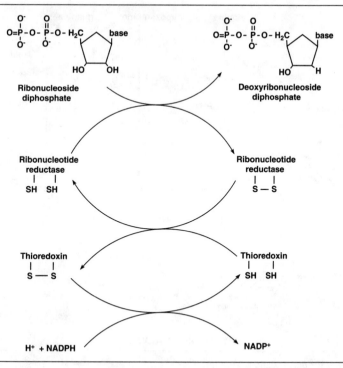

Fig. 26-6. Reduction of ribonucleoside diphosphates to deoxyribonucleoside diphosphates.

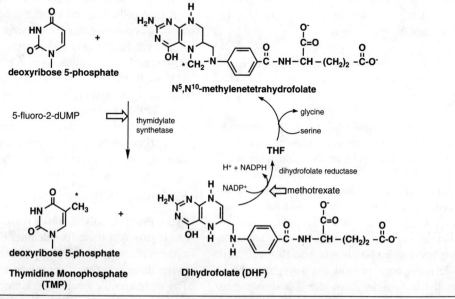

Fig. 26-7. Biosynthesis of thymidine monophosphate and the regeneration of N^5,N^{10}-methylenetetrahydrofolate. The points of inhibition by 5-fluoro-2-dUMP and methotrexate are indicated by open arrows.

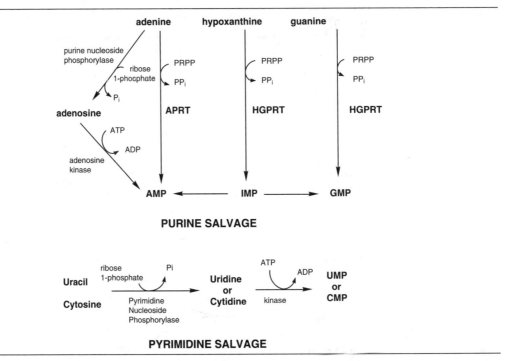

Fig. 26-8. Salvage pathway for purines and pyrimidines.

phosphorylated again to their respective nucleotides or converted to bases by the enzyme **nucleoside phosphorylase.**

nucleoside + P$_i$ → base + ribose 1-phosphate

Purine Salvage

The purine bases can yield nucleotides by a one-step reaction with PRPP. One enzyme, **adenine phosphoribosyl transferase** (APRT), catalyzes the reaction of adenine with PRPP, producing AMP. A different enzyme, **hypoxanthine guanine phosphoribosyl transferase** (HGPRT), reacts with PRPP and hypoxanthine or guanine, producing IMP or GMP.

Adenine can also be salvaged by a two-step route involving nucleoside phosphorylase and a specific kinase. The phosphorylase will use any purine base, forming the nucleoside. Adenosine kinase adds the first phosphate, creating the nucleotide. Because of the low concentration of adenine and ribose 1-phosphate, this pathway does not contribute greatly to the pool of adenine nucleotides unless the major route of salvage

is blocked. Guanosine kinase has never been detected in human cells and, consequently, the two-step route does not operate for guanine. Salvage of guanine must occur by a direct reaction of guanine with PRPP. The existence of inosine kinase is also in doubt in human cells, so the two-step route is questionable for the salvage of hypoxanthine as well. Phosphorylation of mononucleotides is accomplished by kinases showing an increasing level of specificity. These kinases differentiate between the different purine bases and between bases attached to ribose and deoxyribose.

Pyrimidine Salvage

Pyrimidines are salvaged by a two-step route similar to the two-step route for adenine. First, a relatively nonspecific nucleoside phosphorylase converts the pyrimidine bases to their respective nucleosides. The more specific nucleoside kinases react with the nucleosides, forming nucleotides. As with purines, further phosphorylation is carried out by enzymes with increasing specificity. Pyrimidines are not efficiently salvaged by the one-step route because the only available enzyme, orotic acid PRPP transferase, is present in too low a concentration and has too high a K_m for uracil and cytosine to be effective.

Inhibitors

Because growing cells need to synthesize both DNA and RNA, interference with the biosynthesis of purines and pyrimidines or their function has been a major focus in the development of anticancer drugs. Drugs inhibiting nucleotide biosynthesis affect not only tumor cells but normal ones as well. The inhibi-

tion of DNA and RNA synthesis in stem cells inhibits the formation of erythrocytes, lymphocytes, cells of the intestinal epithelium, and hair-forming cells. This inhibition produces the most obvious side effects of cancer chemotherapy. Because most tumor cells possess a more active salvage pathway than do normal cells, drugs entering metabolism via the salvage pathways obtain a higher concentration in tumor cells and thus have a therapeutic advantage. The structures of the drugs discussed here are given in Figure 26-9, and their sites of action are summarized in Table 26-1.

Fig. 26-9. Structures of some inhibitors of purine and pyrimidine metabolism.

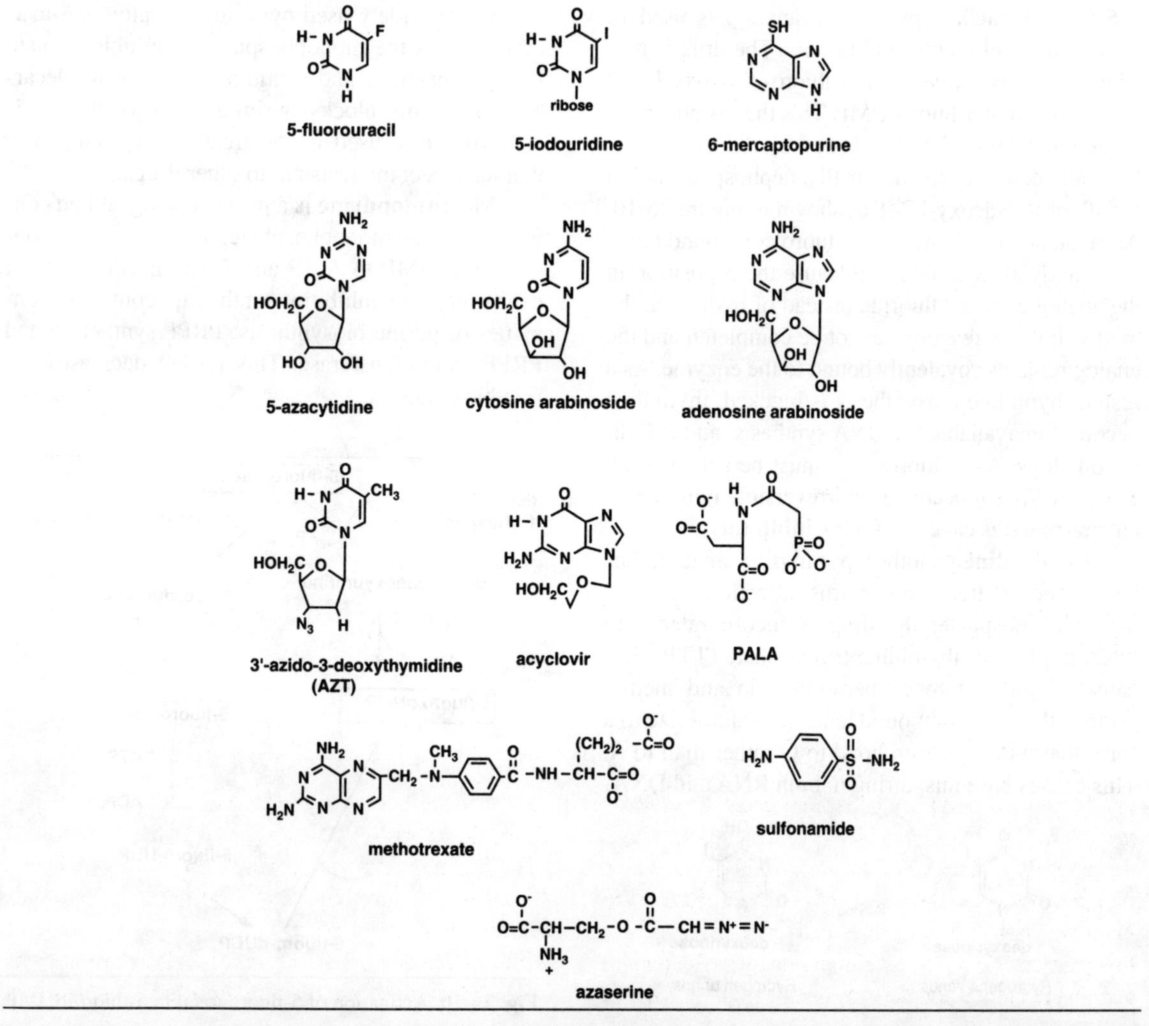

Table 26-1. Site of action of drugs inhibiting nucleotide biosynthesis

Inhibitor	Affect of activated form
5-Fluorouracil	Suicide inhibitor of thymidylate synthetase
5-Iodouridine	Incorporated into DNA, producing miscoding
5-Azacytidine	Inhibits orotate PRPP transferase and OMP decarboxylase
6-Mercaptopurine	Inhibits conversion of IMP to AMP and GMP, PRPP synthetase, PRPP amidotransferase
Cytosine arabinoside	Incorporated into DNA
Adenosine arabinoside	Incorporated into DNA
AZT	Incorporated into HIV viral DNA
Acyclovir	Incorporated into DNA of cells infected with herpesviruses
PALA	Inhibits aspartate transcarbamoylase
Methotrexate	Inhibits dihydrofolate reductase
Sulfonamide	Inhibits folate biosynthesis in bacteria
Azaserine	Inhibits nitrogen transfers from glutamine

PRPP = 5-phosphoribosyl-1-pyrophosphate; OMP = orotidine monophosphate; IMP = inosine monophosphate; AMP = adenosine monophosphate; GMP = guanosine monophosphate; AZT = azidothymidine; PALA = N-phosphonoacetyl-aspartate.

5-Fluorouracil, a pyrimidine analog, is used in the treatment of many solid tumors. The drug is metabolized to its active form, 5-fluoro-2′-deoxy-UMP by first forming 5-fluoro-UMP. This then is phosphorylated to 5-fluoro-UDP, followed by reduction to 5-fluoro-2′-deoxy-UDP and, finally, dephosphorylation to 5-fluoro-2′-deoxy-UMP as shown in Figure 26-10. As an analog of dUMP, the 5-fluoro compound binds to thymidylate synthetase. Because the 5 position in the analog contains fluoride instead of hydrogen, the methyl transfer reaction cannot be completed and the analog remains covalently bound to the enzyme. As a result, thymidine biosynthesis is blocked, thymidine becomes unavailable for DNA synthesis, and cell division stops. As 5-fluorouracil must be activated by the cell before it becomes an irreversible inhibitor of an enzyme, it is called a **suicide inhibitor.**

5-Iodouridine, another pyrimidine analog, has been used to treat herpesvirus infections. As the deoxy triphosphate, the drug is incorporated into DNA in place of thymidine triphosphate (TTP). Because of the difference between iodo and methyl groups, the iodo compound tends to teutomerize to a form that will hydrogen bond to G rather than to A. This causes base mispairing in both RNA and DNA.

Another widely used pyrimidine analog is **5-azacytidine.** As the monophosphate, it inhibits orotate PRPP transferase and orotidine 5-phosphate decarboxylase. This blocks pyrimidine biosynthesis. 5-Azacytidine is used in the treatment of leukemias that have become resistant to other drugs.

6-Mercaptopurine is a purine analog. When converted to the monophosphate, it inhibits the conversion of IMP to AMP and GMP. Furthermore, it exerts feedback inhibition on the rate-controlling enzymes of purine biosynthesis: PRPP synthetase and PRPP amidotransferase. This further decreases purine biosynthesis.

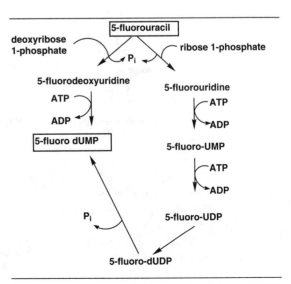

Fig. 26-10. Activation of 5-fluorouracil to 5-fluoro-dUMP.

A different kind of chemotherapeutically effective analog is created by altering the sugar rather than the base. Cytosine and adenosine arabinosides are examples of this type of drug. In their activated forms, both are incorporated into DNA. The mechanism by which these analogs kill cells is unclear. **Cytosine arabinoside** (araC) is used to treat acute leukemias. **Adenosine arabinoside** (araA) is used for viral infections.

Azidothymidine (AZT) is also a sugar analog. It is one of the drugs used to treat infections by HIV, the causative agent for the acquired immunodeficiency syndrome (AIDS). In AZT, the hydroxyl group at position 3′ is replaced with an N_3 (azido) group. AZT can be activated to the triphosphate by the salvage pathway. Then the viral DNA polymerase incorporates the analog into DNA. Because there is no 3′OH group, chain elongation ceases. The reactivity of the azido group allows the drug to bind covalently to the polymerase and inactive it. The human DNA polymerases do not accept AZT as a substrate, so it is not incorporated into replicating human DNA.

Another sugar analog is found in the drug **acyclovir.** In this compound, most of the sugar is missing. This drug can be activated to the triphosphate by an enzyme present in cells infected by herpesviruses. Then it is incorporated into DNA as if it were dGTP. Because there is no 3′OH group, chain elongation ceases. Acyclovir is not activated in uninfected human cells, and so it is used to treat herpesvirus infection.

There are a number of inhibitors that are neither base nor sugar analogs. One of these is **N-phosphonoacetyl-L-aspartate (PALA).** This compound is an analog of the transition state of the substrates for aspartate transcarbamoylase. It is one of the first antimetabolites designed to mimic a transition state.

Any compound that decreases the folate pool will eventually stop purine and pyrimidine biosynthesis and with it DNA and RNA synthesis. **Methotrexate** is an analog of folate that inhibits the reduction of dihydrofolate to tetrahydrofolate. In cells actively synthesizing thymidine, methotrexate rapidly depletes the supply of tetrahydrofolate and stops all one-carbon transfers. It is a potent anticancer drug that is used against both leukemias and solid tumors.

Sulfonamides are analogs of the para-aminobenzoic acid portion of folate and act as competitive inhibitors of folate biosynthesis. Because folate cannot be synthesized by humans, these drugs cause no ill effects on mammalian cells. However, the lack of de novo folate synthesis is lethal to bacteria.

Azaserine, an analog of glutamine, inhibits all reactions involving nitrogen transfers from glutamine. These reactions include the insertion of N3 and N9 into the purine ring and the synthesis of GMP from IMP and CTP from UTP. Because glutamine is required also for the transport of ammonia, azaserine is toxic and is not used clinically.

Inborn Errors of Nucleotide Synthesis

Deficiencies in several of the enzymes of pyrimidine biosynthesis give rise to the excretion of orotic acid in the urine. Lack of both orotate PRPP transferase and orotidylate decarboxylase result in **orotic aciduria type I.** This enzyme defect produces kidney stones of orotic acid, megaloblastic anemia, and an immunodeficiency that if untreated leads to an early death. **Type II orotic aciduria** is due to a deficiency in only the decarboxylase. This defect causes an increase in the excretion of orotic acid and ortidylate and leads to anemia. Both disorders are treatable with uridine. The uridine serves both as a source of cytosine nucleotides and as a feedback inhibitor of further pyrimidine biosynthesis.

A much milder form of orotic aciduria can be caused by a deficiency of the urea cycle enzyme **ornithine transcarbamoylase.** When this enzyme is absent, mitochondrial carbamoyl phosphate cannot be used for urea synthesis. Instead, it leaks into the cytoplasm and is used to make pyrimidines. In addition to excessive pyrimidine production and excretion, this abnormality produces protein intolerance and hepatic enlargement as a result of the inability of the liver to metabolize the nitrogen from amino acid catabolism. Similar symptoms of orotic aciduria occur in Reye's syndrome because of damage to the liver mitochondria.

Inborn errors in the salvage of purines give rise to major medical problems. A lack of APRT results in the inability to salvage adenine by the major one-step route. Adenine is excreted intact and as the 2,8-dihydroxy derivative. Because neither of these compounds is very soluble, they crystallize, forming kidney stones. Mental retardation is another symptom of a lack of APRT.

Lesch-Nyhan syndrome, a rare X-linked recessive trait, is caused by a total lack of HGPRT. The syndrome is characterized by an excessive synthesis of purines, increased levels of PRPP, and increased activity of four of the six enzymes of pyrimidine biosynthesis. Although the reason for the alteration in pyrimidine biosynthesis is not known, the absence of guanine nucleotides for effective feedback on the controlling enzymes of purine biosynthesis is the cause of the increased PRPP and purine levels. The excessive production of purines leads to excessive degradation and gout, which will be discussed in detail in the next chapter.

Lesch-Nyhan patients also show neurological symptoms consisting of mental retardation, extreme aggression, and self-mutilation. The basal ganglia of the brain have one of the lowest levels of PRPP amidotransferase of all cells. Because of this, these cells are almost completely dependent on the salvage pathway for their supply of purine nucleotides. The metabolic relationship between the shortage of guanine nucleotides and the neurological symptoms remains to be elucidated. However, this is one of the few cases in which a known enzymatic alteration results in a specific behavioral change.

Summary

Purines are synthesized as mononucleotides by adding individual atoms or groups to a preformed ribose 5-phosphate. The individual atoms come from glycine, carbon dioxide, aspartate, 10-formyl-THF, glutamine, and N^5,N^{10}- methylidine-THF. The rate-controlling reactions are PRPP synthetase and PRPP amidotransferase. AMP is synthesized from IMP by the transfer of an amino group from aspartate. GMP comes from IMP by oxidation and the addition of a nitrogen from glutamine.

Pyrimidines are synthesized as bases and then added to PRPP to form nucleotides. The first reaction, catalyzed by cytosolic carbamoyl phosphate synthetase, is the rate-controlling reaction in humans. In contrast to the urea cycle enzyme, this enzyme uses glutamine rather than ammonia as the nitrogen donor. The next reaction, aspartate transcarbamoylase, is the rate-controlling reaction in bacteria. CTP is synthesized from UTP by a nitrogen transfer from glutamine. Thymidine is produced from dUMP by a methyl transfer from N^5,N^{10}-methylene-THF. Deoxynucleotides come from the reduction of nucleoside diphosphates by the action of ribonucleotide reductase.

Purines and purine analogs are salvaged by a direct reaction with PRPP. Adenine can also be salvaged by a two-step route. Pyrimidines and their analogs are salvaged only by a two-step route.

5-Fluorouracil, in its activated form, and methotrexate inhibit thymidine biosynthesis. 6-Mercaptopurine blocks purine biosynthesis. 5-Azacytidine and PALA interfere with pyrimidine synthesis. Arabinosides of cytidine and adenosine are incorporated into DNA, as is 5-iodouridine. Sulfonamides interfere with bacterial synthesis of folate. Azaserine inhibits all glutamine-requiring reactions. Azidothymidine and acyclovir are used to treat viral infections caused by HIV and herpes, respectively.

Lack of orotic acid PRPP transferase or decarboxylase causes orotic aciduria and the crystallization of orotic acid in the kidneys. A lack of ARPT also produces kidney stones of adenine and 2,8-dihydroxyadenine. Lack of HGPRT produces Lesch-Nyhan syndrome, which is characterized by excess purine synthesis and by neurological symptoms.

27 Nucleotide Degradation

The nuclear DNA of cells is very stable and does not break down until the entire cell is destroyed. The only cellular DNA degradation that does occur is from repair processes. In addition, any foreign DNA entering the cell, such as a virus, is subject to degradation. These amounts are extremely small and do not have a significant impact on nucleotide metabolism. However, degradation of DNA entering the digestive tract does occur and is important to the overall nucleotide balance.

RNA degradation occurs in the digestive tract and in cells. Messenger, ribosomal, and transfer RNAs are constantly being turned over, with half-lives of hours to weeks. Introns produced from heterogeneous nuclear RNA processing are degraded, as are the nonconserved sequences from the ribosomal and transfer RNA precursors. All these nucleotides either are salvaged for reutilization or are subject to catabolism. Excess catabolism or defects in the enzymes responsible can cause medical consequences ranging from kidney stones to gout to complete immunodeficiency.

Objectives

After completing this chapter, you should be able to

List the major steps in the degradation of polynucleotides.

Diagram the catabolism pathway for purines and pyrimidines, name the major enzymes, and list the excretion products.

Describe the mechanism of action of allopurinol.

Match the inherited diseases of nucleotide degradation with the defective enzyme and the major symptoms.

Degradation of Polynucleotides

Polynucleotides are degraded by the same type of reactions in both the digestive tract and cells. Of those ingested, little degradation occurs in the mouth or stomach. In the intestine, pancreatic enzymes digest large nucleic acids into smaller oligonucleotide units. These reactions are outlined in Figure 27-1. **Pancreatic DNase** hydrolyzes DNA between purine and pyrimidine bases, producing chains averaging four nucleotides in length that have a free 3'OH

group. **Pancreatic RNase** cleaves RNA when the 3' linkage is to a pyrimidine. Because of the need for a 2',3' cyclic intermediate in the hydrolysis reaction, this enzyme cannot hydrolyze DNA or RNA in which the 2'OH group is methylated.

Several exonucleases then act to remove nucleotides from either the 5' or the 3' end of the oligonucleotide. Enzymes acting from the 5' end generally require a free 5'OH group and produce 3' mononucleotides. Enzymes acting from the 3' end require a free 3'OH

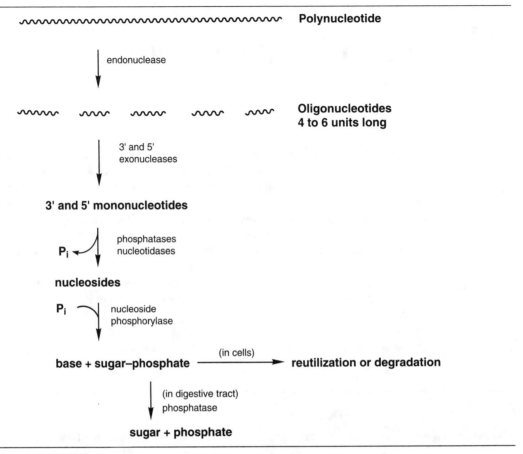

Fig. 27-1. Degradation of polynucleotides.

group and produce 5′ mononucleotides. The result of the combined action of these enzymes is a mixture of 3′ and 5′ nucleotides. Nucleotidases and phosphatases then remove the phosphate, producing nucleosides. The nucleosides are hydrolyzed to the free base and ribose or deoxyribose phosphates. The enzymes that degrade nucleotides are generally more specific than those degrading polynucleotides. These enzymes require a specific base (purine or pyrimidine) and a specific sugar (ribose or deoxyribose).

Purine Degradation

Purines in excess of those needed for polynucleotide synthesis are subject to degradation following the pathway outlined in Figure 27-2. Adenosine is first deaminated by the enzyme **adenosine deaminase** to form inosine. Then, **purine nucleoside phosphorylase** removes the ribose residue, producing hypoxanthine and ribose 1-phosphate. **Xanthine oxidase** converts hypoxanthine to xanthine and then to **uric acid.** Ribose 1-phosphate can be isomerized to ribose 5-phosphate and either phosphorylated to PRPP or oxidized via the pentose phosphate pathway.

Guanosine follows a similar pathway. Purine nucleoside phosphorylase produces guanine and ribose 1-phosphate. The base then is deaminated by guanase, yielding xanthine.

Sodium urate is normally the major excretion product of purine degradation in humans, although small amounts of xanthine, hypoxanthine, and guanine are excreted as well. Sodium urate is relatively

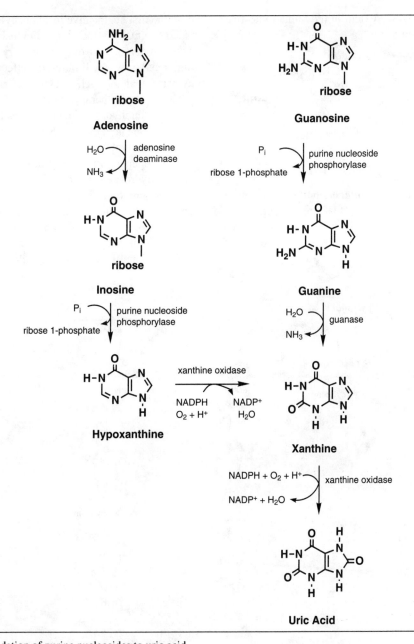

Fig. 27-2. Degradation of purine nucleosides to uric acid.

insoluble in aqueous solutions. If the pH of urine is below 5.7, the pK of uric acid, then the major excreted form is uric acid. This is even less soluble than sodium urate. If the concentration exceeds the solubility, uric acid precipitates. In most species other than primates and Dalmatian dogs, the problems caused by the low solubility of urate are bypassed by oxidizing uric acid to more soluble compounds such as **allantoin, allantoic acid,** urea, or carbon dioxide, ammonia, and water. These latter reactions occur to some extent in the human intestine by bacterial oxidation.

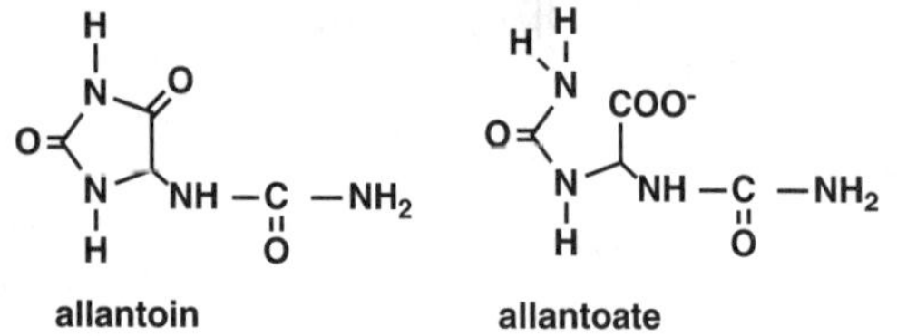

allantoin   allantoate

Pyrimidine Degradation

The degradation of pyrimidines, outlined in Figure 27-3, produces compounds that are all soluble in aqueous solution. Because of this, no extensive medical problems will arise if excessive amounts of pyrimidines are degraded.

In the first degradative reaction, cytosine is deaminated to uracil by the enzyme **cytosine deaminase.** The activity of this enzyme is very high in many tumor cells, thereby limiting the effective half-life of cytosine analogs, such as cytosine arabinoside, used in cancer chemotherapy. Inhibition of cytosine deaminase by tetrahydrouridine decreases the rate of drug degradation and increases its therapeutic potential.

Uracil is first oxidized to dihydrouracil. Then the ring is opened, forming β-ureidopropionate. This can be hydrolyzed to form β-alanine, carbon dioxide, and ammonia. The β-alanine can be excreted or converted to malonic semialdehyde and oxidized to carbon dioxide, ammonia, and water.

Thymidine degradation follows a similar pathway, eventually forming β-ureidoisobutyrate. This

Fig. 27-3. Degradation of pyrimidine bases.

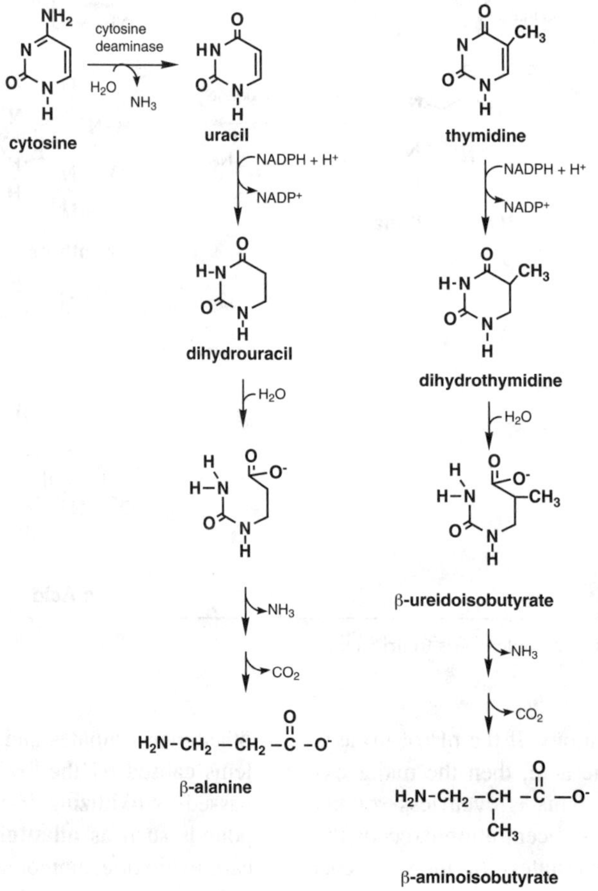

compound loses NH_4 and carbon dioxide, forming β-aminoisobutyrate. This can be excreted or metabolized to form methylmalonic semialdehyde. After conversion to the CoA derivative, it eventually enters metabolism as succinyl-CoA.

Minor bases such as pseudouridine and the methylated bases from the breakdown of transfer RNA have no metabolic pathway for further degradation. They are excreted intact.

Enzymatic Defects

Gout

In humans, other primates, and Dalmatian dogs, the normal route of purine oxidation stops at uric acid. Excess purines—whether caused by overproduction, inadequate salvage, inefficient excretion, massive cell destruction, or high purine intake in the diet—can lead to excessive uric acid production and hyperuricemia. If the sodium urate concentration exceeds the solubility, crystallization occurs, producing **tophi.** Crystallization can occur in joints, especially the big toe, and in the kidney as sodium urate stones. If the pH of the urine is less than 5.7, the stones will be composed of the even less soluble uric acid. Crystallization of sodium urate in the joints causes **gout.** Lymphocytes infiltrate at the sites of the tophi and produce inflammation, fever, and pain.

More than 90% of the clinical cases of gout are due to insufficient renal excretion of uric acid. However, primary gout can also be due to an overproduction or underutilization of PRPP, a precursor to the purine nucleotides. High levels of PRPP stimulate excessive purine biosynthesis. Because the purines are not needed for RNA or DNA biosynthesis, they are degraded. This excessive degradation produces urate levels above the solubility, and gout. Overproduction of PRPP can be caused by a lack of feedback inhibition by purine nucleotides on PRPP synthetase, the first enzyme in purine biosynthesis, or by an excessive amount of the enzyme. Underutilization occurs with defects in either of the purine salvage enzymes, APRT or HGPRT. Similarly, a lack of feedback inhibition on PRPP amidotransferase, a key enzyme of purine biosynthesis, also can cause excessive purine synthesis and gout. Finally, an overproduction of ribose 5-phosphate, which occurs in von Gierke's disease (type I glycogen storage disease), can produce gout. The excess ribose 5-phosphate is converted to PRPP and then to purines, which must be degraded.

Some secondary conditions can also give rise to gout symptoms. Lactate and ketone bodies, produced during heavy exercise, compete with urate for renal tubular resorption. This decreases the excretion of urate and increases its concentration in the blood. These metabolic acids also decrease the pH of the urine and increase the proportion of urate that is in the form of the less soluble uric acid, thereby increasing the possibility of the formation of uric acid kidney stones. Dehydration, another possible consequence of heavy exercise, also increases the relative concentration of sodium urate in the circulation. Alcohol inhibits the antidiuretic hormone, producing dehydration and a rise in the concentration of urate. Other conditions that may give rise to gout symptoms include a diet rich in organ meats, such as kidney and liver, and successful cancer chemotherapy, which produces a massive cell kill and results in the degradation of large amounts of nucleic acids.

Gout usually is treated with **allopurinol,** an analog of hypoxanthine.

allopurinol   **alloxanthine**

At low concentrations, allopurinol is a competitive inhibitor of xanthine and hypoxanthine for oxidation by **xanthine oxidase.** At high concentrations, allopurinol is oxidized by xanthine oxidase to form **alloxanthine,** which irreversibly inhibits the enzyme. Therefore, allopurinol acts as a suicide inhibitor.

In the presence of allopurinol, xanthine is not oxidized to urate. The problem of crystallization is eliminated as the new excretion products, xanthine and hypoxanthine, do not cocrystalize. Allopurinol does not decrease the amount of purines entering the degradative pathway. It merely alters the proportions of excreted products so that none exceed their solubility. Allopurinol treats the gout symptoms of Lesch-Nyhan syndrome (discussed in Chapter 26) but does not affect the mental symptoms of the disease.

Allopurinol is used for long-term control of gout. Acute attacks can be treated with the drug **colchicine.** This compound inhibits the formation of the mitotic spindle. In the presence of colchicine, all cell division is inhibited and the inflammation is eliminated but the crystals of urate remain. By preventing cell division, colchicine also inhibits the immunological response to bacterial infection, an undesirable side effect.

Immunodeficiency

Severe immunological defects are caused by a deficiency of two of the enzymes of purine degradation. A lack of **purine nucleoside phosphorylase** produces a T-cell immunodeficiency, characterized by a much-increased susceptibility to infection. It appears that in the absence of purine nucleoside phosphorylase, guanosine and its nucleotides accumulate. This produces excessive amounts of dGTP which, in some manner, inhibits the ability of the T cells of the immune system to divide and function. Why T cells are especially affected is not known.

Severe combined immunodeficiency is associated with a lack of **adenosine deaminase.** Because adenosine cannot be catabolized in severe combined immunodeficiency, adenosine and its nucleotides accumulate. This produces an increased level of dATP, which serves to inhibit **ribonucleotide reductase.** Reduction of all ribonucleotide diphosphates to their respective deoxyribonucleotides is inhibited. Without deoxynucleotides, DNA cannot be synthesized and neither T nor B lymphocytes can divide. The affected individual has no functioning cell-mediated immune response. Except for a bone marrow transplant, there is no effective long-term treatment for severe combined immunodeficiency. A patient with this disease was the first person given a human gene transplant. The gene transplant was effective in providing a measurable level of adenosine deaminase and some restoration of immune function, and so several other gene transplants have followed. Because of its unique effect on the division of lymphocytes, inhibition of adenosine deaminase by the drug deoxycoformycin has proved to be effective in treating hairy cell leukemia.

Summary

Adenosine is degraded to uric acid by the sequential action of adenosine deaminase to produce inosine, purine nucleoside phosphorylase to yield hypoxanthine, and xanthine oxidase to produce xanthine and uric acid. Guanosine degradation requires purine nucleoside phosphorylase to give the guanine, guanase to produce xanthine, and xanthine oxidase to produce uric acid. Excess urate can precipitate in joints, causing gout. Sodium urate or the less soluble uric acid can precipitate in the kidney, forming kidney stones. Gout and urate stones are treated with allopurinol, a xanthine oxidase inhibitor. Acute attacks of gout can be treated with colchicine, an antimitotic agent. In most species other than primates, urate is further oxidized to allantoin and allantoate. A defect in purine nucleoside phosphorylase is associated with T-cell immunodeficiency. Severe combined immunodeficiency is caused by a lack of adenosine deaminase.

Cytosine and uracil are metabolized to β-alanine. Thymidine is converted by a similar pathway to β-aminoisobutyrate. All these are soluble excretion products.

Part VII Questions: Nucleotide Metabolism

1. The rate of purine biosynthesis is controlled by the concentration of all of the following *except*
   A. inorganic phosphate.
   B. ADP.
   C. PRPP.
   D. ribose 1-phosphate.
   E. AMP.

2. The atom indicated by the arrow comes directly from

   A. glutamine.
   B. glycine.
   C. aspartate.
   D. ammonia.
   E. N-acetylglutamate.

3. The rate-controlling reaction for pyrimidine biosynthesis in humans
   A. is carbamoyl phosphate synthetase.
   B. is aspartate transcarbamoylase.
   C. requires N-acetylglutamate as a cofactor.
   D. is inhibited allosterically by ATP.
   E. is located in the mitochondrial membrane.

4. Deoxyribonucleotide production
   A. involves the reduction of ribonucleoside triphosphates.
   B. is directly inhibited by PALA.
   C. uses FAD as the final reducing agent.
   D. requires the function of the protein thioreductase.
   E. is inhibited directly by methotrexate.

5. All the following compounds enter metabolism by way of the salvage pathway *except*
   A. 5-azacytidine.
   B. 6-mercaptopurine.
   C. methotrexate.
   D. adenosine arabinoside.
   E. 5-fluorouracil.

6. The major excretion products of purine degradation in humans include
   A. uric acid and adenosine.
   B. ammonia, carbon dioxide, and water.
   C. allantoin and allantoic acid.
   D. xanthine and uric acid.
   E. β-alanine and ammonia.

7. All the following are excretion products from pyrimidine degradation *except*
   A. uric acid.
   B. β-alanine.
   C. carbon dioxide.
   D. β-aminoisobutyric acid.
   E. ammonia.

8. Gout
   A. is produced by the crystallization of orotic acid in the joints.
   B. can be caused by a deficiency in PRPP synthetase.
   C. is associated with a T-cell immunodeficiency.
   D. is due to an inability to salvage adenine.
   E. is usually treated with an inhibitor of xanthine oxidase.

9. Severe combined immunodeficiency is associated with a lack of
   A. adenosine deaminase.
   B. guanase.
   C. hypoxanthine guanine phosphoribosyl transferase (HGPRT).
   D. cytosine deaminase.
   E. xanthine oxidase.

10. Which of the following statements is true about the degradation of polynucleotides?
    A. Pancreatic DNase hydrolyzes DNA from the 5′ end, producing 5′ mononucleotides.

B. Mononucleotides and small oligonucleotides are the final digestion products.

C. Pancreatic RNase is an exonuclease that
can hydrolyze both RNA and DNA into
segments four to six nucleotides long.

D. Nucleotide phosphatases generally are specific for both the base and the sugar.

E. Degradation of polynucleotides occurs only
in the small intestine.

Part
VIII Integration of Metabolism

28 Amino Acid and Protein Hormones

The metabolic activities of different tissues are coordinated by hormones. Hormones are released by the endocrine glands into the circulation. The hormones then travel to receptor cells and alter enzymatic activity by either modifying existing proteins or stimulating or inhibiting the synthesis of new proteins. The same hormone can produce very different effects in different types of receptor cells.

Hormones can be classified into two large groups: those derived from amino acids and those derived from lipids. This chapter will discuss the amino acid and protein hormones. The next chapter will cover those derived from lipids. The amino acid and protein hormones usually are stored in the cells that produce them. They are released in response to hypothalamic or pituitary hormones or changes in the levels of circulating metabolites or ions. In general, the amino acid and protein hormones do not enter their target cells. Instead, they bind to specific receptors on the plasma membrane and induce the production of internal second messengers that alter cellular metabolism. These second messengers include cAMP, phosphoinositol, diacylglycerols, and calcium.

Objectives

After completing this chapter, you should be able to

Describe the general mechanism by which amino acid and protein hormones produce their metabolic effects.

List the hormones of the anterior pituitary, their major metabolic effects, the hypothalamic factors controlling their release, and the results of excess or deficiency.

Recognize the structures of the thyroid hormones and describe the process of synthesis.

List the characteristics and metabolic effects, including second messengers produced, for the following hormones: thyroxine, triiodothyronine, parathyroid hormone, calcitonin, insulin, glucagon, the catecholamines, and the gastrointestinal hormones. Describe the results of deficiency or excess of these hormones.

Hypothalamic and Anterior Pituitary Hormones

The hypothalamus produces nine releasing and inhibiting factors that affect the anterior pituitary. Their relationships are summarized in Table 28-1. In addition to the releasing and inhibiting factors, the hypothalamus produces two hormones, oxytocin and vasopressin, that are released to the general circulation. All the hypothalamic hormones are modified amino acids or small peptides.

The anterior pituitary produces eight hormones in re-

Table 28-1. Relationship between hypothalamic and pituitary hormones and their target cells

Hypothalamic hormone	Effect on pituitary hormone	Target cell
Thyrotropin-releasing hormone (TRH)	Stimulates thyroid-stimulating hormone (TSH)	Thyroid
Somatostatin	Inhibits growth hormone (GH)	Bone
Growth hormone–releasing hormone (GHRH)	Stimulates GH	Bone
Prolactin-inhibiting hormone (PIH)	Inhibits prolactin	Mammary gland
Prolactin-releasing hormone (PRH)	Stimulates prolactin	Mammary gland
MSH-inhibiting hormone	Inhibits melanocyte-stimulating hormone (MSH)	Central nervous system (?)
MSH-releasing hormone	Stimulates MSH	Central nervous system (?)
Gonadotropin-releasing hormone (GRH)	Stimulates luteinizing hormone (LH)	Testis, ovary
	Stimulates follicle-stimulating hormone (FSH)	Testis, ovary
Corticotropin-releasing hormone (CRH)	Stimulates adrenocorticotropic hormone (ACTH)	Adrenal cortex
Vasopressin		Kidney, blood vessels
Oxytocin		Uterus, mammary gland
Endorphins		Central nervous system

sponse to stimulation by hypothalamic hormones. Three of these—**follicle-stimulating hormone** (FSH), **luteinizing hormone** (LH), and **ACTH**—affect the synthesis or release of steroid hormones. The others are **thyroid-stimulating hormone** (TSH), **growth hormone, prolactin, melanocyte-stimulating hormone** (MSH), and the **endorphins.** The anterior pituitary hormones are larger than the hypothalamic hormones. FSH, LH, and TSH are glycoproteins with two subunits. The α-subunit is identical in all three hormones. The β-subunit is different and gives the hormones their specificity. ACTH, MSH and β-lipotropin (the precursor to the endorphins) are all derived from a single precursor molecule, **proopiomelanocortin.** Gamma-MSH is on the N-terminal end of proopiomelanocortin, followed by ACTH and β-lipotropin. After release from the precursor, ACTH can be cleaved to α-MSH and β-lipotropin to γ-lipotropin and β-endorphin (Fig. 28-1).

Lack of all the pituitary hormones produces **panhypopituitarism,** in which there is hypoglycemia because the lack of ACTH causes a lack of the glucocorticoids. In addition, there are few, if any, sex

Fig. 28-1. Conversion of proopiomelanocortin to γ-MSH, ACTH, and β-lipotropin. ACTH can be cleaved to form α-MSH. Beta-lipotropin can be converted to γ-lipotropin and β-endorphin.

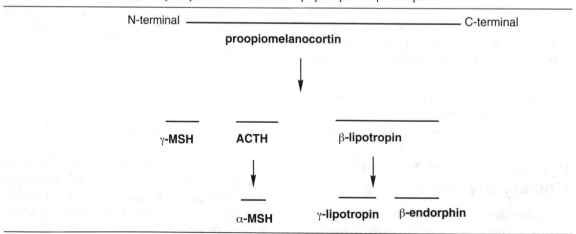

hormones. This syndrome occurs only with complete destruction of the pituitary gland.

Thyroid-Releasing and Thyroid-Stimulating Hormones

Thyroid-releasing hormone (TRH) is a three–amino acid peptide released by the hypothalamus into the portal circulation. It enters the anterior pituitary and stimulates the release of **thyroid-stimulating hormone.** TSH then acts on the thyroid gland and stimulates the synthesis and release of the thyroid hormones and stimulates growth of the gland. TSH also facilitates uptake of iodine into the thyroid. Its mechanism of action is a cAMP cascade. Lack of TSH produces **primary myxedema,** which shows the same symptoms as a lack of thyroid hormone.

Growth Hormone

Growth hormone (GH) or **somatotropin** is an anterior pituitary protein of 191 amino acids. Its release by the pituitary is inhibited by hypothalamic **somatostatin,** a 14–amino acid peptide. The release of somatotropin is stimulated by another hypothalamic factor, **growth hormone–releasing hormone.** Growth hormone increases cell growth, protein synthesis, and plasma glucose and calcium levels. Because many individual events are involved, hours or days are required to observe the effects of growth hormone on the body. Growth hormone levels increase with exercise, stress, and diets rich in amino acids. Because of its effect on cell and muscle growth, growth hormone has become a drug that is abused by some athletes.

The placenta produces **chorionic growth hormone** or **placental lactogen,** which has a structure and function similar to the pituitary growth hormone. It is the level of this hormone that is measured in home pregnancy tests.

Lack of growth hormone in infants produces **dwarfism,** which can be treated only by injection of the missing hormone. Human growth hormone is now available as a recombinant protein and is used in treating children of small stature. Excess growth hormone in infants produces **gigantism.** If an excess occurs in adults, the bones of the jaw and nose grow, as do those of the fingers and toes. This result is a deformity known as **acromegaly.**

Prolactin

A second set of hypothalamic inhibiting and releasing factors control **prolactin** release from the pituitary. **Prolactin-inhibiting hormone,** which is dopamine, inhibits the secretion of prolactin. The peptide **prolactin-releasing hormone** stimulates prolactin release. Prolactin is a peptide hormone containing 198 amino acids, and it induces lactation in the mammary gland.

Melanocyte-Stimulating Hormone

Melanocyte-stimulating hormone (MSH) is the third pituitary hormone controlled by a set of stimulating and inhibiting hypothalamic hormones: **MSH-releasing hormone** and **MSH-inhibiting hormone.** The exact function of MSH in humans is unclear. In frogs, it promotes melanin pigment formation.

Gonadotropins

Gonadotropin-releasing hormone is a 10–amino acid peptide that induces the release of LH and FSH. **Follicle-stimulating hormone** induces the early stages of sperm production, initiates maturation of the ovarian follicles, and stimulates the synthesis of estrogens. Its concentration rises 10-fold at ovulation.

Luteinizing hormone stimulates the production of testosterone, estrogens, and progesterone. It also induces ovulation. These effects occur through an increase of cAMP and the phosphorylation of specific proteins. The placenta produces an LH-like hormone, **chorionic gonadotropin.** The amino acid sequences and activities of the placental and pituitary hormones are very similar.

Adrenocorticotropic Hormone

Adrenocorticotropic hormone (ACTH) is a 39–amino acid peptide that stimulates the synthesis of cortisol in the adrenal gland and the proliferation of the cells of that gland. Its release is inhibited by the adrenocortical steroids. The effects of ACTH will be discussed in more detail in the next chapter, in which the adrenocortical hormones will be covered.

Vasopressin and Oxytocin

The two hypothalamic hormones that enter the general circulation are vasopressin and oxytocin. Like the other hypothalamic hormones, these are secreted by the hypothalamus and then are taken up by the anterior pituitary. However, instead of producing or inhibiting the release of another hormone, the pituitary stores and releases these hormones unchanged.

Vasopressin is also called **antidiuretic hormone** (ADH). It contains nine amino acids and functions to increase water resorption in the kidney. In addition, vasopressin is a powerful vasoconstrictor. Lack of vasopressin causes **diabetes insipidus,** which is characterized by a high output of very dilute urine.

Oxytocin contains nine amino acids, only two of which are different from those in vasopressin. Oxytocin produces uterine contractions and milk ejection. It is used clinically to induce labor.

Endorphins

β-**lipotropin** serves as a precursor to the endogenous analgesic peptides, the **endorphins.** These hormones inhibit the perception of pain and hunger. Because the endorphins mimic morphine or, more accurately, morphine mimics them, one of the endorphins may be responsible for the "high" experienced by long-distance runners and other endurance athletes. Inhibition of the endorphin receptors is one method of treating those addicted to morphine and other related drugs.

Thyroid Hormones

The hormones of the thyroid control the overall metabolic rate by a mechanism that is not yet understood. Unlike other amino acid and peptide hormones, the thyroid hormones enter their target cells and are transported to the nucleus. Whether a specific cytosolic binding protein is required is a matter of dispute. Once in the nucleus, the thyroid hormone binds to a specific sequence on DNA and induces the synthesis of RNA and new protein.

The thyroid hormones are **thyroxine** (T_4) and the more active **triiodothyronine** (T_3). Both these hormones are synthesized as part of a large protein **thyroglobulin** that contains 115 tyrosine residues. Io-

dine is actively transported into the thyroid gland and then attached to the tyrosine residues. After stimulation by THS, the thyroglobulin is broken down by lysosomal enzymes, and T_4 and T_3 are released into the circulation.

thyroxine

triiodothyronine

Lack of sufficient thyroid hormone produces retardation of physical and mental development. In infants, this is **cretinism.** In childhood, lack of thyroid hormone produces **juvenile myxedema,** which is characterized by decreased bone growth, slowing of mental development, and predisposition to atherosclerosis. In adults, hypothyroidism results in a decreased metabolic rate, decreased body temperature, and extreme cold sensitivity. **Hashimoto's disease,** another hypothyroid condition, is due to autoimmune destruction of the thyroid gland.

Insufficient dietary iodine will produce a hypothyroid condition and enlargement of the thyroid gland. This enlargement is caused by lack of feedback by T_4 and T_3 on the production of TSH, a condition known as **simple goiter.** It can be treated by increasing the amount of iodine in the diet, usually by the use of iodized salt. In the past, simple goiter was common in areas of the United States with low amounts of iodine in the soil, specifically the Great Lakes area. The widespread use of iodized salt has made simple goiter a much less common condition, although it can still be observed in rural communities in Europe.

In **toxic goiter,** the thyroid gland is enlarged and excessive amounts of T_4 and T_3 are produced. The symptoms of toxic goiter are nervousness, fatigue, increased body temperature, weight loss, and increased heart rate. There is also a characteristic protrusion of the eyes. Toxic goiter can be treated by surgical removal of part of the thyroid or, more frequently, by destruction of some of the gland with radioactive iodine.

Graves' disease or diffuse goiter is due to the existence in the serum of a novel long-acting TSH that is different from the TSH produced by the pituitary. Graves' disease also can be treated by destruction of the thyroid gland or by drugs that inhibit the uptake and utilization of iodine. Propylthiouracil and methimazole are drugs that act in this way.

Parathyroid Hormone and Calcitonin

Parathyroid hormone increases the calcium concentration in the serum. The function of parathyroid hormone is antagonized by the thyroid hormone calcitonin. Together these hormones maintain the calcium level in the blood within a very narrow concentration range.

Parathyroid Hormone

Parathyroid hormone is a protein hormone of 84 amino acids. It is synthesized as a prehormone and released when the calcium level of the serum decreases. Parathyroid hormone increases the resorption of calcium from bone, increases calcium absorption in the gastrointestinal tract, decreases urinary excretion of calcium, and increases excretion of potassium. These effects are mediated through a cAMP cascade. In addition, parathyroid hormone stimulates the enzymes necessary for the conversion of native vitamin D to the active form, which is necessary for the response of bone and gastrointestinal tract to parathryoid hormone.

Hypoparathyroidism produces muscular weakness, tetany, cataracts, and calcification of the basal ganglia. This condition usually occurs with autoimmune destruction of the parathyroid gland or its accidental removal during thyroid gland surgery. Tumors of the parathryoid produce **hyperparathyroidism.** Symptoms include decalcification of the bones with resulting fractures and calcium stones in the kidney.

Calcitonin

Calcitonin is a 32–amino acid peptide that has effects opposite those of parathyroid hormone. It stimulates calcium deposition in bone and renal excretion of calcium. Unlike parathyroid hormone, these effects are not mediated through a cAMP cascade. Calcitonin is released in response to an increase in serum calcium. Calcitonin is used to treat **Paget's disease,** which is characterized by rapid turnover of bone and increased urinary excretion of calcium.

Pancreatic Hormones

The pancreas produces two hormones with antagonistic effects on glucose metabolism: insulin and glucagon. The effects of these hormones are summarized in Figure 28-2. Insulin is produced by the B cells and glucagon by the A cells of the pancreas.

Insulin

Insulin is composed of two peptide chains connected by three disulfide bonds. It is synthesized as a single peptide, **preproinsulin.** Then, 23 amino acids are split off, producing **proinsulin.** This is the stored form of the insulin. An increase in the glucose level of the blood induces cleavage of proinsulin and the release of insulin into the circulation. The sulfonylurea drugs **tolbutamide, tolazamide,** and **chlorpropamide** also promote the release of insulin by a mechanism that is not understood. The catecholamines have an opposite effect and inhibit insulin release by the pancreas.

Insulin produces both rapid and delayed effects on target cells. The rapid effects occur in seconds to minutes after binding of insulin to its plasma membrane receptor. The identity of the intracellular second messenger mediating the effects of insulin is not currently known. The rapid effects of insulin in liver include an increase in glucose and amino acid uptake, a decrease in gluconeogenesis, and an increase in glycogen and protein synthesis. In adipose tissue, lipid synthesis also is increased. After a longer period of time, new RNA is synthesized and new protein made. These effects may be due to internalization of the insulin and binding it or a fragment of it to specific DNA sequences.

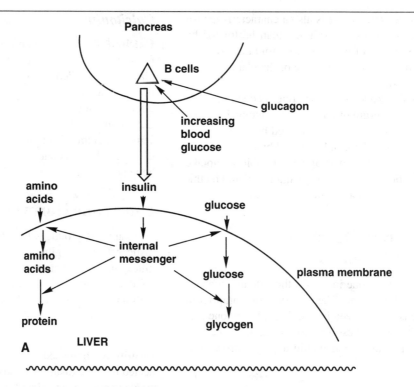

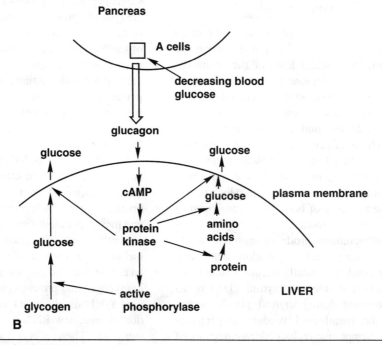

Fig. 28-2. Effects of insulin (A) and glucagon (B) on liver.

Lack of sufficient insulin or appropriate response to insulin produces **diabetes mellitus.** The juvenile form, called **type I** or **insulin-dependent diabetes mellitus,** is due to destruction of the B cells of the pancreas. This appears to be brought on by an autoimmune reaction following a viral infection. In type I diabetes, blood glucose levels increase while cells dependent on insulin for glucose uptake starve. Fatty acids then become the primary cellular fuel. Ketone bodies are produced in excess, which results in metabolic acidosis. The excess glucose is excreted in the urine along with large amounts of water. Persons with juvenile diabetes require daily injections of insulin for survival.

Maturity onset, **type II** or **non-insulin-dependent diabetes mellitus,** shows a high level of circulating insulin but insufficient response to it. This lack of response may be due to circulating antibodies to insulin or to a defect in the plasma membrane insulin receptor. The symptoms of type II diabetes are similar to but less severe than those of type I. Type II diabetes is associated with obesity and frequently can be controlled by weight loss and a low-carbohydrate diet. An excess of insulin from either a tumor or overinjection of insulin can produce severe hypoglycemia, coma, and death.

Glucagon

Glucagon is a peptide of 28 amino acids. Like insulin, it is synthesized as a prohormone that is larger than the final active compound. The metabolic effects of glucagon are opposite those of insulin: increased blood glucose levels, gluconeogenesis, and glycogen breakdown in liver and increased proteolysis in adipose tissue and muscle. These metabolic changes are caused by an increase in the intracellular concentration of cAMP. Compounds that decrease the breakdown of cAMP enhance the effect of glucagon and increase the blood level of glucose. Caffeine, theobromine, and theophylline—all methylated xanthines found in coffee, tea, and chocolate—have this effect.

Glucagon is released in response to hypoglycemia, exercise, stress, and cortisol. Its release is inhibited by increased amounts of blood glucose, fatty acids, ketone bodies, and insulin.

Hormones of the Adrenal Medulla

The adrenal gland is located on top of the kidney and is composed of two metabolically distinct areas. The outer area or cortex produces the **steroid hormones** cortisol, corticosterone, and aldosterone. The effects of these hormones will be covered in the next chapter. The chromaffin cells of the inner medulla synthesize the **catecholamines** epinephrine and norepinephrine. These are produced from tyrosine, using the same pathway as is used in the parasympathetic nerves (Fig. 28-3).

Epinephrine and **norepinephrine** are the "fight or flight" hormones. Their effects are summarized in

Fig. 28-3. Synthesis of epinephrine from tyrosine.

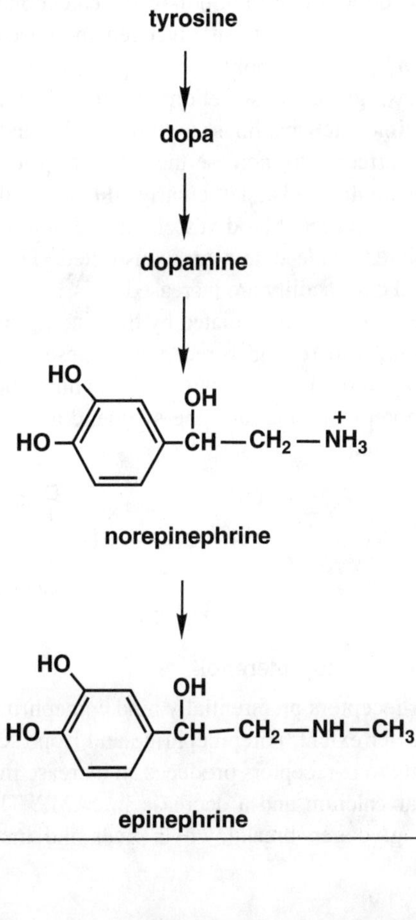

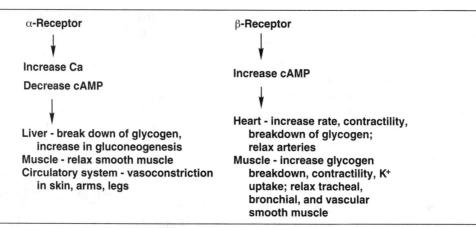

Fig. 28-4. Multiple effects of epinephrine and norepinephrine.

Figure 28-4. In combination with the glucocorticoids and the thyroid hormones, the catecholamines induce the breakdown of glycogen in muscle and liver and promote lipolysis in adipose tissue. In the pancreas, glucagon secretion is stimulated at the same time that insulin secretion is inhibited. The overall effect is to increase the circulating levels of the metabolic fuels, particularly glucose and fatty acids. In addition, blood vessels in liver and muscle are dilated while others are constricted. The heart rate and contractility are increased.

These events are mediated by the binding of catecholamines to α- and β-receptors. These receptors can be differentiated by their ability to bind epinephrine, norepinephrine, and the synthetic agonist **isoproterenol.**

$$HO-\langle\rangle-CH-CH_2-NH-CH\begin{array}{c}CH_3\\|\\CH_3\end{array}$$

isoproterenol

Alpha-receptors preferentially bind epinephrine and, to a lesser extent, norepinephrine and isoproterenol. Binding to α-receptors produces an increase in intracellular calcium and a decrease in cAMP. This induces glycogen breakdown in liver and increases

gluconeogenesis, vasoconstriction in the skin, dilation of the pupil of the eye, and relaxation of intestinal smooth muscle. Beta-receptors have a preference for isoproterenol. Binding to β-receptors produces an increase in intracellular cAMP. This causes breakdown of stored lipids, an increase in heart rate and contractility, relaxation of coronary arteries, increase in potassium uptake, and relaxation of tracheal and bronchial smooth muscles. Because of the two different types of receptors, the same hormone can produce different effects on different tissues.

Gastrointestinal Hormones

The gastrointestinal system synthesizes and releases at least eight different hormones that promote or inhibit the release of digestive enzymes and HCl. The effects of these hormones are summarized in Table 28-2.

Cholecystokinin (CCK) is a 39–amino acid peptide that induces the contraction of the gallbladder and the release of pancreatic enzymes. **Secretin,** a 27–amino acid peptide, also promotes the release of pancreatic enzymes. The effect of these two hormones is opposed by **somatostatin,** a 14–amino acid peptide. Release of intestinal enzymes is promoted by **vasoactive intestinal peptide** (VIP) and inhibited by **gastric inhibitory peptide** (GIP). **Gastrin** and **motilin** both stimulate the secretion of HCl in the stomach.

Table 28-2. Gastrointestinal hormones

Hormone	Where produced	Function
Cholecystokinin (CCK)	Small intestine	Contract gallbladder, release pancreatic enzymes
Secretin	Duodenum, jejunum	Release pancreatic secretions
Somatostatin	Small intestine	Inhibit release of pancreatic enzymes
Vasoactive intestinal peptide (VIP)	Small intestine	Release intestinal H_2CO_3, inhibit HCl secretion
Gastric inhibitory peptide (GIP)	Duodenum, jejunum	Inhibit HCl secretion, inhibit insulin release
Gastrin	G cells of antrum, duodenum	Stimulate HCl secretion
Motilin	Duodenum, jejunum	Stimulate HCl secretion, increase gastric motility
Enterogastrone	Small intestine	Inhibit HCl secretion

Summary

With the exception of thyroid hormone and perhaps insulin, the peptide hormones do not enter their target cells. Instead they bind to specific plasma membrane receptors and induce the production of internal second messengers that alter metabolic activity. The hypothalamus produces nine inhibiting and releasing hormones that affect the pituitary hormones FSH, LH, TSH, ACTH, GH, MSH, and prolactin. The tropic hormones control other endocrine glands. Prolactin, GH, and two other hypothalamic hormones, oxytocin and vasopressin, act directly on cells.

The thyroid hormones increase the overall metabolic rate. A lack of these hormones in infants produces cretinism and in children results in juvenile myxedema. Simple goiter is an enlargement of the thyroid gland due to a lack of iodine. In toxic goiter, the gland enlarges and produces excess thyroid hormone. Graves' disease is due to the existence of a nonpituitary long-acting TSH.

Parathyroid hormone increases serum calcium by increasing resorption of bone, increasing calcium absorption in the gastrointestinal tract, and decreasing calcium excretion by the kidney. Calcitonin has the opposite effects.

Blood glucose levels are also controlled by antagonistic hormones. Insulin from the B cells of the pancreas decreases the glucose level by increasing the uptake of glucose and decreasing gluconeogenesis and glycogen breakdown. Glucagon produced by the A cells has the opposite effects. A lack of insulin produces type I diabetes mellitus. Lack of response to insulin produces type II. Both are associated with hyperglycemia, glucose excretion in the urine, excessive ketone body production, and polyuria.

The adrenal medulla synthesizes epinephrine and norepinephrine. These hormones increase the levels of glucose and fatty acids in the blood. Their effects are mediated through binding to α-receptors, which increase calcium influx and decrease the concentration of cAMP and β-receptors, which increase the cAMP levels.

The gastrointestinal system produces at least eight hormones that promote or inhibit the secretion of HCl and the intestinal and pancreatic digestive enzymes.

29 Steroid Hormones

The lipid-derived hormones include the eicosanoids and the steroids. The eicosanoids, which include the prostaglandins, thromboxanes, and leukotrienes, were discussed previously in Chapter 20. The other group of lipid-derived hormones are the steroids. These compounds are synthesized from cholesterol in the gonads and the adrenal cortex. The sex steroids are essential for reproductive function and the secondary sex characteristics. The adrenocortical hormones are required for mobilization of stored fat, production of glucose from amino acids, and maintenance of salt and water balance.

Objectives

After completing this chapter, you should be able to

Describe the general mechanism by which steroids influence the metabolic activity of their target cells.

Recognize the structures of cortisol, corticosterone, aldosterone, progesterone, estradiol, estrone, and testosterone. Outline the synthesis and degradation of these steroids.

List the tissues synthesizing these steroids, their target cells, and their major metabolic effects.

Mechanism of Action

Steroid hormones are synthesized in response to pituitary tropic hormones which, in turn, are released in response to hypothalamic-releasing hormones, as outlined in Figure 29-1. As a general rule, steroids are not stored in the tissues synthesizing them. Instead, they are released to the circulation immediately after synthesis. While in the circulation, steroids are bound to albumin or specific transport proteins, such as **transcortin** for cortisol and **gonadal steroid-binding protein** for estradiol and testosterone.

On reaching the target cell, the steroid enters the cell and binds to a specific receptor protein located in the cytosol. The steroid-protein complex then en-ters the nucleus and binds to a specific DNA sequence. As a result, genes are activated and new RNA and new protein are synthesized. If genes are inactivated as a result of steroid binding, RNA synthesis from this gene stops, and the resulting protein is no longer produced. Because several steps are involved, there is a delay of 30 minutes to several hours between initial binding of the steroid hormone and a change in the metabolic activity of the cell.

Adrenocortical Steroids

The outer area or cortex of the adrenal gland produces the steroid hormones cortisol, coticosterone, and aldosterone. The inner medulla synthesizes the catecholamines epinephrine and norepinephrine.

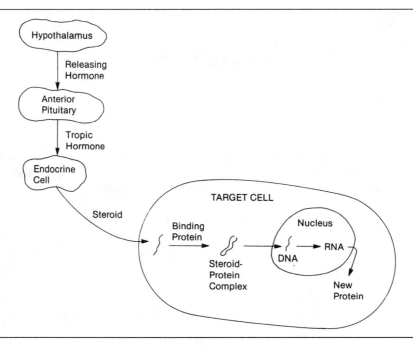

Fig. 29-1. The mechanisms of release and action of the steroid hormones.

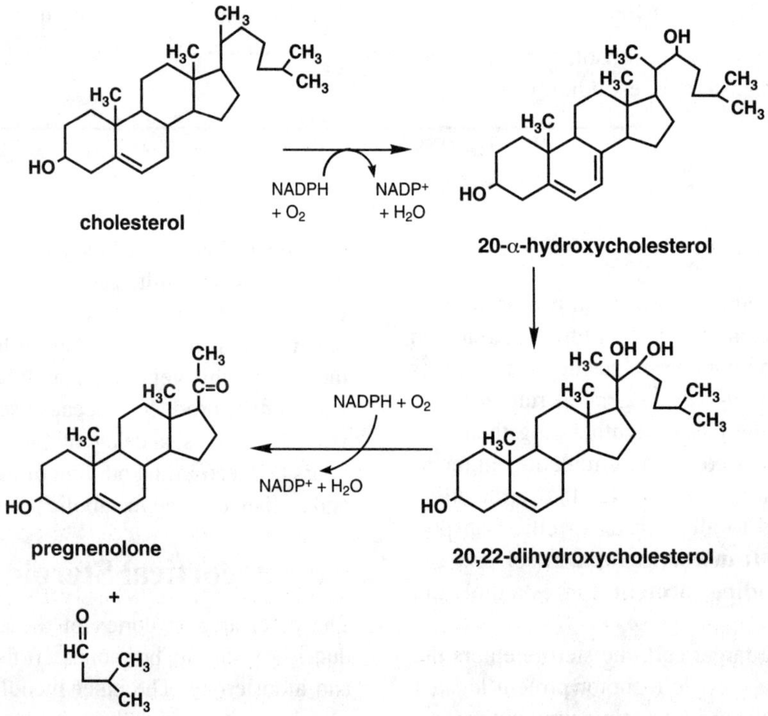

Fig. 29-2. Synthesis of pregnenolone from cholesterol.

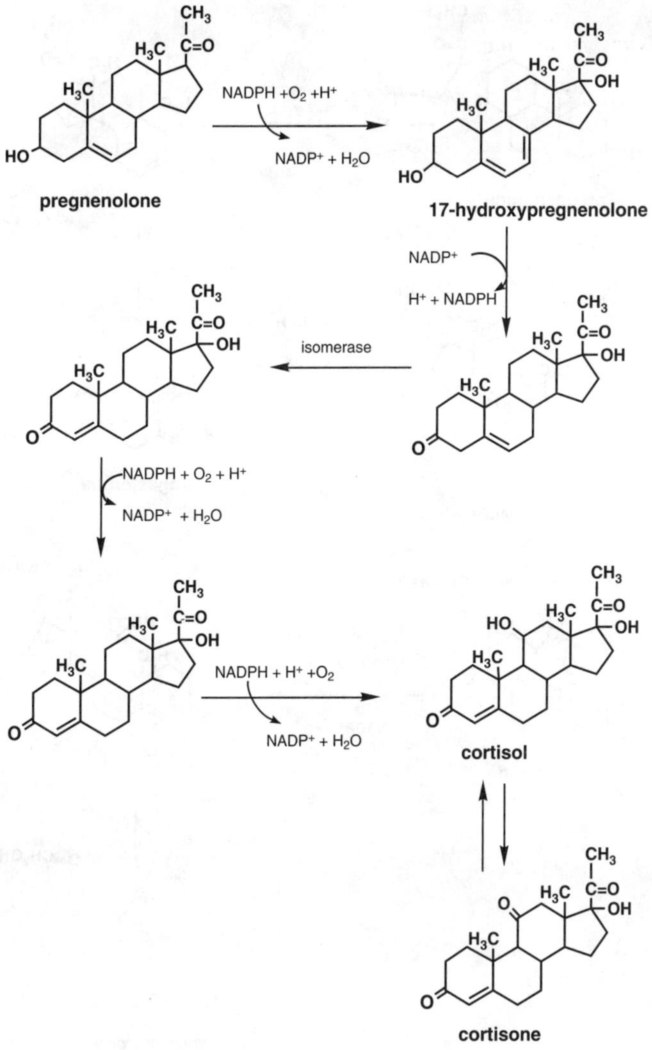

Fig. 29-3. Synthesis of cortisol from pregnenolone.

The starting point for the synthesis of the adreno-cortical hormones is **pregnenolone.** This is derived from cholesterol by the pathway outlined in Figure 29-2. The conversion of cholesterol to pregnenolone involves hydroxylation at positions 20 and 22 and cleavage of the side chain. All the hydroxylation enzymes are mixed-function oxidases that utilize NADPH and molecular oxygen as substrates. These enzymes are bound to the membranes of the mitochondria or the endoplasmic reticulum.

Glucocorticoids

The major glucocorticoid of humans is **cortisol,** also called **hydrocortisone.** Corticosterone is a minor glucocorticoid for humans but is the major one in other species. The pathway for synthesis of cortisol is given in Figure 29-3 and that for corticosterone in Figure 29-4. Both these compounds are synthesized in response to **adrenocorticotropic hormone** (ACTH). ACTH induces the synthesis of the adrenocortical hor-

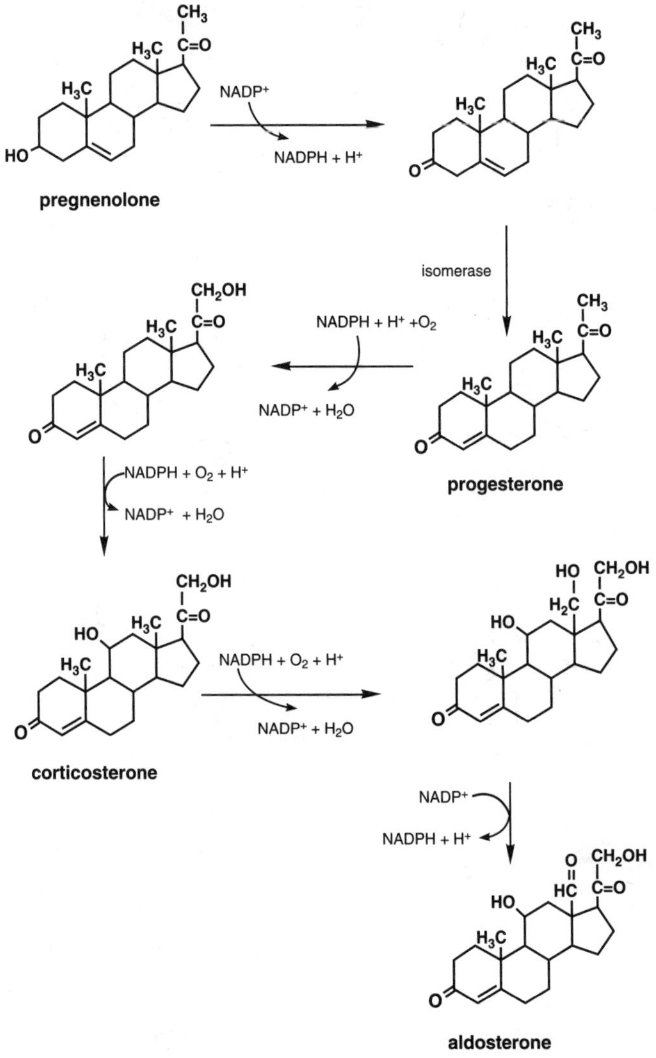

Fig. 29-4. Synthesis of progesterone, corticosterone, and aldosterone from pregnenolone.

mones, the enzymes responsible for their synthesis, and the growth and division of the adrenocortical cells. These events are triggered by a cAMP cascade that occurs after the binding of ACTH to the plasma membrane of the adrenocortical cells.

As discussed in the previous chapter, ACTH is released in response to the hypothalamic peptide hormone corticotropin-releasing hormone. Corticotropin-releasing hormone is released into the pituitary portal circulation, taken up by the anterior pituitary, and induces the release of ACTH. ACTH release is inhibited by feedback from circulating glucocorticoids.

Glucocorticoids produce a catabolic effect in muscle, connective tissue, adipose tissue, and lymphocytes. In these tissues, protein and fat are degraded while glucose uptake and utilization are inhibited. Amino acids, fatty acids, and glycerol are released into the circulation. These compounds are taken up by the liver and converted into metabolic fuel—glucose, glycogen, and ketone bodies. In addition to stimulating enzyme activity, cortisol also stimulates the biosynthesis of degradative enzymes in muscle, adipose tissue, connective tissue, and lymphocytes. In liver, it stimulates production of the synthetic enzymes responsible for gluconeogenesis, glycogen synthesis, and amino acid degradation.

Excessive amounts of glucocorticoids inhibit lymphocyte proliferation and the inflammatory response. Because of this effect, cortisol and its synthetic analog **dexamethasone** are used clinically as local and topical antiinflammatory drugs.

dexamethasone

An excess of glucocorticoids produces **Cushing's syndrome,** which is usually due to a tumor of the pituitary or the adrenal gland. Persons with Cushing's syndrome have a characteristic round moon-shaped face caused by edema. In addition, they show obesity of the trunk, protein depletion, and wasting in the muscles of the arms and legs, hyperglycemia, glucosuria, and mental retardation. Because the mineralocorticoids are also found in excess in Cushing's syndrome, there is hypertension and a loss of potassium as well.

An insufficiency of adrenocortical hormones results in **Addison's disease.** This occurs in tuberculosis or as a result of autoimmune destruction of the adrenal cortex. The symptoms in Addison's disease are opposite those of Cushing's syndrome. Addison's disease is characterized by hypoglycemia, weight loss, weakness, and hyperpigmentation. The latter is due to the melanocyte-stimulating effect of excessive amounts of ACTH, present because there are no adrenocortical steroids available for feedback inhibition. Because of the lack of mineralocorticoids, hypotension and loss of sodium occur. Complete loss of adrenocortical function is fatal owing to dehydration and circulatory collapse.

Mineralocorticoids

The major mineralocorticoid in humans is **aldosterone.** It is made from corticosterone by the pathway given in Figure 29-4. The main function of aldosterone is to increase the resorption of sodium ions in the distal tubules of the kidney. This increases the resorption of water, resulting in increasing blood volume and blood pressure. Aldosterone also increases the excretion of potassium ions. These events occur by stimulation of the synthesis of the proteins involved in the sodium channels, not by stimulating the sodium-potassium pump.

Release of aldosterone is stimulated by an increase of potassium in the blood or a decrease of blood pressure in the renal arteries. The mechanism involves the cleavage of circulating **angiotensinogen** by the renal hormone **renin.** This cleavage produces **angiotensin I,** which is further cleaved to **angiotensin II,** which then stimulates the release of aldosterone (Fig. 29-5).

An excess of aldosterone produces **primary aldosteronism** or **Conn's syndrome.** The symptoms include increased sodium resorption, increased water resorption, hypertension, and loss of potassium.

A loss of one or more of the hydroxylation enzymes in the adrenal cortex produces a lack of both glucocorticoids and mineralocorticoids. Adrenocortical hyperplasia also occurs if there is a lack of feedback inhibition by the adrenocortical hormones on the release of ACTH. Defects in hydroxylation produce the **adrenogenital syndromes,** in which unused precursors to the cortical hormones are released into the circulation and converted into compounds with androgenic activity.

A defect in the **21-hydroxylase** allows no production of aldosterone, cortisol, or corticosterone. Instead, the Δ3,4,17-dione is released into the circulation. This is a precursor of testosterone, which in young boys causes **pseudopuberty.** In young girls it causes **pseudohermaphroditism** and ambiguous genitalia. Older females show **virilization.** In all these cases, there is a loss of sodium and water and a decrease in blood pressure. Lack of the **11-hydroxylase** causes accumulation of 11-deoxycorticosterone, a compound that has mineralocorticoid properties. As a result, potassium decreases, sodium increases, and blood pressure rises. Lack of the **17-hydroxy-**

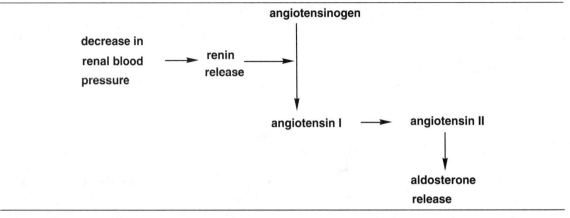

Fig. 29-5. The pathway connecting a decrease in renal blood pressure to the release of aldosterone. A decrease in renal blood pressure triggers the release of renin from the kidney. Renin converts circulating angiotensinogen to angiotensin I, which in turn is converted to angiotensin II, which stimulates aldoesterone release from the adrenal gland.

lase produces persons with no male or female hormone production. A female appearance results, regardless of the genotype.

Gonadal Hormones

The major male steroid hormone is testosterone, and the major female steroids are estradiol, estrone, and progesterone. These hormones are responsible for the production of ova and sperm and the maintenance of the secondary sexual characteristics.

The gonadal steroids are synthesized in response to an increase in cAMP caused by pituitary luteinizing hormone. Release of LH is inhibited by feedback from circulating gonadal steroids.

Androgens

Testosterone, the major androgen, is produced in the Leydig cells of the testis. It is synthesized from pregnenolone by the pathway outlined in Figure 29-6. In some target cells, testosterone is converted to the more potent androgen **dihydrotestosterone.**

dihydrotestosterone

Testosterone is necessary for the induction and maintenance of the secondary male sexual characteristics, including facial and body hair, deep voice, and the growth of the sex organs. Testosterone also is responsible for anabolic or growth effects in muscle. In addition, it produces an increase in the number and respiratory rate of mitochondria. Because of these effects, testosterone analogs have been used and abused by athletes training for strength and endurance sports.

Estrogens

The ovary produces **estradiol** and **estrone** from testosterone by the pathways outlined in Figure 29-7. As you can see, the estrogens are the only steroids that have an aromatic ring. Another estrogen, **estriol,** is produced by the placenta from estrone. Estriol is the major urinary steroid in pregnant women.

Like testosterone in males, estradiol and estrone are responsible for the maintenance of the secondary sexual characteristics in females. These include enlarged breasts, increased body fat, maintenance of the lining of the vagina, and control of the menstrual cycle. These hormones inhibit the production of the pituitary hormone FSH and, at high levels, stimulate the secretion of LH.

Because of their slower rate of degradation and greater stability with respect to the action of the digestive tract enzymes, estrogen analogs have been

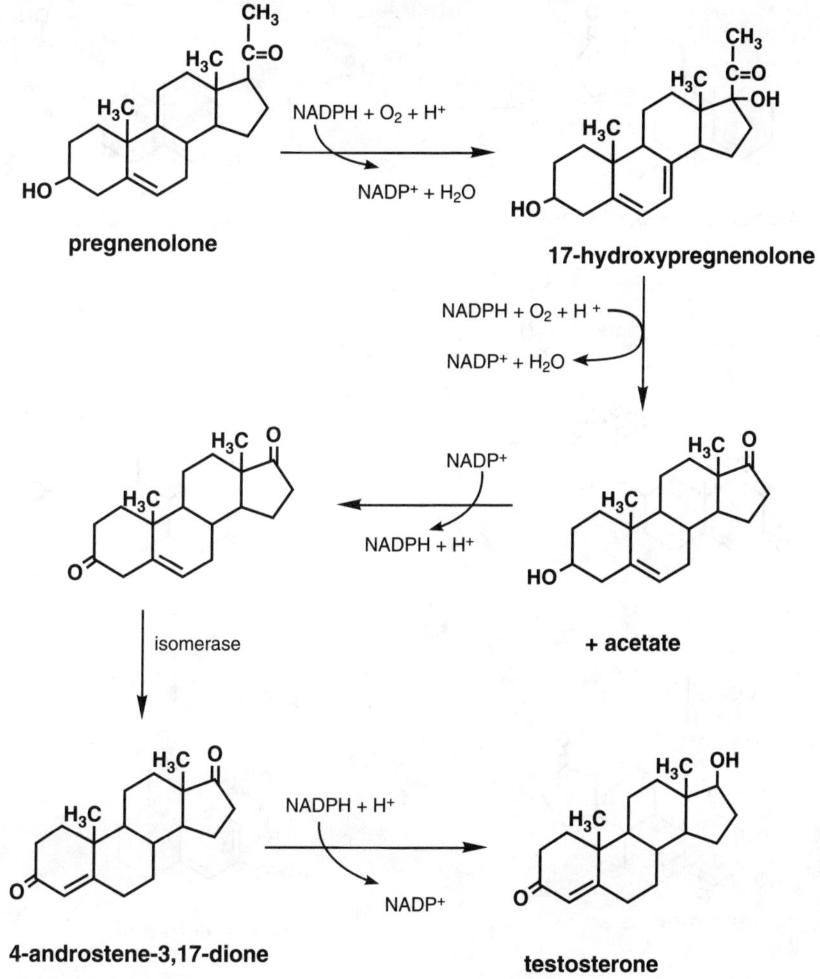

Fig. 29-6. Synthesis of testosterone from pregnenolone.

found clinically useful in the contraceptive pill and as treatment for osteoporosis, dysfunctional vaginal bleeding, and prostate cancer. **Ethinyl estradiol** and **mestranol** are two compounds used in this way.

Another estrogen analog that is not a steroid is **diethylstilbestrol** (DES). This compound has been used as a "morning-after" contraceptive pill. Its use during the 1950s to prevent miscarriage resulted in an increased rate of vaginal cancer in the daughters and testicular cancer in the sons of women who took the drug.

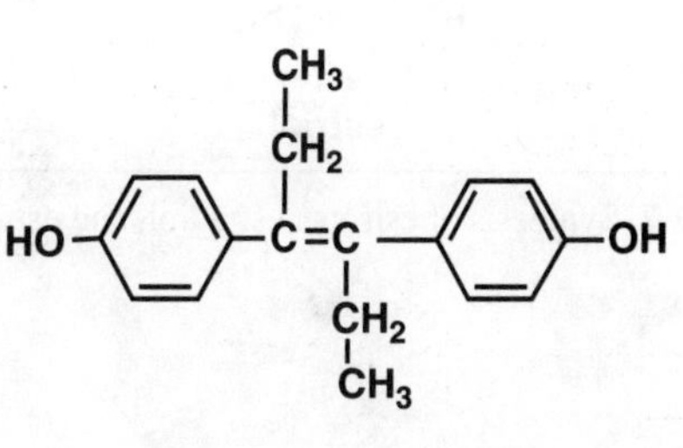

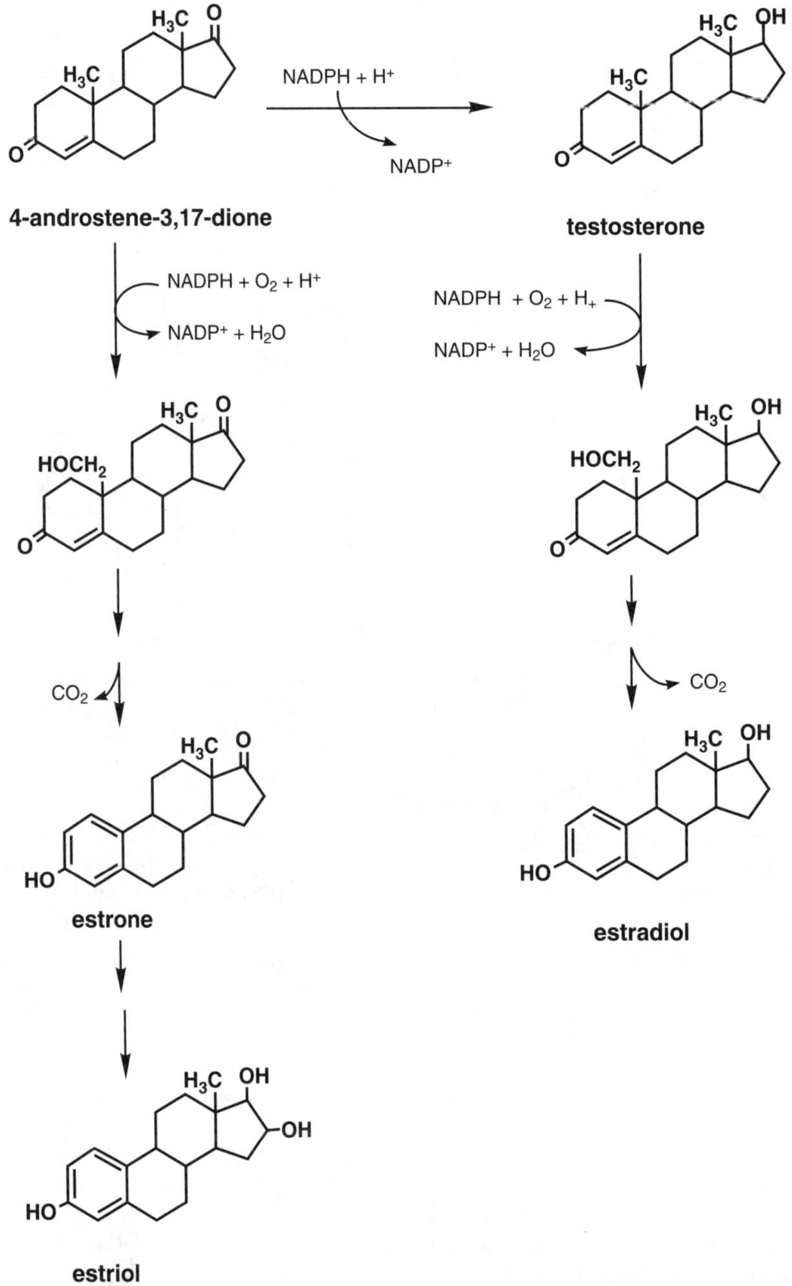

Fig. 29-7. Synthesis of estrone, estradiol, and estriol from testosterone.

The antiestrogen **tamoxifen** is used in treatment of breast cancer.

Progesterone

Progesterone is made by the corpus luteum and the placenta. As seen in Figure 29-4, progesterone also is made in the adrenal cortex where it serves as a precursor to corticosterone and aldosterone. Progesterone induces the development of the endometrium and the mammary glands. Unlike the estrogens, it inhibits the release of LH.

Several synthetic progesterones have been used in contraceptive pills. Among these are **norethindrone** and **norgestrel.** Norgestrel is being used as a subcutaneous implant for long-term contraception.

Degradation of Steroids

The steroid ring structure is not broken down during catabolism of the steroid hormones. Instead, the keto groups are reduced to hydroxyl groups and then the compound is conjugated with glucuronate or sulfate. These conjugates are excreted in the urine or the bile. Measurement of conjugated 17-keto steroids is a method of determining the level of androgens, especially testosterone.

Summary

Steroid hormones are synthesized by the adrenal cortex and the gonads. After release into the circulation, they are taken up by their target cells, bind to a specific cytosolic protein, and are transported into the nucleus. There they bind to specific nucleotide sequences on the DNA, resulting in alterations in gene expression. Steroids are degraded by reduction and conjugation with sulfate or glucuronate.

The adrenal cortex produces cortisol and corticosterone, which are the glucocorticoids, and aldosterone, which is a mineralocorticoid, The glucocorticoids are released in response to ACTH. They produce a breakdown of fats and protein in muscle, connective tissue, lymphocytes, and adipose tissue. The glucocorticoids increase the synthesis of glucose and glycogen from amino acids in liver. Aldosterone is released in response to angiotensin II. It increases the renal resorption of sodium and the excretion of potassium. An excess of adrenal steroids produces Cushing's syndrome, whereas a deficiency results in Addison's disease. Conn's disease is due to excessive amounts of aldosterone. The adrenogenital syndromes are caused by defects in one or more of the hydroxylating enzymes necessary for the synthesis of the adrenal steroids.

Testosterone is the major male gonadal steroid. It is responsible for maintenance of the male secondary sexual characteristics and for the anabolic or growth-promoting effects. Estradiol and estrone are the two major female gonadal steroids. They maintain the female secondary sexual characteristics and control the menstrual cycle. Progesterone is produced by the corpus luteum and the placenta. It is important for the menstrual cycle and for maintenance of pregnancy.

30 Vitamins

Vitamins are small organic compounds that are essential for life but cannot be made by the organism. As you have seen in previous chapters, many vitamins serve as functional parts of enzymes. For humans, there are four fat-soluble vitamins and ten water-soluble vitamins. Two compounds, vitamin D and niacin, are frequently listed as vitamins for humans. However, both these compounds can, in fact, be synthesized. Vitamin D is made from cholesterol, and niacin is made from tryptophan. In addition, inositol and pantothenic acid occur so widely in foods that no natural deficiency has ever been identified. Although specific deficiency diseases are known for many of the vitamins, it is unusual to observe the deficiency of a single vitamin in developed countries unless the person is consuming a very restricted diet.

Objective

After completing this chapter, you should be able to

Recognize the structure, list the dietary sources, describe the metabolic functions, and identify the effects of excess and deficiency for each vitamin.

Fat-Soluble Vitamins

The four fat-soluble vitamins are A, D, E, and K. Of these, only vitamin K serves as a coenzyme. The rest have other metabolic functions. All the fat-soluble vitamins are synthesized from isoprene units and can be stored in the liver.

Vitamin A

Vitamin A is produced by oxidation of β-**carotene,** a constituent of most green and yellow vegetables.

The active form is **retinal** (Fig. 30-1). Vitamin A also occurs as the alcohol, retinol, and the acid, retinoic acid. These forms are interconvertible in the liver. The visually active form of vitamin A is 11-*cis*-retinal. The storage form is the palmitate ester.

In the retina, **11-*cis* retinal** combines with the protein **opsin** to make **rhodopsin.** Different kinds of opsin proteins occur in rod and cone cells and are sensitive to different wavelengths of light. When exposed to light, the double bond in the side chain of retinal changes from the *cis* to the *trans* configuration. After this occurs, *trans*-retinal dissociates from opsin. In the dark, *trans*-retinal can be isomerized back to the *cis* form and reassociate with opsin, as illustrated in Figure 30-2. Illumination of rhodopsin also produces an increased synthesis and degradation of cGMP. This turnover appears to amplify the light signal.

A mild deficiency in vitamin A produces an inability to see well in dim light or **night blindness.** A more severe deficiency results in dryness of the eye, followed by necrosis, ulceration and, finally, blindness. This is **xerophthalmia keratomalacia.** It is un-

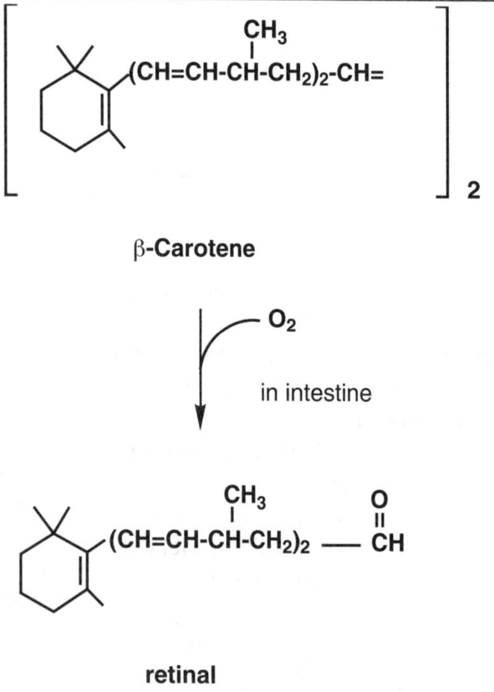

CH

$$CH_3$$
$$(CH=CH-CH-CH_2)_2-CH=$$

β-Carotene

O_2

in intestine

$$CH_3 \qquad\qquad O$$
$$(CH=CH-CH-CH_2)_2 \longrightarrow CH$$

retinal

Fig. 30-1. Conversion of β-carotene to retinal in the intestine.

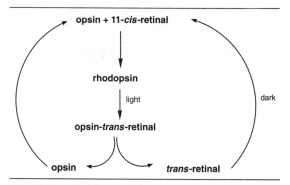

Fig. 30-2. Light-induced isomerization of rhodopsin and its resynthesis.

common in the United States because of vitamin enrichment of many staple foods. However, worldwide, xerophthalmia keratomalacia is the leading cause of blindness.

In addition to its role in vision, vitamin A is involved also with the maintenance and differentiation of epithelial cells. Although the mechanism is not fully understood, it involves binding of vitamin A to a binding protein and then to a specific sequence in the DNA that alters gene expression. Because of the effect of vitamin A on cell differentiation, retinoids and analogs of retinoids are being explored as possible dietary cancer preventive agents.

Vitamin A is not excreted but is stored in the liver, and so symptoms of excess can occur. These include headache, nausea, abdominal pain, lethargy, painful joints, muscular weakness, and skin lesions. In addition, yellow carotene may be deposited on the soles of the feet and the palms of the hands.

Vitamin D

Vitamin D is not truly a vitamin as it is derived from a steroid that can be synthesized from cholesterol. The pathway is shown in Figure 30-3. In the biosynthesis pathway, cleavage of ring B of 7-dehydrocholesterol requires ultraviolet light. The product, **cholecalciferol** or **vitamin D₃**, is further metabolized in the liver and kidney to **25-hydroxycholecalciferol** and **1,25-dihydroxycholecalciferol,** the active forms of vitamin D.

Active vitamin D, in conjunction with parathyroid hormone, promotes the absorption of calcium and phosphorus from the gastrointestinal tract. Because vitamin D is synthesized by the liver and acts on other cells to promote the synthesis of new proteins, the mechanism of action is more like that of a hormone than that of a traditional vitamin.

Vitamin D can be found in eggs, irradiated dairy products, and fish oils. Infants and young children whose environments lack sufficient exposure to sunlight or whose diets are deficient in vitamin D can develop a vitamin D deficiency. Although they may appear plump and well-fed, these children show slow development, poor muscle tone, and weak bones. When vitamin D–deficient children begin to walk, their poorly calcified bones bend, producing bowed legs and spinal deformities characteristic of **rickets.**

In the elderly, a lack of vitamin D leads to bone demineralization (osteomalacia) and fractures of the pelvis and femur after falls. This condition is especially common in women with light skin color and

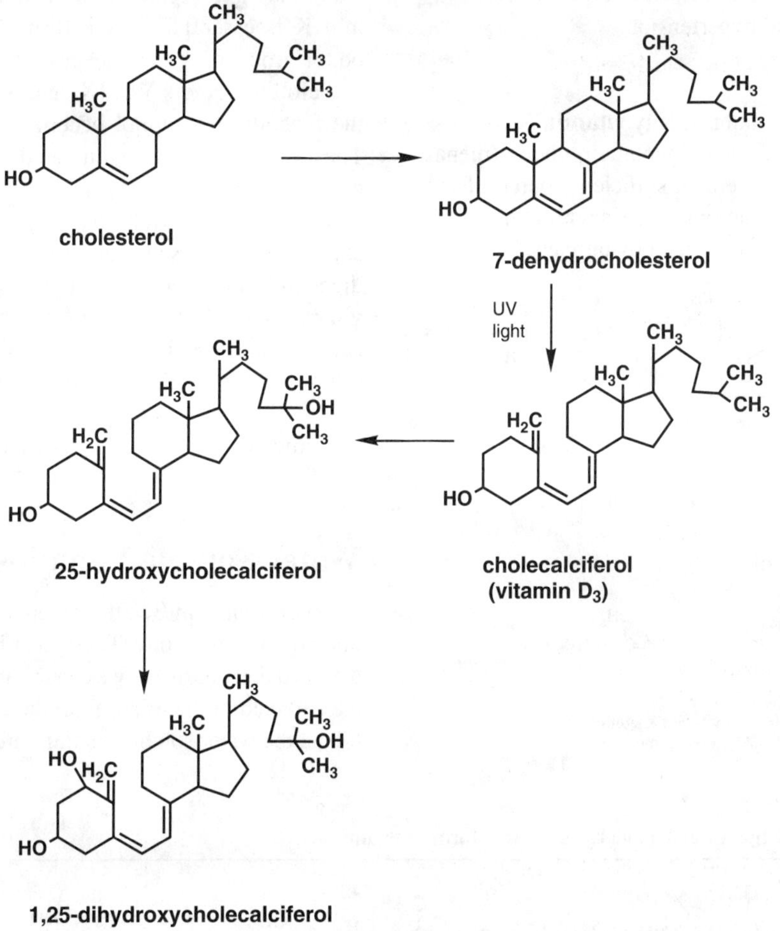

Fig. 30-3. Synthesis of vitamin D from cholesterol.

slender build who have a minimal intake of milk products or other sources of calcium and minimal exposure to sunlight.

An excess of vitamin D promotes demineralization of bone and calcium deposits in the soft tissue. Infants show anorexia, vomiting, extreme thirst, and polyuria. Excess vitamin D usually occurs when mothers give their infants overdoses of fish liver oils or commercial vitamin D preparations.

Vitamin E

No specific metabolic function for **vitamin E** has been identified in humans. It appears to act as an antioxi-dant, protecting the double bonds of polyunsaturated and essential fatty acids in the plasma membrane.

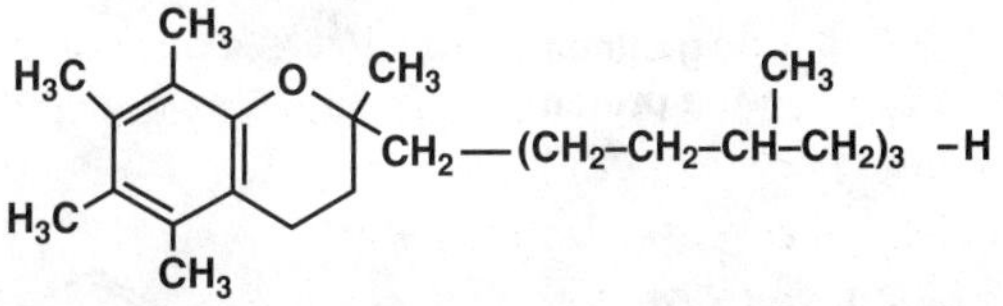

The active form of vitamin E is α-**tocopherol.** Vitamin E can be found in most plants but is especially prevalent in seed oils and wheat germ. Experimental deficiency produces skin lesions in humans and infertility in rats. Experimental excess destroys vitamin K and prevents the formation of retinal from β-carotene. In addition, excess vitamin E produces fatigue, swelling, and hypertension.

Vitamin K

Of the fat-soluble vitamins, only **vitamin K** acts as a coenzyme. Because methylnaphthylquinone (menadione) alone can serve as a sufficient source of the vitamin, it appears that only the ring structure of vitamin K cannot be synthesized by humans.

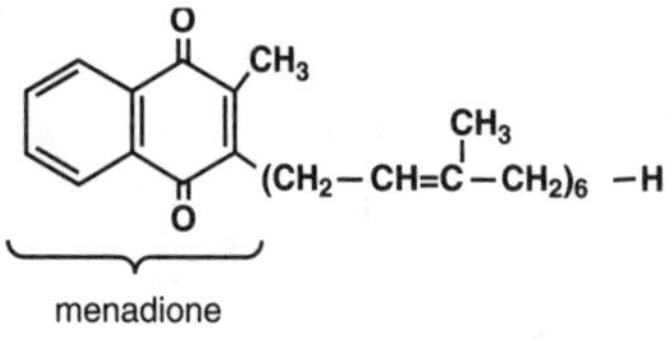

menadione

(quinone form)

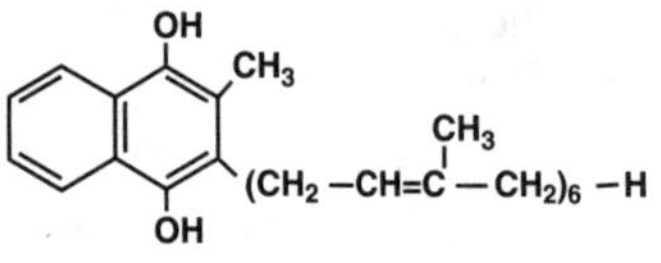

vitamin K quinol
(coenzyme form)

As shown in Figure 30-4, the quinol form of vitamin K is essential for the formation of the γ-carboxyl group on glutamate residues of prothrombin and clotting factors VII, IX, and X. In the course of the reaction, the quinol becomes an endoperoxide. The quinol must be regenerated before vitamin K can act again as a coenzyme.

Vitamin K is synthesized by intestinal bacteria and requires bile salts for absorption. Defects in fat digestion and absorption can produce a deficiency of vitamin K and the other fat-soluble vitamins. Because vitamin K is necessary for the synthesis of several of the clotting factors, a deficiency produces a lack of normal clotting and hemorrhage. There is no known effect of an excess of vitamin K.

Water-Soluble Vitamins

Except for vitamin B_{12}, the water-soluble vitamins are not stored in the body. They must be supplied in the diet on a daily basis. Any excesses above requirements are excreted in the urine. With the exception of inositol, all the water-soluble vitamins are coenzymes.

Fig. 30-4. Carboxylation of glutamate by the quinol form of vitamin K.

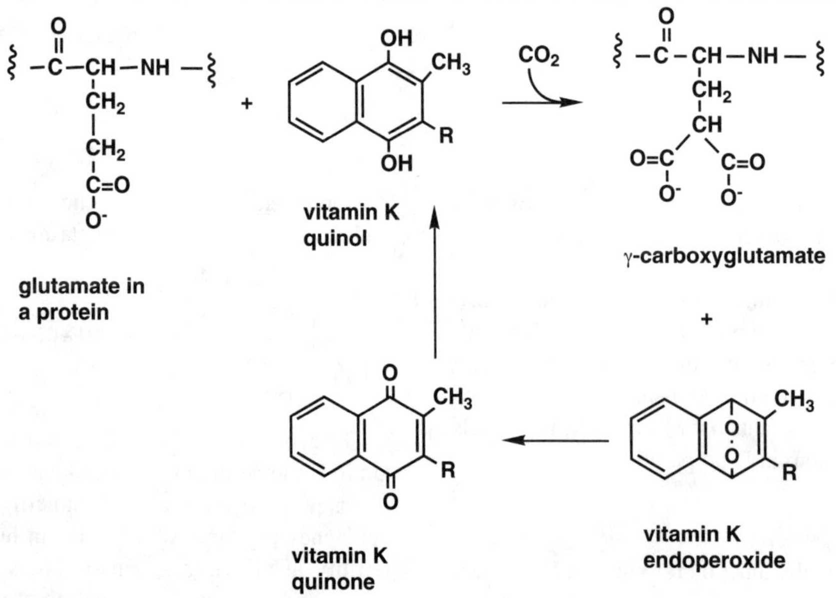

Ascorbic Acid

Ascorbic acid is **vitamin C.** It is found in most fresh fruits and vegetables, not just citrus fruits. The vitamin is easily destroyed by cooking. Of all animals, only primates and guinea pigs are unable to synthesize vitamin C.

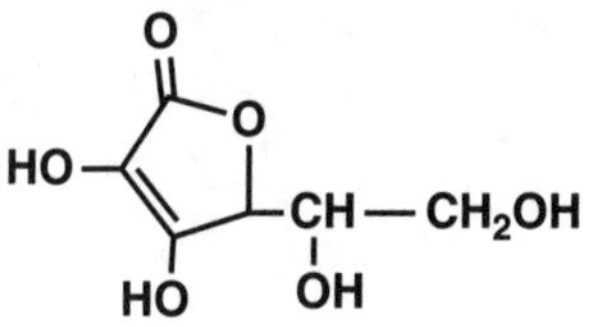

**vitamin C
(ascorbic acid)**

Ascorbic acid is necessary for hydroxylation of proline and lysine in the synthesis of collagen (Fig. 30-5). A deficiency of vitamin C causes **scurvy,** which is characterized by malformation of connective tissue. The symptoms of scurvy include muscular weakness, sore gums, tooth loss, skin hemorrhage, slow wound healing, and anemia. In children, ascorbic acid deficiency is known as **Barlow's disease.** It occurs in infants whose diets consist entirely of heat-sterilized milk in which the little vitamin C normally present has been destroyed. Such children show muscle soreness and hemorrhage and lie in a peculiar "pithed frog" position. In adults, a vitamin C deficiency is most frequently found in elderly men who live alone and consume few fresh fruits and vegetables.

Although there is no convincing scientific evidence, large doses of vitamin C have been recommended by some for the prevention and treatment of diseases ranging from the common cold to cancer. Such large doses of vitamin C are not without hazard. They cause the urine to be very acidic, which converts urate into uric acid and oxalate into oxalic acid, thereby promoting the formation of kidney stones in susceptible individuals.

Biotin

Biotin is also called **vitamin H.** It is required for the carboxylation of propionyl-CoA to methylmalonyl-CoA and of pyruvate to oxaloacetate.

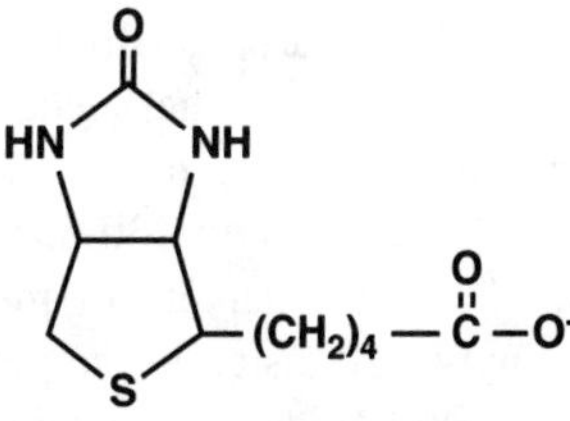

biotin

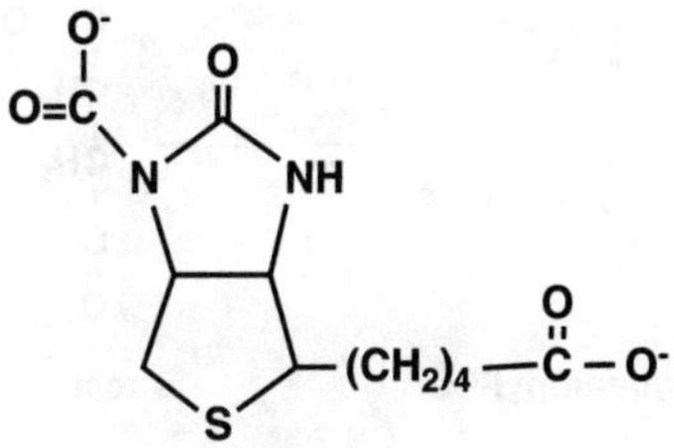

carboxybiotin

Fig. 30-5. Synthesis of hydroxylysine from lysine residues in collagen.

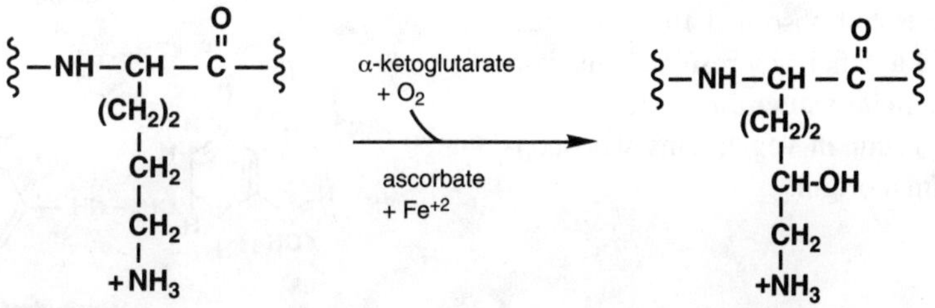

**lysine residue
in collagen**

**hydroxylysine
in collagen**

A deficiency of biotin is rare except in persons consuming large amounts of uncooked egg white. Egg white contains the protein **avidin,** which binds biotin and prevents its absorption. Cooking egg whites denatures avidin, prevents binding of biotin, and alleviates the deficiency. Biotin deficiency produces dermatitis, fatigue, malaise, nausea, and neuropathy. There appears to be no toxicity even when large doses of biotin are consumed. The excess is merely excreted in the urine.

Cobalamin

Cobalamin or **vitamin B_{12}** usually is isolated as the cyano derivative, although the adenosyl derivative appears to be the coenzyme form. The structure of vitamin B_{12} is given in Figure 30-6. Vitamin B_{12} is essential for the rearrangement of methylmalonyl-CoA to succinyl-CoA, a step in the pathway for the degradation of fatty acids that contain an odd number of carbon atoms.

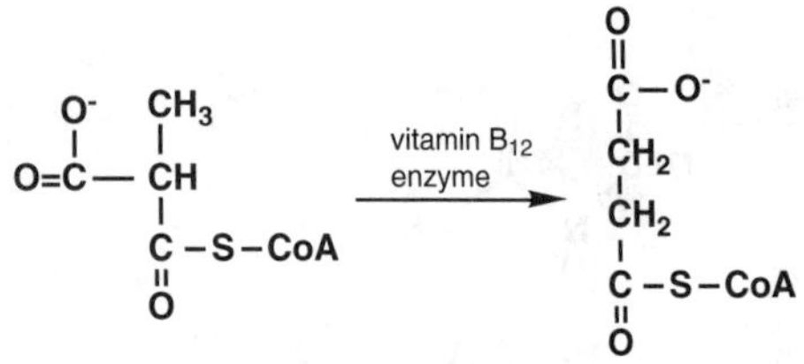

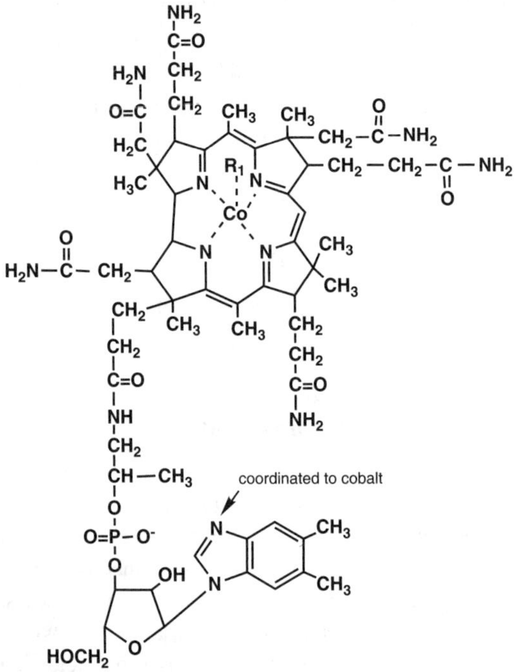

R$_1$ = CN in cyanocobalamin

R$_1$ = adenosine in the coenzyme form
    (attached at 5' position)

Fig. 30-6. Structure of vitamin B_{12}.

methylmalonyl-CoA   succinyl-CoA

Although vitamin B_{12} can be synthesized only by bacteria, it is found in most animal tissues. A deficiency in vitamin B_{12} usually is due to an inability to absorb the vitamin rather than a dietary deficiency. Absorption of vitamin B_{12} requires the function of a protein called **intrinsic factor.** When this factor is absent, vitamin B_{12} is not absorbed from the gastrointestinal tract and a deficiency results. This deficiency produces **pernicious anemia.** A vitamin B_{12} deficiency can also occur in vegetarians who consume no food of animal origin.

Folic Acid

The coenzyme form of **folic acid** is **tetrahydrofolate,** which is necessary for one-carbon metabolism.

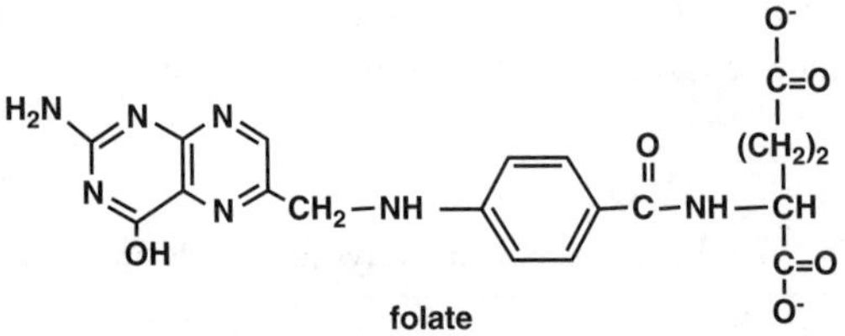

folate

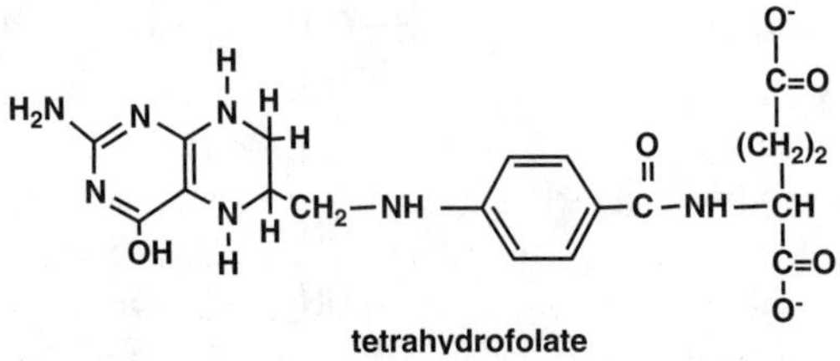

tetrahydrofolate

Folate can be found in green leafy vegetables, kidney, and liver. A deficiency of both folate and vitamin B_{12} produces **megaloblastic anemia.** A deficiency of folate alone results in **macrocytic anemia.** These conditions are rare except in malnourished pregnant women, infants on diets consisting of only milk, and chronic alcoholics. A deficiency of folate and most of the other water-soluble vitamins occurs in persons with malabsorption syndromes such as **sprue.**

Inositol

Inositol is a component of phospholipids. As the 1,4,5-triphosphate, inositol functions as a second messenger produced in response to a variety of hormones, nerve transmitters, and growth factors.

inositol

inositol triphosphate

In the inositol triphosphate cascade, diagrammed in Figure 30-7, the extracellular factor binds to a plasma membrane receptor. This leads to the hydrolysis of GTP and the cleavage of diphosphoinositol phosphatidic acid to diacylglycerol and triphosphoinositol. Both compounds act as intracellular messengers. The diacylglycerol activates protein kinase C. The triphosphoinositol mobilizes intracellular calcium which, in turn, activates other protein kinases. The phosphoinositol cascade has been linked to the growth factor–related oncogenes erbB, sis, ras, and src.

Inositol cannot be synthesized by humans but is so widely distributed in foods that no natural deficiency has been described. Experimental deficiencies in animals produce failure to grow.

Niacin

Niacin also is called **nicotinic acid.** The amide **nicotinamide,** is part of NAD and NADP, which are frequent cofactors in oxidation-reduction reactions. Nicotinamide can be synthesized from tryptophan, an essential amino acid, by the pathway described in Chapter 24, Figure 24-6. Both nicotinamide and tryptophan are found in high concentrations in whole grains and legumes.

nicotinate

NAD

A deficiency of both tryptophan and nicotinamide produces **pellagra.** Pellagra is rare except in persons consuming corn as the major part of the diet. Corn is deficient in both tryptophan and nicotinamide and appears to prevent their absorption when they are present in other foods. The symptoms of pellagra include dermatitis, diarrhea, and neural disorders. Death occurs if proper treatment is not given.

Pantothenic Acid

Pantothenic acid is part of **coenzyme A** (Fig. 30-8) and is essential for the metabolism of fatty acids and the function of the Krebs cycle. Pantothenic acid is found so widely in foods that no disease associated with its deficiency is known.

Pyridoxine

Pyridoxine is **vitamin B$_6$**. The coenzyme form is **pyridoxal phosphate,** which is essential for many reactions of amino acid metabolism, including transamination, decarboxylation, and oxidative deamination.

A deficiency of vitamin B$_6$ in infants produces hyperirritability, convulsions, and anemia. In adults, there is a loss of appetite and a resulting weight loss.

Riboflavin

Riboflavin is **vitamin B$_2$** and functions as part of FAD and FMN in oxidation-reduction reactions. Riboflavin can be found in milk, green vegetables, liver, fish, and eggs. When a deficiency of riboflavin occurs, it is usually in conjunction with a deficiency of several other vitamins.

Fig. 30-7. The inositol triphosphate cascade.

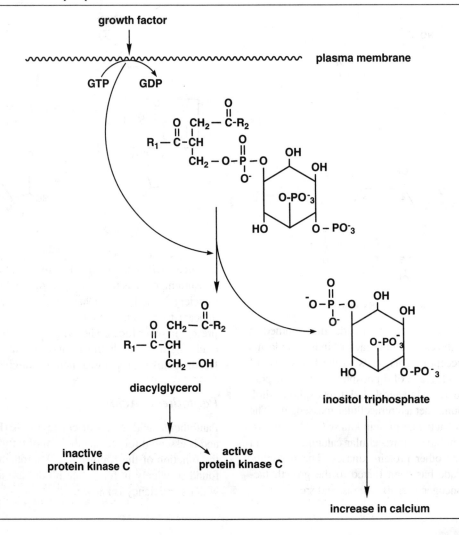

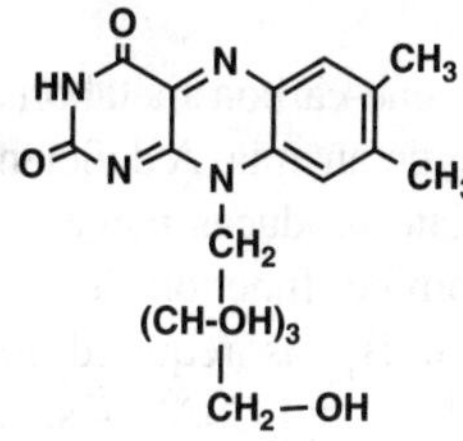

riboflavin

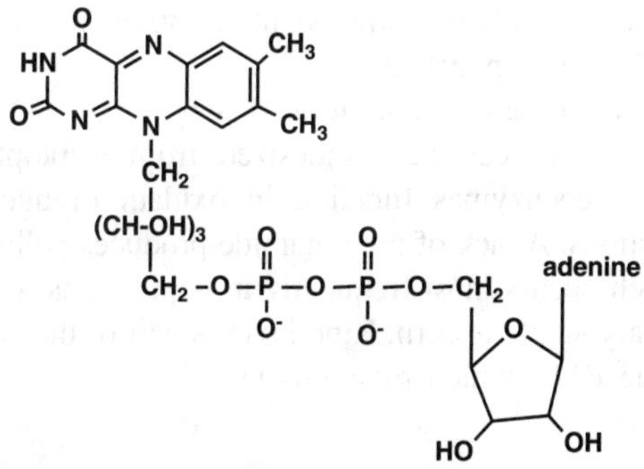

FAD

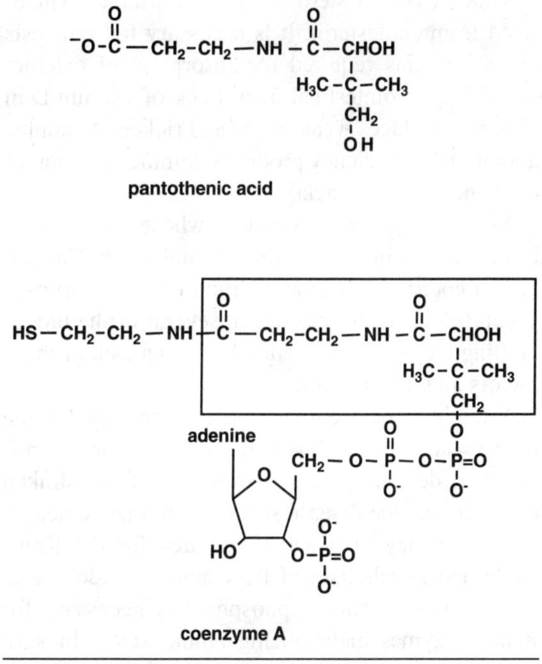

pantothenic acid

coenzyme A

Fig. 30-8. Structure of coenzyme A. The vitamin pantothenic acid, one of the components of coenzyme A, is shown in the box.

Thiamine

Thiamine is **vitamin B$_1$**. As the pyrophosphate, thiamine functions as a coenzyme for pyruvate dehydrogenase, α-ketoglutarate dehydrogenase, and transketolase. Thiamine is found in meat, whole grains, and legumes.

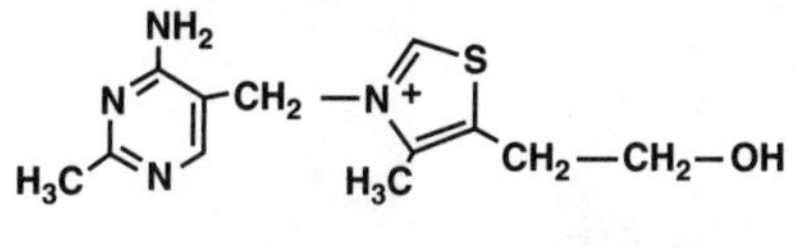

thiamine

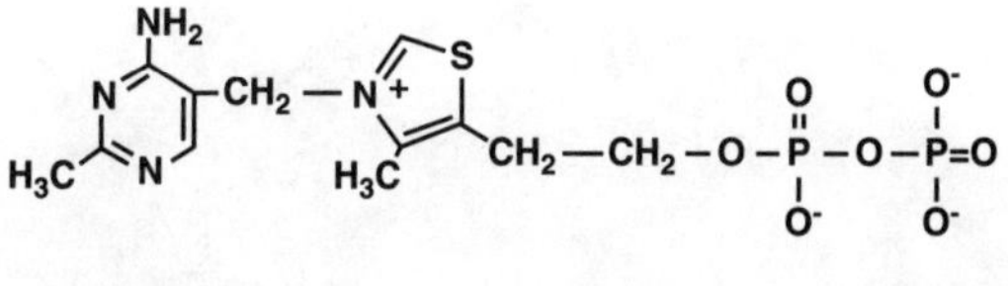

The classic thiamine deficiency disease is **beriberi,** which occurs in countries whose staple food is polished rice. The early symptoms include muscular weakness, fatigue, and swelling and numbness of the legs. Later, extreme edema may develop. In some cases, muscular weakness is extensive enough to cause the affected individual to become bedridden.

Chronic alcoholics also show signs of thiamine deficiency. In these cases, it is called **Wernicke-Korsakoff syndrome.** The symptoms are memory loss, ocular palsy, and ataxia. Because of a chronically poor diet, alcoholics usually show the effects of deficiencies of more than one vitamin.

Summary

Vitamins A, D, E, and K are the fat-soluble vitamins. They are all synthesized in part from isoprene units. Vitamins A and D are stored in the liver, and excesses are toxic. The main metabolic function of vitamin A is as part of the visual pigment rhodopsin. A deficiency produces night blindness and, eventually, permanent blindness as a result of ulceration and scarring of the cornea.

Vitamin D is a steroid-derived hormone synthesized from cholesterol. It is necessary for synthesis of the proteins required for absorption of calcium from the gastrointestinal tract. Lack of vitamin D in children produces weak bones and rickets. In adults, a vitamin D deficiency produces demineralization of the bones (osteomalacia).

Vitamin E is an antioxidant whose exact metabolic function in humans remains unknown. Vitamin K is a coenzyme for the addition of the γ-carboxyl group to glutamate residues in several of the blood-clotting factors. A deficiency leads to a lack of these factors and hemorrhage.

Vitamin C or ascorbic acid is necessary for the hydroxylation of proline and lysine residues in collagen. A deficiency results in a lack of crosslinked collagen and the disease scurvy. Pantothenic acid is part of coenzyme A and is required for the Krebs cycle and metabolism of fatty acids. Pyridoxine in the form of pyridoxal phosphate is necessary for many enzymes metabolizing amino acids. Inositol forms part of phospholipids and serves as a second messenger in cells.

Folate is necessary for one-carbon metabolism. A deficiency causes macrocytic anemia. A deficiency of both vitamin B_{12} and folate produces megaloblastic anemia. Biotin and thiamine function in carboxylation reactions. Vitamin B_{12} is required for the rearrangement of methylmalonyl-CoA to succinyl-CoA. A deficiency of vitamin B_{12} occurs when absorption is blocked by a lack of intrinsic factor. This results in pernicious anemia. A deficiency of thiamine produces beriberi, common in countries where the staple food is polished rice.

The amide of nicotinic acid is part of NAD and NADP and can be synthesized from tryptophan. These coenzymes function in oxidation-reduction reactions. A lack of nicotinamide produces pellagra, which occurs most frequently in populations whose dietary staple is corn. Riboflavin is part of the coenzyme FAD, which functions in oxidation-reduction reactions.

31 Minerals

In addition to vitamins, a large number of minerals are also necessary for human metabolic processes. Some of these, such as calcium and phosphorus, are present in relatively large quantities. The trace minerals, such as zinc, selenium, molybdenum, and cobalt, are present in only minute amounts. Although specific deficiency states have been recognized for some of the minerals, most are so widespread in foods from both animal and vegetable sources that clinical problems due to natural deficiencies are rare.

Objective

After completing this chapter, you should be able to

List the metabolic function, major dietary sources, and effect of deficiency and excess, if known, for each of the minerals.

Calcium

Calcium is the most abundant of the minerals of the body. More than 95% of it is found in the bones and teeth as hydroxyapatite, a complex of calcium, phosphate, and hydroxyl ions. In addition, calcium is found in the serum, where it is essential for the function of the blood-clotting proteins. Intracellular calcium is required for muscle contraction, transmission of nerve impulses, and enzyme reactions. It is found in many foods, including milk, dark green vegetables, and legumes.

The level of calcium is controlled by parathyroid hormone, calcitonin, and vitamin D. A deficiency of serum calcium produces depolarization of cells and muscle tetany. An excess of calcium in the serum results in hyperpolarization of cells, muscle weakness, and cardiac failure. Both excess and deficiency are more likely to be caused by faulty hormonal regulation than by an improper diet. Only the elderly, especially women, are likely to suffer from a dietary calcium deficiency. This is usually due to a decreased intake of milk and other dairy products and possibly decreased calcium absorption. Such a calcium deficiency can lead to **osteoporosis,** in which resorption of calcium from bone is a major symptom. This loss of calcium results in a tendency of the bones to fracture under mild stress, weakness of the spine, stooped posture, and loss of height. All these symptoms are frequently observed in elderly women.

In addition to its roles in bone structure, blood clotting, and muscle contraction, calcium also serves as an important intracellular messenger, as discussed previously in Chapter 8. When bound to the protein calmodulin, calcium modulates a number of different reaction pathways by activating protein kinases. One of these pathways is glycogen metabolism. A calcium-activated protein kinase phosphorylates phosphorylase kinase. This, in turn, activates phosphorylase, which breaks down glycogen (Fig. 31-1). At the same time, calmodulin binds to glycogen syn-

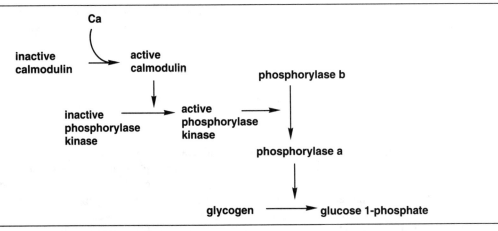

Fig. 31-1. The pathway connecting an increase in intracellular calcium with an increase in glycogen breakdown.

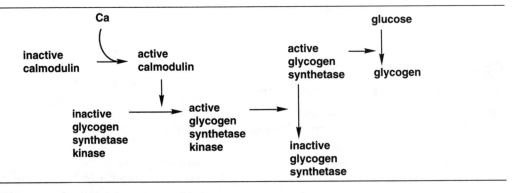

Fig. 31-2. The pathway connecting an increase in intracellular calcium with a decrease in glycogen synthesis

thetase kinase. This enzyme phosphorylates glycogen synthetase, inactivates it, and stops glycogen synthesis (Fig. 31-2). Other enzyme pathways modulated by calmodulin include the synthesis and degradation of cAMP and cGMP.

Phosphorus

Most **phosphorus** in the body is found with calcium as hydroxyapatite in the bones and teeth. It also occurs inside cells as phosphorylated sugars, nucleotides, DNA, and RNA. Parathyroid hormone decreases the resorption of phosphorus in the kidney and increases the serum level. Calcitonin has the opposite effect.

A phosphorus deficiency is rare in persons on a normal mixed diet, although it can occur in those taking large quantities of aluminum hydroxide antacids. This is because aluminum forms insoluble phosphorus salts that cannot be absorbed from the gastrointestinal tract.

Magnesium

Like calcium and phosphorus, **magnesium** is found primarily in bone. In addition, it occurs intracellularly and is required for a number of enzymatic reactions, especially those utilizing ATP.

A deficiency of magnesium tends to occur with deficiencies in other minerals because of faulty absorption or excessive excretion. A purely dietary deficiency is prevalent only in alcoholics because of their grossly inadequate diet.

Iron

Although the amount of iron in the body is small, approximately 4 g, it is essential for life. Iron is found in hemoglobin, myoglobin, and cytochromes of the electron transport chain. The amount of iron in the body is controlled at the level of uptake from the gastrointestinal tract. Because there is no mechanism for excretion of iron, excesses can occur.

Iron is found in high concentration in meat, egg yolk, beans, peas, and spinach. Milk, on the other hand, is poor source of iron. Only 10% of the iron in the diet is absorbed, although this percentage can increase with iron deficiency. After absorption, iron is transported bound to the serum protein **transferrin.** Then it is stored in the liver as **ferritin,** a protein-iron complex.

In **primary hemochromatosis,** too much iron is absorbed and too much is stored. The ferritin is fully saturated, and iron is stored in another protein-iron complex, **hemosiderin.** This iron excess causes cirrhosis of the liver and damage to the pancreas, which produces diabetes. **Secondary hemochromatosis** can occur with multiple transfusions as the excess iron, which enters the body as hemoglobin in the transfused blood, cannot be excreted sufficiently.

Iron deficiency produces anemia. This is most common in infants on a diet consisting only of milk and in young women at puberty and during pregnancy. Because of the blood lost during menstruation, the iron requirement for women is higher than for men. However, because of the iron enrichment of many foods, iron-deficiency anemia is not as common as some advertisements would lead one to believe.

Sodium

Sodium is the major ion of the extracellular fluid. It is involved in the maintenance of water balance and blood pH and the conduction of nerve impulses. Sodium levels are controlled by aldosterone, which regulates the resorption of sodium in the kidney.

The primary source of sodium in the diet is table salt. An excess of sodium is associated with increased blood pressure. A deficiency can result from sodium loss by extreme sweating and in diabetes insipidus. The symptoms of deficiency include muscle cramps, headache, and nausea.

Potassium

Potassium is the major intracellular ion. It is involved in the conduction of nerve impulses and in many enzymatic reactions. Potassium is found in milk, meat, cheese, and grains, but is especially prevalent in bananas and oranges.

An excess of potassium is found in renal failure, dehydration, and shock. Because of a decrease in blood volume and inhibition of nerve impulse conduction, potassium excess can lead to cardiac failure and central nervous system dysfunction. A deficiency of potassium can occur with the prolonged use of diuretics.

Chloride

Chloride in the diet occurs predominantly with sodium in table salt. Because they are ingested together, a sufficient amount of sodium provides a sufficient amount of chloride. Chloride ions are involved in the maintenance of water balance and blood pH. A deficiency in chloride can occur with the loss of HCl in prolonged vomiting.

Sulfur

Sulfur is found in the amino acids methionine and cysteine as well as in biotin and thiamine. It usually is ingested in one of these forms. Sulfur occurs as sulfate in chondroitin sulfate in cartilage and bound in ester linkage to serine in the peptide hormone cholecystokinin.

Copper

Copper is essential for the function of cytochrome oxidase and for heme synthesis. The ion is actively transported in the gastrointestinal tract. A lack of this transport system produces **Menke's kinky-hair**

syndrome. In the plasma, copper is bound to a specific protein, **ceruloplasmin.** Like iron, copper is stored in the liver.

In **Wilson's disease,** free copper rather than bound copper is the predominant form in the plasma. The copper is deposited in the liver, brain, and eye, where it disrupts the function of these tissues.

Trace Minerals

The trace minerals are required for specific enzymatic reactions. With the exception of iodine, they are found only rarely as individual deficiencies. Only fluoride results in toxicity when ingested in excess.

Almost all the **iodine** in the body is associated with the thyroid gland. Lack of iodine produces an enlargement of the gland and goiter. In the diet, iodine is found in fish, seafood, and especially iodized salt.

Fluoride ions substitute for the hydroxyl ions in hydroxyapatite and produce stronger bones and teeth that are less subject to fracture and decay. Fluoride is added to the drinking water of most cities and towns at a concentration of approximately 1 part per million (ppm). Fluoride can be toxic if consumed in excess. Excess fluoride produces mottled teeth and calcium deposits in muscle. It is not possible to ingest an excess of fluoride by drinking properly fluoridated water. Excesses occur with water that contains a natural high level (10 to 45 ppm) of fluoride.

Selenium is required for the function of glutathione peroxidase, which catalyzes the destruction of hydrogen peroxide (Fig. 31-3). Removal of peroxide prevents the oxidation of the double bonds in the fatty acids found in biological membranes.

The other trace minerals are found associated with specific enzymes. **Zinc** is found in poly-A polymerase, carboxypeptidase, carbonic anhydrase, and the zinc-

finger transcription factors. **Manganese** is essential for the function of arginase, one of the enzymes of the urea cycle. Xanthine oxidase, the last enzyme in the degradative pathway for purines, contains **molybdenum. Tin** is found in heme oxidase. **Cobalt** is part of vitamin B_{12} and usually is ingested with the vitamin. As more enzymes are isolated in pure form, it is likely that additional trace minerals will be identified.

Summary

Calcium is the most abundant of the minerals in the human body. It is necessary for the structure of bones and teeth, blood clotting, muscle contraction, and transmission of nerve impulses. Inside cells, calcium acts as a second messenger by binding to calmodulin, which then stimulates particular protein kinases. A dietary deficiency of milk products can cause a calcium deficiency. Excesses and deficiencies of calcium are usually due to faulty regulation by parathyroid hormone, calcitonin, and vitamin D.

Phosphorus also is found predominantly in bone. Inside cells, it is an integral part of RNA, DNA, phospholipids, and phosphorylated sugars. Aluminum hydroxide antacids form insoluble phosphate salts and can cause a phosphate deficiency.

Magnesium is located in bone and also inside cells. There it is associated with enzymes, especially those utilizing ATP.

The level of iron is controlled at the absorption step. It is transported to the liver bound to transferrin and is stored as ferritin. Excess iron intake in the form of transfusions or excessive absorption produces hemochromatosis and liver damage. Iron deficiency results in anemia.

The dietary source of both sodium and chloride is table salt. Both ions are involved in maintaining blood pH and water balance. In addition, sodium is necessary for the transmission of nerve impulses. Diabetes insipidus produces a sodium deficiency; prolonged vomiting causes a chloride deficiency.

Potassium is the major intracellular ion and is involved in the conduction of nerve impulses and enzymatic reactions. An excess of potassium can occur with renal failure, dehydration, or shock. Continual use of diuretics can produce a potassium deficiency.

Sulfur is found in methionine, cysteine, thiamine,

Fig. 31-3. Conversion of reduced glutathione to oxidized glutathione using an enzyme containing selenium.

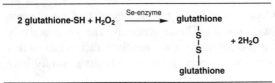

and biotin. It is part of chondroitin sulfate and the enzyme cholecystokinin.

The trace minerals are necessary for the function of particular enzymes. Copper is found in cytochrome oxidase; iodine in thyroxine; fluoride in hydroxyapatite; selenium in glutathione peroxidase; zinc in carbonic anhydrase, carboxypeptidase, and some of the proteins interacting with DNA; manganese in arginase; molybdenum in xanthine oxidase; tin in heme oxidase; and cobalt in vitamin B_{12}.

32 Diet and Nutrition

Barely a day passes without a major article about some aspect of nutrition appearing in the local newspaper or in a weekly news magazine. People assume that anyone with training in the medical sciences is an expert on such subjects. Because of this, it is necessary for you to understand the concepts of good nutrition and to be able to explain what metabolic changes occur when a restricted diet is consumed or when dietary needs change.

Objectives

After completing this chapter, you should be able to

Define the terms *specific dynamic action, basal metabolic rate, biological value,* and *respiratory quotient.*

Calculate the number of calories in a diet from a known number of grams of fat, carbohydrate, and protein.

Discuss the following relative to diet, caloric requirements, and possible deficiencies: obesity, anorexia nervosa, bulimia, kwashiorkor, marasmus, alcoholism, vegetarianism, pregnancy, and total parenteral nutrition.

Energy Yield

The amount of usable energy found in dietary foods varies. When burned in a bomb calorimeter, 1 g of fat yields 9 kcal of energy. Carbohydrate gives 4 kcal/g and protein yields between 5 and 6 kcal/g. For protein, however, the results from the bomb calorimeter do not represent the equivalent of biological oxidation. In the body, nitrogen from protein is converted to ammonia or urea. In the bomb calorimeter, it is burned to nitrogen oxide. Because extra energy is required to produce ammonia or urea from protein, the biological caloric yield from protein averages only 4 kcal/g.

Another way to estimate the energy consumed and its source is by the **respiratory quotient** (RQ).

The RQ is the ratio of moles of carbon dioxide produced per mole of oxygen used.

$$RQ = CO_2/O_2$$

For carbohydrate, the RQ is 1.

$$C_6H_{12}O_6 \text{ (glucose)} + 6\,O_2 \rightarrow 6\,CO_2 + 6\,H_2O$$

For fat, the RQ is approximately 0.7.

$$C_{18}H_{32}O_2 \text{ (stearic acid)} + 23\,O_2 \rightarrow 16\,CO_2 + 16\,H_2O$$

The RQ for protein is approximately 0.8. The RQ can be greater than 1 if carbohydrate is being converted to stored fat—for example, if glucose is converted into stearic acid.

$$3 (C_6H_{12}O_6) \rightarrow C_{18}H_{32}O_2 + 2 H_2O + 8\frac{1}{2}O_2$$
$$\text{glucose} \qquad\qquad \text{stearic acid}$$

By measuring the amount of urinary nitrogen produced, the amount of protein being oxidized can be estimated. The rest of the metabolic energy must be a combination of fat and carbohydrate. The percentage of each can be determined from the protein-corrected RQ.

At rest, the major fuel for the human body is fat (RQ approximately 0.8). On beginning exercise, glycogen is burned (RQ > 0.95). As the glycogen is exhausted, the percentage of fat increases and the RQ decreases.

Because the body does not function like a bomb calorimeter, not all the energy in food is available for biological activities. Some energy is expended in digesting the food and absorbing, transporting, and finally activating the resulting products. The energy input required before ATP can be synthesized is the **specific dynamic action.** It is highest for protein and lowest for carbohydrate. In general, the specific dynamic action amounts to approximately 10% of the calories ingested.

Basal Metabolic Rate

The **basal metabolic rate** (BMR) is the amount of energy required for basic life functions. It is measured in a fasting person who is resting but still awake, lightly clothed, and in a warm room. Under these conditions, primarily fat is being utilized as metabolic fuel and the RQ is close to 0.8. If the amount of oxygen consumed per hour is measured, the number of calories being burned can be calculated. The BMR is reported as kilocalories per square meter of surface area per hour.

The BMR decreases with age in adults. It is higher in men than in women, due to the larger muscle mass in men and the larger amount of insulating fat in women. Caffeine, thyroxine, and catecholamines all increase the BMR. For a lean adult man of 150 lb (approximately 70 kg), the BMR is around 1700 kcal/day.

Dietary Requirements

The number of calories required per day include those for the BMR plus those for daily activities. These extra calories can vary from 120 kcal/hr for reading this book to more than 600 kcal/hr for running a marathon. In addition to total calories, a balanced diet must supply adequate protein, fat, carbohydrate, vitamins, minerals, water, and fiber. The US government publishes a list of the recommended daily allowances (RDAs) of the major nutrients required for healthy persons. It is adjusted for age, sex, height, and weight.

The protein intake in a balanced diet must be sufficient to achieve nitrogen balance and to provide all the essential amino acids. For this, animal protein is better than vegetable protein. In egg, all the essential amino acids are present and at the proper ratio for human needs. Because of this, egg protein has been assigned a **biological value** of 100. Meat and milk also contain protein with high biological value. Corn has a low biological value because it is low in the essential amino acid tryptophan. Wheat is also low because of a small amount of lysine. However, a mixture of these two grains would provide all the essential amino acids and a diet of high biological value. The minimum amount of protein necessary for an adult is approximately 20 g/day. The RDA varies from 20 g/day (< 1 oz) for children to 56 g/day (approximately 2 oz) for adult men. A typical diet of persons in the United States contains much more protein than is required.

An adequate diet must also contain the two essential fatty acids, linolenic and linoleic acids. Fat is also important for the taste and palatability of food, as a source of fat-soluble vitamins, and for calories. It is recommended that less than 30% of the calories in the diet should come from fat.

There is no RDA for carbohydrate. However, carbohydrate reduces the amount of protein needed by providing the intermediates for the Krebs cycle, which would otherwise come from the degradation of amino acids. In a balanced diet, approximately 50 to 60% of the calories should come from carbohydrate.

Water is the major constituent of the body. It is lost in sweat, respiration, urine, and feces. The recommended water intake for infants and children is

10 to 15% of the body weight per day. In adults, it is 2 to 4% of the body weight per day, or six to eight 8-oz glasses of water daily.

Although there is no RDA for fiber, it is an important part of the diet. Dietary fiber comes from the undigestible cellulose of plants. Fiber speeds the passage of food through the intestine and produces more frequent and softer stools. A lack of sufficient fiber in the diet is associated with diseases of the intestine, including cancer.

Special Nutritional Problems

Obesity

If the number of calories ingested exceeds the number expended, the excess calories—whether ingested in the form of protein, fat, or carbohydrate—will be converted to fat and stored. The person will gain weight. If the intake over expenditure continues over time, the person eventually will become obese. **Obesity** is defined as a body weight 15% or more in excess of the reference weight for a person of the same sex, height, and build. Obesity is associated with cardiovascular disease, adult-onset diabetes, and renal disease.

The only way to lose weight is to ingest fewer calories than are needed for daily activities. A dietary deficiency of 3500 kcal is necessary for loss of 1 lb of body weight. This amounts to a 500-kcal deficiency per day to lose 1 lb weekly. In the process of losing weight, there is a tendency to consume fewer dairy products and less red meat. This practice can lead to a deficiency of both calcium and iron. A drastic decrease in the intake of fats can produce a deficiency in the essential fatty acids and the fat-soluble vitamins. This is more likely to happen if the diet is a particularly restrictive one. Furthermore, strange and restrictive diets do not encourage the person to learn about good nutrition, which will be needed to maintain proper weight and function when the desired weight loss is complete.

Starvation

The opposite of obesity is **starvation.** It can occur in developed countries in the form of anorexia nervosa or bulimia and in underdeveloped countries in the form of kwashiorkor or marasmus. **Anorexia nervosa** is most common in young women who starve themselves sometimes to death. This extreme underweight condition is treated by increasing the food intake slowly, sometimes forcibly, under hospital care. A slow increase of food intake is necessary because the enzymes necessary to digest and process foods may be in low supply or entirely absent in these individuals. Thus, they cannot digest large quantities of food taken at one time. In **bulimia,** another condition that occurs predominantly in young women, the patient binges, consuming large quantities of food. Weight gain is avoided by vomiting or using laxatives after the binge. This type of behavior can be fatal due to the loss of chloride ion and eventual circulatory collapse.

Kwashiorkor and marasmus are associated with protein and calorie malnutrition. **Kwashiorkor** occurs in weaned children who are fed a diet lacking in sufficient amounts of protein. It is characterized by growth failure, muscular wasting (often concealed by edema), apathy, anorexia, and changes in the skin and hair. In classic kwashiorkor, there are sufficient calories but insufficient protein. In **marasmus,** there are not enough calories of any kind. This condition is characterized by growth failure, wasting without edema, and anemia. In contrast to those with kwashiorkor, a child with marasmus is not anorexic but is ravenously hungry. This is true unless the child is so weakened that eating requires the expenditure of more energy than is available. It is rare to see either one of these conditions as a separate entity. They tend to merge with a deficiency of both calories and protein.

Alcoholism

Alcohol provides approximately 7 kcal/g, so alcoholics are not starving for calories. However, their diet provides few other essential nutrients. Because of this, alcoholics generally lack protein, calcium, and the water-soluble vitamins, especially thiamine, folate, and pyridoxine. Liver damage caused by excessive alcohol intake can make it difficult to correct the protein deficiency. This is because the enzymes necessary to synthesize urea from catabolized amino acids are in low supply. A sudden influx of amino acids can produce high levels of ammonia in the blood, which can be fatal.

Vegetarians

Persons who consume vegetarian diets containing only foods from plant sources tend to lack vitamin B_{12}, which is available only from animal sources. A deficiency of essential amino acids can also occur, as plant proteins have a lower biological value than do animal proteins. Mixing of proteins from different plant sources at the same meal can relieve this dietary deficiency. Iron is frequently in low supply in a vegetarian diet because no red meat is consumed. Omission of dairy products can produce a calcium and phosphorus deficiency. Because of these deficiencies, a strict vegetarian diet is inadequate for infants and growing children. Adults on vegetarian diets may need vitamin and mineral supplements to provide the RDA of necessary nutrients.

Pregnancy

Pregnancy requires an increase of 15% in the number of calories and 50% in the amount of protein consumed. Substantial increases are also necessary for folate, calcium, phosphorus, and magnesium. For a woman consuming a standard American diet, an iron supplement is necessary to provide for the extra iron requirements of the increasing blood supply of mother and growing fetus. Smaller increases are necessary in other major nutrients as well.

Total Parenteral Nutrition

Persons unable to take food by mouth must be maintained by intravenous intake of all nutrients. A solution of 5% glucose, 4.25% amino acids, plus vitamins and minerals is sufficient for nitrogen balance. However, it will not provide enough calories for the BMR, much less for the repair of wounds or the effects of fever. Higher concentrations of nutrients produce a hypertonic solution that can damage tissues at the site of entry and increase the risk of thrombosis. Because dilution is necessary to prevent these complications, only 600 kcal/day can be given by total parenteral nutrition.

If this type of total nutritional support is to be prolonged, the addition of higher concentrations of nutrients and emulsified fats is required. A feeding of this type must be given through an indwelling catheter placed in a major vein.

Summary

The caloric yield from biological oxidation of 1 g of fat is 9 kcal, 1 g of carbohydrate yields 4 kcal, 1 g of protein yields 5 to 6 kcal. The energy that must be added before ATP can be captured from food is the specific dynamic action. The ratio of fat and carbohydrate being burned can be calculated from the respiratory quotient, the ratio of carbon dioxide produced to oxygen consumed. The basal metabolic rate measures the amount of energy required for the essential metabolic needs. This plus the calories required for activity determines the daily caloric needs.

Any calories consumed in excess of the amount needed will be stored as fat. Weight loss requires a reduction in the number of calories or an increase in activity. A balanced diet for normal activity or weight reduction must contain sufficient protein of a high biological value to achieve nitrogen balance. It must also contain the essential fatty acids, carbohydrates, and the required vitamins and minerals. Starvation occurs in developed countries in the form of anorexia nervosa and bulimia and in underdeveloped countries as kwashiorkor or marasmus. Serious nutritional deficiencies may also occur in alcoholism, pregnancy, total parenteral nutrition, and in diets containing only food from plant sources.

Part VIII Questions: Integration of Metabolism

1. The effects of all the following hormones are mediated through a cAMP cascade *except*
   A. thyroid hormone.
   B. parathyroid hormone.
   C. glucagon.
   D. norepinephrine.
   E. thyroid-stimulating hormone.

2. Thyroid hormone
   A. is synthesized as a small prohormone.
   B. is released from the thyroid in response to a decrease in serum iodine.
   C. is increased in simple goiter.
   D. enters the target cell and is transported into the nucleus.
   E. synthesis is stimulated by thiourea.

3. Insulin
   A. secretion is from the A cells of the pancreas.
   B. secretion increases with increasing serum glucose levels.
   C. acts on the liver by increasing cAMP levels.
   D. stimulates gluconeogenesis in muscle.
   E. requires catecholamines for activity.

4. Serum calcium levels
   A. are increased by calcitonin.
   B. if above normal, stimulate the secretion of parathyroid hormone.
   C. are low in hypoparathyroidism.
   D. are controlled by the rate of uptake of calcium by the liver.
   E. are decreased in the presence of isoproterenol.

5. ACTH
   A. is a glycoprotein produced by the pancreas.
   B. stimulates aldosterone release.
   C. is released by the hypothalamus.
   D. is inhibited in its release by cortisol.
   E. is synthesized in the parathyroid gland.

6. Iron is
   A. transported to the liver bound to ceruloplasmin.
   B. stored in the liver as hemosiderin.
   C. controlled at the level of excretion by the kidney.
   D. in excess after hemorrhage.
   E. found only in the liver, bone marrow, and erythrocytes.

7. All the following statements about calcium are true *except* calcium
   A. levels in serum are regulated by aldosterone.
   B. is the primary mineral in bone.
   C. is necessary for muscle contraction.
   D. can stimulate the phosphorylation of proteins by binding to calmodulin.
   E. deficiency in the diet can lead to osteoporosis in older women.

8. Cobalt is found in
   A. xanthine oxidase.
   B. heme oxidase.
   C. glutathionine reductase.
   D. vitamin B_{12}.
   E. α-ketoglutarate reductase.

9. Which of the following statements is true about trace minerals?
   A. Potassium is required for blood clotting.
   B. The level of copper is controlled by parathyroid hormone.
   C. Sulfur is necessary for the conduction of nerve impulses.
   D. Most of the iodine in the body is found in the thyroid gland.
   E. Most of the body's selenium is found in the pancreas.

10. Hydroxyapatite
   A. is a complex of magnesium, phosphate, and sulfur.
   B. is a major constituent of bone and teeth.
   C. synthesis is stimulated by epinephrine.
   D. is weakened if fluoride is substituted for some of the hydroxyl ions.
   E. degradation is stimulated by insulin.
11. The basal metabolic rate
   A. is measured while the person is sleeping.
   B. increases with age in adults.
   C. is increased by thyroxine.
   D. is higher in women than in men.
   E. determines the number of calories required for weight reduction.
12. Specific dynamic action
   A. is the amount of calories per day needed for essential metabolic activities.
   B. amounts to approximately 25% of the calories in the diet.
   C. is highest for carbohydrates.
   D. is the amount of energy needed for processing food before it can produce ATP.
   E. is lowest for amino acids.
13. A balanced diet should include
   A. protein from predominantly plant sources.
   B. linoleic and linolenic acids.
   C. sufficient protein to produce a negative nitrogen balance.
   D. at least 25% of the calories from carbohydrate.
   E. at least 50% of the calories from fat.
14. A daily food intake of 90 g of egg protein, 200 g of fat, and 150 g of carbohydrate would provide
   A. approximately 2500 calories of usable energy.
   B. insufficient protein for most healthy adults.
   C. approximately 50% of the calories as carbohydrate.
   D. a small enough number of calories for weight reduction in a sedentary person.
   E. a diet lacking some of the essential amino acids.
15. A deficiency of calories occurs in all the following *except*
   A. alcoholism.
   B. anorexia nervosa.

C. bulimia.
D. marasmus.
E. total parenteral nutrition.

For questions 16 through 20, select the letter of the structure that correctly completes the sentence.

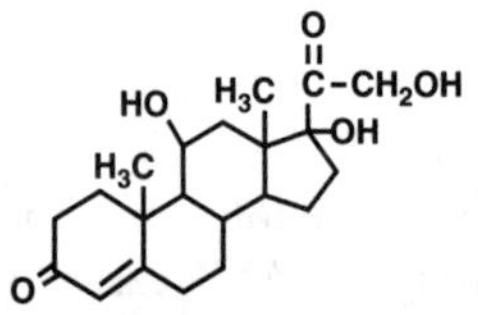

A

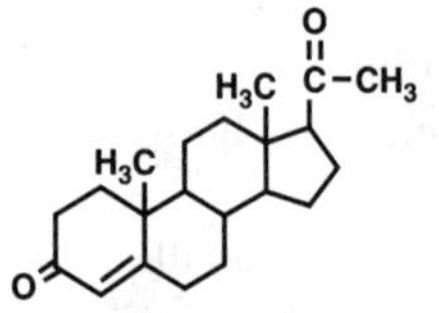

B

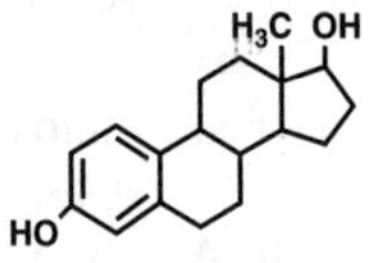

C

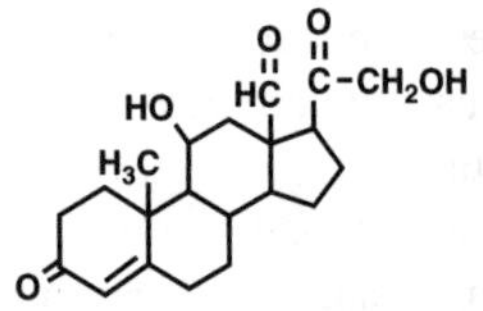

D

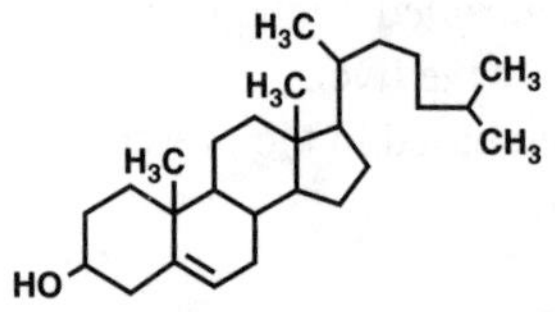

E

16. _____ causes resorption of sodium in the kidney.
17. _____ is a glucocorticoid.
18. _____ has antiinflammatory properties.
19. _____ is the major steroid in humans.
20. _____ is synthesized in greater amounts in women than in men.

For questions 21 through 25, match the letter of the structure to the definition.

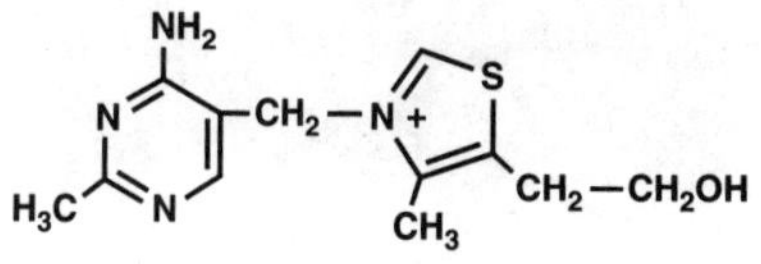

A

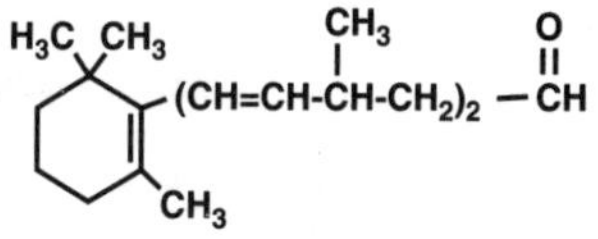

B

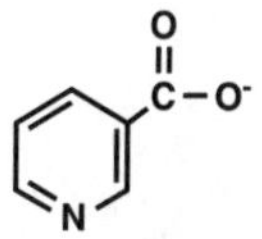

C

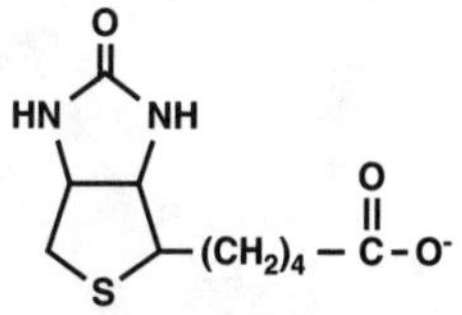

D

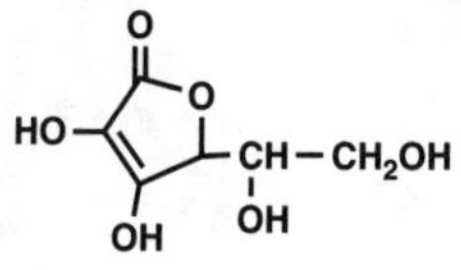

E

21. _____ is necessary for the hydroxylation of proline in collagen synthesis.
22. _____ deficiency produces pellagra.
23. _____ is a constituent of rhodopsin.
24. _____ functions in carboxylation reactions.
25. _____ is a coenzyme for pyruvate dehydrogenase and α-ketoglutarate dehydrogenase.

Answers

Part I: The Cell

1. D
2. C
3. A
4. B
5. A

Part II: Molecular Biology

1. D
2. C
3. D
4. C
5. A
6. E
7. A
8. C
9. D
10. B
11. B
12. C
13. A
14. E

15. B
16. E
17. D
18. C
19. D
20. B
21. B
22. A
23. D
24. A
25. E
26. C
27. D

28. B
29. A
30. C
31. D
32. A
33. A
34. C
35. B
36. E
37. B
38. B
39. B
40. C

Part III: Proteins

1. A
2. A
3. D
4. A
5. C

6. A
7. B
8. B
9. D
10. B

11. D
12. D
13. E
14. C
15. B

Part IV: Metabolism of Carbohydrates

1. D
2. D
3. C
4. B
5. A
6. C
7. D
8. C
9. E

10. A
11. C
12. B
13. C
14. E
15. C
16. D
17. D

18. A
19. A
20. C
21. D
22. A
23. B
24. E
25. B

Part V: Lipid Metabolism

1. C
2. A
3. D
4. A
5. C
6. B
7. A

8. C
9. E
10. A
11. A
12. D
13. C
14. D

15. E
16. D
17. C
18. A
19. B
20. D

Part VI: Amino Acid Metabolism

1. A
2. C
3. B
4. D
5. E
6. B
7. D

8. D
9. A
10. E
11. C
12. D
13. B
14. C

15. A
16. E
17. B
18. A
19. C
20. B

Part VII: Nucleotide Metabolism

1. D	5. C	8. E
2. B	6. D	9. A
3. A	7. A	10. D
4. D		

Part VIII: Integration of Metabolism

1. A	10. B	18. A
2. D	11. C	19. E
3. B	12. D	20. C
4. C	13. B	21. E
5. D	14. A	22. C
6. C	15. A	23. B
7. A	16. D	24. D
8. D	17. A	25. A
9. D		

Index

Index

Note: Page numbers in italic type indicate figures; page numbers followed by t indicate tables.